Hot Stamping Advanced Manufacturing Technology of Lightweight Car Body

Ping Hu · Liang Ying · Bin He

Hot Stamping Advanced Manufacturing Technology of Lightweight Car Body

Science Press
Beijing

Springer

Ping Hu
School of Automotive Engineering
Dalian University of Technology
Dalian
China

Bin He
School of Automotive Engineering
Dalian University of Technology
Dalian
China

Liang Ying
School of Automotive Engineering
Dalian University of Technology
Dalian
China

ISBN 978-981-10-2400-9 ISBN 978-981-10-2401-6 (eBook)
DOI 10.1007/978-981-10-2401-6

Jointly published with Science Press, Beijing, China

Library of Congress Control Number: 2016947930

Printed on acid-free paper

This Springer imprint is published by Springer Nature
The registered company is Springer Science+Business Media Singapore Pte Ltd.

Preface

Since the reform and opening up more than 30 years ago, automobile industry, as the pillar industry of national economy in our country, has played an important role in improving the national standard of living and the quality of travel. The support from all fields has also contributed to the unprecedented development of our country's automobile industry. Two oil crises in 1970s made automobile industry face three big challenges: safety, energy saving and environmental protection. How to design and manufacture cars featured with energy saving and safety is the eternal pursuit goal of automobile industry in the future. The automobile in new ages not only needs to meet all kinds of collision safety laws and regulations such as front crash, side impact, offset collision, etc., but also has to reduce their own curb weight in order to get better fuel economy and achieve energy conservation and emission reduction. Studies have indicated that using new materials with high specific strength and good lightweight effect, such as high-strength steel, aluminum alloy, magnesium alloy and carbon fiber enhanced composite material, in automobile lightweight design and manufacture is the most effective way to achieve this goal. However, new materials such as high-strength steel plate, aluminum alloy and magnesium alloy all have disadvantage of poor toughness and plasticity, which has limited its application in car body, and new technology and new method therefore must be adopted to realize the design and manufacture. High-strength steel hot stamping technology emerges in this new situation.

Hot stamping technology is a new manufacturing technology combining the new material and new technology effectively to manufacture automobile parts. In hot stamping, the original steel plate of boron alloy steel is heated to a temperature of about 950 °C, then transferred to the water-cooling tools for stamping, quenching and forming, finally obtaining the lightweight components with satisfying performance. Hot stamping technology not only can solve the problem of poor formability, unmanageable springback and manufacture precision of high-strength steel sheet, but also can obviously improve the strength and hardness of steel during the forming and quenching process, and obtain ultrahigh-strength hot stamping car body structural parts with tensile strength as high as 1500 MPa. In addition, car body parts with hardness gradient composite properties based on the optimization of forming process can also effectively improve the characteristics of anti-collision and energy absorption, to improve the safety of the car. Based on the background of automobile lightweight and the advantages introduced above, high-strength steel hot stamping technology is booming in the global automotive body manufacturing industry. From the perspective of making China the world's biggest automobile producer with annual production up to 20 million in 2015, the development prospect of this technology is extremely broad, the corporate demand is also very big.

At present, the study of high-strength steel hot stamping technology abroad is very mature. The hot stamping technology was used in the aviation industry such as United States NASA (National Aeronautics and Space Administration), and nuclear industry at the beginning of the last century. The hot stamping technology suitable for auto parts production was first developed by N. Jernverkin 1973, and opened up its industrialization tour in the 1990s. It has been gradually applied and popularized on a global scale in big companies such as BMW and Volvo. As the mature hot stamping technology has been strictly closed in China, it had to be researched and developed from the very beginning in China. The AMT (Advanced Manufacture of Technology) research team led by Prof. Ping Hu in Dalian University of Technology have been studying on the hot stamping technology for more than 10 years, starting from the research field of mechanics for manufacturing process, focusing on the establishment of basic mechanics theory and the constitutive equation and the finite element algorithm for hot stamping. They have developed the KMAS_HF hot stamping sheet forming software with independent intellectual property rights, being the first to break the foreign monopoly and successfully develop a complete set of hot stamping process database and complete sets of production line with completely independent intellectual property rights.

The related scientific research achievements have been published in journals at home and abroad under the premise of not leaking the core technology. The research results have also been successfully applied and demonstrated in the industry in Japan KOBELCO Steel Company, China's Chery Automobile Co., Ltd., FAW Technology Center, JiLin VAFT Lightweight Technology Co., Ltd., and other related units.

As the industry's first monograph that systematically introduces the hot stamping technology from aspects of experiment, theory, method, and industrial application, this book comprehensively introduces the developing situation, equipment and process mechanism of the hot stamping technology. This book mainly introduces the related basic theory about multi-field coupled relationship among heat, stress and phase transformation, the finite element simulation technology and the actual engineering application of hot stamping products in automotive lightweight, together with the theoretical background for sheet metal hot stamping technology and its engineering significance in the field of auto parts. The book also provides a useful reference for other new technology related temperature and phase transformation, such as aluminum–magnesium alloy hot stamping. We sincerely hope the book will be beneficial for advanced manufacturing engineers, automotive design engineers, and researchers in other related fields.

The latest achievements and progress of hot stamping technology in the last 5 years are included in this book, which is organized in ten chapters. The contents include the research achievements and patents of the author and the AMT group for years, and have referred to the related scientific papers published in recent years. Chapter 1 introduces the basic knowledge of sheet metal stamping, including the development of stamping technology, the core basic knowledge such as process, tools, press machine, the production process, as well as the basic requirement of stamping process for material property, which lays a foundation for the subsequent introduction of hot stamping technology. Chapter 2 provides a systematic and concise introduction about the high-strength steel hot stamping technology and the main single equipment technology based on mass production line to make the readers have a macrolevel understanding of the technology. Chapter 3 expounds the process factors that affect the performance of high-strength steel and the original results of process optimization by authors' team in recent years. It also puts emphasis on illustrating the process mechanism to produce auto body parts with tailored properties. Chapter 4 mainly elaborates the hot stamping mechanical theory and constitutive equation for high-strength steel plate from a phenomenological level. Through experiments and theoretical analysis, quantitative research on multi-field coupling heat, stress, and phase transformations in hot stamping process is introduced, together with the stress–strain relationship derived from the law of mixture, and the hot stamping constitutive model of total strain theory and incremental theory, which have established the basic mechanics theory of hot stamping based on phenomenological significance. Chapter 5 establishes the single crystal and polycrystalline finite deformation constitutive integration algorithm under the condition of variable temperature based on the finite element algorithm by taking elastic–plastic deformation gradient and stress as basic variables. Combined with the thermal tensile curves, numerical simulation and experimental verification under thermal coupling are carried out. Chapter 6 focuses on the heat transfer theory in hot stamping process, including the mixed heat transfer theory between blank-tools and tools-channel in the process of transfer, punching, and quenching. The heat transfer coefficient between blank-tools and tools-channel is measured by inverse calculation and experiments. The factors such as high-temperature oxidation and the steel blank surface roughness are also studied in this chapter. Chapter 7 discusses the factors influencing the plasticity and deformation resistance of hot stamping materials, and establishes the high-temperature material constitutive model, which is suitable for multi-field coupling analysis, based on high-temperature material mechanics performance. Combined with the first set high-temperature forming limit TFLD test equipment developed independently in China, the 3D forming limit surface 3D-TFLD suitable for high-temperature formability prediction is obtained. In Chap. 8, the high-strength steel hot stamping FEM simulation algorithm is discussed from the four key problems of numerical simulation: the discussion of the variational equation in temperature field modeling and simulation, cell division, transient spatial domain, and discrete time domain. Meanwhile the fundamental equation and the solving method of the hot stamping phase transformation are analyzed and described. Based on the hot stamping multi-field coupling numerical simulation needs, this paper expounds the static explicit algorithm and dynamic explicit algorithm. Chapter 9 mainly introduces the application of hot stamping components and hardness gradient parts in lightweight car body. According to practical engineering, the hot stamping technology is applied to the typical body bearing parts such as door anti-collision beam, side wall, body beam frame, and the school bus pillars. And the application and optimization of typical body structures such as B Pillar are conducted with the hardness gradient composite properties of hot stamping. Chapter 10 mainly introduces the key technologies involving tool optimization design and manufacture in hot stamping technology. It also analyzes the fatigue and life of hot stamping die.

The relevant research work in this book is strongly supported by projects such as the Key Project of the National Natural Science Foundation of China, "973" National Basic Research Project of China and "04" Great Project of the Ministry of Industrialization and Information of China. After years of interdisciplinary collaboration research, from scientific theory to process practice, from the scientific problems to product research and development, the systemic research progress has been made. To promote the new technology of hot stamping automotive components manufacturing combining new material, new process and new equipment and to guide the innovation and development of auto parts manufacturing industry and then provide a new train of thought for the design and development of new cars are the purpose and motivation for the author to write the book. Errors are inevitable in this book due to the continuous development of hot stamping technology and the limitations of the author. Any comments from readers will be appreciated.

Dalian, China

Ping Hu
Liang Ying
Bin He

Acknowledgment

This book is funded by the Key Project of the National Natural Science Foundation of China (No. 10932003). This support is gratefully acknowledged.

Special thanks go to Dr. Ning Ma, Minghua Dai, Dan Zhao, Dongyong Shi, Wenquan Liu, Ye Yu, Xi Zhao, and others for their significant contribution to the research and development of hot stamping technology. Without their effort, it is impossible to complete this book. The author also wishes to thank VAFT, Changchun auto Parts Co., Ltd., who provided special hot stamping laboratory equipment and relevant experience. Thanks to Dantong Wang, Yang Liu, Xianda Zhang, Fubo Zhang and other students, who gave unstinted effort in editing this book.

In addition, Science Press has done a lot of productive work in editing the manuscript, we sincerely thank all the people and units who have made contribution to the publication of the book.

Contents

Chapter 1
The Basis of Sheet Metal Forming Technology

Metal pressure processing, also known as metal plastic working, is a kind of manufacturing method, which makes use of metal plastic deformation caused under external force to obtain raw materials, blanks, or components with a certain shape, size, and mechanical property.

Stamping is a kind of manufacturing technology that deforms sheet metal in the stamping tool by the deformation force supplied by the power of regular or special stamping equipment to obtain product components with certain shape, size, and mechanical property. Sheet metal, stamping tool, and stamping equipment are three major factors for stamping.

Along with the new compulsive policy about front impact, side impact, and emissions in automotive field, the automobile body parts made of thin and high strength sheet metal blanks have become the main trend of automobile industry. But for cold stamping technology, thinning and high strength are the double factors to worsen formability, which will make it easy to craze and produce excess springback in the forming process for body parts that affect the assembly of car's body subsequently. Especially, when the steel strength is higher than 1000 MPa, the traditional cold stamping technology is difficult to produce autobody parts with relatively complex structure and shape. To solve this problem, the hot stamping technology [2] arises at the historic moment.

Stamping can be divided into cold stamping and hot stamping according to the working temperature. The former is commonly used for metal sheet stamping at room temperature while the latter is suitable to process a kind of sheet which has high resistance to deformation and low plasticity.

Hot stamping, also called hot forming [1, 4, 5], is a plate processing technology that combines heat treatment process and cold stamping technology of sheet metal, aims at producing complex stamping parts with strength greater than 1000 MPa. For parts that have relatively complex shape and cannot be formed in one (direct) hot stamping, however, need an additional cold stamping before the hot stamping process. Thus, it is necessary to know the basic knowledge of cold stamping in consideration of the inheritance of hot stamping for cold stamping.

P. Hu et al., *Hot Stamping Advanced Manufacturing Technology of Lightweight Car Body*, DOI 10.1007/978-981-10-2401-6_1

1.1 The Development of Stamping Technology

Plastic working is a kind of metal processing method that has already existed in ancient times. The ancients struck the preformed metal blocks with hammers to make necessary tools, such as food containers and hunting forks. In China, the application of metal plastic working method can be traced back to 4000 years ago. At that time, forging was the main processing method, including cold forging and hot forging. Besides, there were some other processing methods applied, such as foil stacking forging, wire drawing, sheet metal forming, and stamping. The products made through plastic working method include weapons, production tools, daily necessities, leisure goods, etc. [3].

Nowadays, sheet metal forming is experiencing a fully automatic mass production in the field of automotive industry, household commodity, beverage cans, and other industries, while ancient sheet metal forming is made using hammer and anvil for piece production. Figure 1.1 shows the ancient sheet metal forming process: extension, thickening, projection, and hollowing. The wall thickness of the unearthed copper pottery of Yan State in late Spring and Autumn period (from the fifth century B.C to the sixth century B.C) is only 1–2 mm, which was excavated in Beixingbao, Huailai, Hebei province, China. The copper pottery is consisted of two

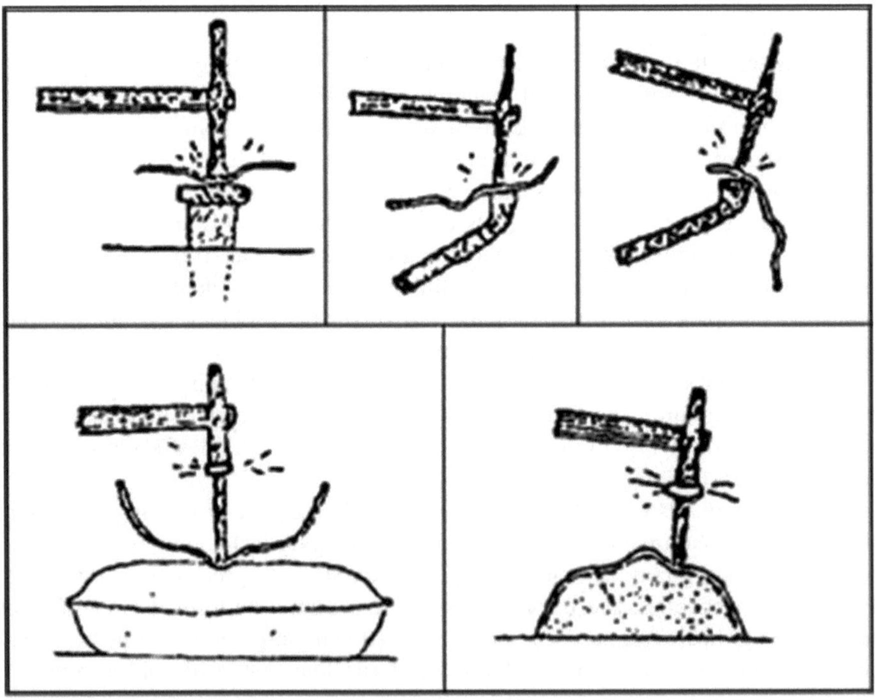

Fig. 1.1 Ancient sheet metal forming process

parts, the superstructure and infrastructure, which is hammer-shaped separately and then combines together in crimping snapped way.

It is generally considered that stamping era began during the European Industrial Revolution about 200 years ago. At that time, the development of modern iron technology contributed to the emergence of stamping machines. It means that the structure of the original machine that is controlled first by hands and then by feet (commonly known as pedaled machine) has changed. New stamping machine moves up and down, and forms products directly by feeding raw materials into the stamping tool installed in the middle of punching machine.

Subsequently, the world witnessed an era of great changes that the car originated in Europe crossed over the Atlantic. It brought a great convenience to Americans who used to be on horseback in the wilderness. However, the price was a problem in the process of car popularization. No matter how hard they tried, they could not afford to buy a car with such a high price, which led to the reduction of its value in use. As a result, using stamping to produce automotive parts became popular because of its contribution to lower the production cost. In the modern society, 60–65 % of car parts are made by stamping. In other words, the car can be described as the outcome of stamping. Stamping is proved to be the best way of mass production especially in this society where car is becoming more and more popular.

In the twentieth century, stamping technology has experienced five stages in developed countries [11].

Stage I: Before 1950s, the stamping line was composed of a double-action drawing press and several single-action drawing presses, feeding and unloading by hands, which caused the low production efficiency, poor security environment, and poor product quality.

Stage II: In the 1960s, the stamping line was still composed of a double-action drawing press and several single-action drawing presses. However, each press was equipped with an unloading robot hand in order to reduce the manual labor and the number of operators.

Stage III: In the 1970s, the automatic stamping line was established and could be operated only by one or two persons. This automatic stamping production was composed of several presses which were equipped with unstacking devices, feeding—unloading robots, middle turns-over or transmission mechanisms, and electronic control systems.

Stage IV: In the 1980s, owing to the emergence of multi-station presses, a double-action drawing press and a multi-station press were used for composing a stamping line. The main plywood of the multi-station press penetrated the mold area of the double-action press, while certain turns-over structures were equipped between the double-action press and the multi-station press to complete the transfer of the workpieces.

Stage V: Since the 1990s, with the advent of numerical controlled (NC) hydraulic cushion, the double-action drawing press was no more used as the leading equipment of stamping lines. Instead, hydraulic cushion was installed into the first station of the large multi-station press, which made this press form a flexible production unit independently. The large multi-station press equipped with such a NC hydraulic cushion as to produce flexibly becomes the development direction of large automobile cover stampings. Besides, it represents the most advanced technology level of stamping, and is also the most advanced stage of large automotive cover stamping development.

At present, China has become a big country of manufacturing and consumption. The metal plastic forming industry has also been developed rapidly [6] in China that the yearly output value has increased by 30 % on average during the period of 2001–2007. Metal forming industry involved in the whole manufacturing industry, the representative industry including [10] automobile, motorcycle, household refrigerators, washing machines and other household appliances, micro computers, integrated circuit, and mobile phones, among which cars, trucks, and buses are the main products in auto industry.

Automobile industry is the pillar industry of national economy in many countries with developed industry or emerging industrialization. In a car, the amount of sheet metal parts takes up more than 75 % of the total number of parts. In Japan, a country with developed automobile industry, the sales of automobile sheet metal parts account for 68 % of total sales in sheet metal industry, implying that the automobile sheet metal parts play an important role in the sheet metal industry, also declaring that the sheet metal forming technology plays an important role in a nation's industrial development. Since the twenty-first century, China's auto industry has developed thriftily and gradually into line with the international community, which has directly led to the progress of sheet metal forming technology and the plastic processing industry [8].

1.2 The Basics of Sheet Metal Forming

1.2.1 The Process of Traditional Cold Stamping

Stamping is a pressure processing method that puts pressure on the sheets or profiles at room temperature using stamping tools to produce plastic deformation or separation and then obtain parts with certain shape, size, and performance. The application range of stamping technology is so wide that it may not only process sheet metal and bar metal, but also a variety of nonmetallic materials. Since the process is usually carried out at room temperature, it is also called cold stamping.

The traditional cold stamping process of sheet metal is completed using stamping tools and stamping equipment. Compared with other processing methods, it has the following characteristics:

(1) The workpieces with complex shape, such as the shell parts, can be obtained by cold stamping method, which can be hardly formed by other machine processing methods;
(2) The dimensional accuracy of cold stamping is determined by the stamping tools. Therefore, it has the advantages of dimensional stability and good interchangeability;

(3) Because of its simple operation and low labor intensity, it is easy to realize mechanization, automation, and high productivity;
(4) Because of its high material utilization ratio, and its workpieces with light weight, good rigidity, high strength, and low energy consumption, the cost of the workpiece is rather low;
(5) The structure of the stamping tool used in the stamping process is relatively complex and the stamping tool has a long production cycle and high cost. So the stamping process is mainly used for mass production, and its application for single-piece and small batch production is subject to the restrictions. Nevertheless in recent years, the development of simple dies, combination dies, and zinc-based alloy dies offers the opportunity to the use of stamping process in single-piece and small batch production.

There are many different types of stamping process to meet the requirements on the shape, size, internal and external quality, and amount of workpieces. Generally, a multi-channel stamping process is needed for a stamping part. Due to the variety of shape, size, precision, production volume, and raw materials, there are also different processing methods which are used in cold stamping. To sum up, it can be divided roughly into separating process and shaping process.

Separating process is a stamping process through which the sheet can be separated by a certain contour to obtain stampings (also known as blankings) with certain shape, size, and cut surface quality. It includes blanking, piercing, incision, slice, and other processes.

Shaping process is a stamping process in which plastic deformation is produced under the condition of no material cracking and then the stamping parts with a certain shape, size, and precision are obtained. It includes bending, deep drawing, flanging, distortion, bulging, necking, etc.

1.2.2 The Cold Stamping Tool

Stamping tool is an important equipment in stamping production and is a kind of technology-intensive product. The quality, productivity, and production costs of stampings are directly related to the stamping tool design and manufacture. The level of stamping tool design and manufacture is one of the most important symbols to measure a country's level of manufacturing. It largely determines the quality and efficiency of products and the developmental capability of new products.

The type of stamping parts is varied with the type of stamping die. For the convenience of study, the stamping die is classified according to different characteristics. The following classifications are generally adopted:

(1) According to stamping process, it can be divided into blanking dies, bending dies, drawing dies, forming dies, etc.;
(2) According to the combination of process, it can be divided into single-process mode, composite mold, continuous mold, etc.

(3) According to the orientation mode of the upper and lower die, it can be divided into non-oriented opening dies, oriented guide plate dies, oriented guide pillar dies, etc.
(4) According to the way of guide pin or position, it can be divided into fixed guide pin die, movable guide pin die, pilot pin oriented die, side blade oriented die, etc.

The die can be also divided into fine blanking die and ordinary blanking die based on the size, quality, and precision of blanking. In addition, the die is also divided into small die, medium die, and large die based on the size of the die. Sometimes the die can be classified according to the type of punching machine, feeding method, and reclaiming method.

The kind of stamping methods determines the type and working conditions of relevant dies, and the requirements of die material. The blanking die is mainly used for the cutting of various sheet metals. Its edge is suffered from strong friction and impacted in the working process, so the materials with high wearing resistance, impact toughness and fatigue resistance are required for its working parts. The bending die is mainly used for the bending of the sheet metal, which has low workload and some friction. Thus the materials with high wearing resistance and fracture resistance are required for its working parts. The drawing die is mainly used for deep drawing of the sheet metal. The working stress is not very high, but its entrance has to withstand strong friction. Thus the materials with high hardness and wearing resistance are required for its working parts and the surface roughness of cavity die should be relatively small.

The selection of die material not only concerns the life of the die, but also directly influences the cost of die manufacture. Therefore, it is an important task in the process of die design. In the stamping process, the die has to withstand shock load and continuous work, which makes its punch and matrix work under great pressure and intense friction in poor working conditions. Therefore, the selection of die material should follow the following principles: (1) according to the type of the die and working conditions, the selection of materials has to meet the requirements, such as high strength, hardness, wearing resistance, impact resistance, fatigue resistance, etc.; (2) the selection of the materials is made according to press materials and production quantities of stampings; (3) to meet the processing requirements, the materials should have good processing performance to ease of machining, good harden ability, and low heat treatment deformation; (4) the materials should meet the economical requirements.

After the design of stamping die, the die's manufacturing is also an important process. Nowadays, the international mold and die manufacturing industry is seeking for product specialization and manufacturing digitization, while the mold and die manufacturing enterprises are also gradually turned to the develop direction of specialization, such as mold factory of Volkswagen, mold factory of Audi company, SCHULLER, Italy COMAU, American AUTODIE, SECKLY, mold factory of TOYOTA, Japan Fuji and Hyundai auto mold center, etc. Some of them are specialized in manufacturing dies for the overall side surrounding parts, dies of four doors, while some are specialized in dies of floor and structural parts. All the dies are moving to the direction of multi-stage, automotive, and progressive. In the meantime, the 3D entity design is widely used in die manufacturing, and the three-dimensional DL figure transferred from two-dimensional DL figure can be directly used in FM analysis and the CNC programming, realizing 100 % CAE analysis. The 3D parametric design technology applied to the tool design can realize the design of dies for different thickness sheet metal forming.

In China, mold was first listed in the catalogue of mechanical and electronic products mold in 1987 when the total GDP in mold industry reached 3 billion RMB. By 2004, the mold output value in China reached 53 billion RMB, ranking the third in the world. The total output value of 15 mold companies such as FAW mold manufacturing co., LTD. and Tianjin automobile mold co., LTD. became more than 1.5 billion RMB in 2004, and the national total sales of mold industry in 2005 is as high as 61 billion RMB. At present, the mold enterprise for auto body has spread throughout the country, and 50 of them have formed a certain scale, while half have an annual production of 10–50 million RMB.

1.2.3 Stamping Press

Stamping press is a kind of general equipment with exquisite structure and can drive dies to form steel sheet. To processing part, it makes the metal to plastically deform and fracture by applying powerful pressure, which can be widely used in cutting, punching, blanking, bending, riveting, and forming process. Stamping press, also called press machine, is characterized with widely used and high production efficiency, etc.

There are two kinds of commonly used stamping press, the mechanical press and hydraulic press. When the mechanical press works, the big belt pulley is driven by motor through the triangle belt, then drive the slider-crank mechanism straight up and down through gear pair and clutch. When the forging work is completed, the slider moves upward, clutch releases automatically and the automatic device on crank shaft connects at the same time, making the slider stop near the top dead center. Mechanical press can be divided into crank press and friction press according to the type of driven slider mechanisms, or be divided into single-action press and double-action press according to the number of slider blocks, or be divided into frame press and straight side press according to the structure of machine tool bed, or be divided into general press and high speed press according to the degree of automation, etc.

And hydraulic press [7] is a machine used to transfer energy and achieve various techniques. It is produced according to the PASCAL's principle and takes liquid as working medium. Hydraulic press usually consists of the main engine, power system, and the hydraulic control system. According to the working medium, the hydraulic press can be divided into oil hydraulic press and water press. The working principles and features of commonly used stamping presses are shown in Table 1.1.

Table 1.1 Working principles and features of commonly used hot stamping equipments

Type	Device	Working principle	Feature
Mechanical press	Friction press	It transmits the motion by using the friction engagement between the flywheel and the friction disks, and works under the function of the screw and the nut with the principle of relative motion	The structure is simple. Only sliding between the flywheel and the disks will occur under the condition of overload. However, the abrasion to the rim of the flywheel is serious and the productivity is low. It is suitable for middle and small workpiece stamping, especially for the procedures of calibration, coining and forming, etc.
	Crank press	It works with the use of crank-link mechanism. The crankshaft is driven by the electric motor with the pulleys and gear to make the slide reciprocate in a straight line through the connecting rod. Crank press is divided into eccentric press and knuckle-joint press. The main difference between the two presses lies in the spindle, as the former has an eccentric shaft and the latter has a crankshaft. Eccentric presses are generally open presses, while knuckle-joint presses are divided into open and close presses	The productivity is high, so it is applicable to all kinds of stamping processes
	High speed press	It is a kind of special crank press developed due to the need of mass stamping production. The working principle of this press is the same as that of general crank press. But relatively speaking, its rigidity, accuracy, stroke times are all higher. Its stroke times are generally 5–9 times than that of the general crank press. The application of high speed press must be equipped with auxiliary devices (like automatic feeding device, safety testing device, etc.) to achieve high speed machining for stamping	The productivity is very high; it is suitable for mass production. Multi-station progressive die is generally used
Hydraulic press	Oil hydraulic press and Water press	Using Pascal's principle and taking water or oil as working medium, hydraulic press works with the transmission of static pressure to make the slide travel up and down	The pressure is high and static, but the productivity is low. It is suitable for the forming processes such as drawing, extrusion, etc.

Furthermore, for the small batch production of flat sheet, it is generally machined by numerical control (NC) stepping press in order to reduce production cost, shorten production cycle, reduce labor intensity, and improve production efficiency. NC stepping press is a type of press which fulfills punching and nibbling on the blanks with NC technology, so it is an efficient and sophisticated single stamping device of sheet. Punching sheet is fixed on the platform, and then driven to move from left to right, up and down, and positioned according to the prescribed procedure. The molds are installed on the turret for automatic shifting, or installed on the die adapter for manual rapid shifting. The holes and parts are punched into different shapes and sizes with single punching or nibbling punching.

At present, the development of world's large stamping press are going to two directions, one is large and multi-station, the other is focusing on flexible production equipped with automation manipulator. In the recent 10 years, the press has contained many key technologies after continuous development as follows: fully automatic system for die changing, perfect automatic monitoring system, and good human–machine interface, high stroke frequency for high production efficiency and the necessary high precision for high-quality stamping parts. Nowadays, the automobile companies in United States, Japan, and German have equipped higher proportion of transfer press. In China, a majority of the production line is for single wire press and the character for key stamping equipment is large tonnage, large stroke, large table-board, large-tonnage air cushion and the slider air cushion, manipulator system for loading and unloading, high speed, high precision, automatic technology for die changing and function perfect touch screen monitor technology.

So far, the most advanced press adopts servo motor, the tonnage in a few years ago is up to 250 and 500 t, respectively, but now is up to 1000 t. Press of this type controlled by servo motor can realize the control of press speed and working stroke, has attracted attention of many famous automobile company, together with their bulk order. This press can further reduce the stamping process and improve the accuracy of parts, having obtained a much bigger application range. Combined with the application of the nitrogen cylinder, there must be a very broad developmental space for this kind of press.

1.2.4 The Production Process of Stampings

The general production process of stampings is shown in Fig. 1.2. Stamping technology includes stamping process design, die design, and die manufacture. Though the contents of the three aspects are different, they are interrelated, influenced, and interdependent. The basic contents and requirements of them are described in Table 1.2.

It should be pointed out that die design and manufacture must be taken into comprehensive consideration according to the actual situation of enterprise and products' production batch to find the optimal economical technology and

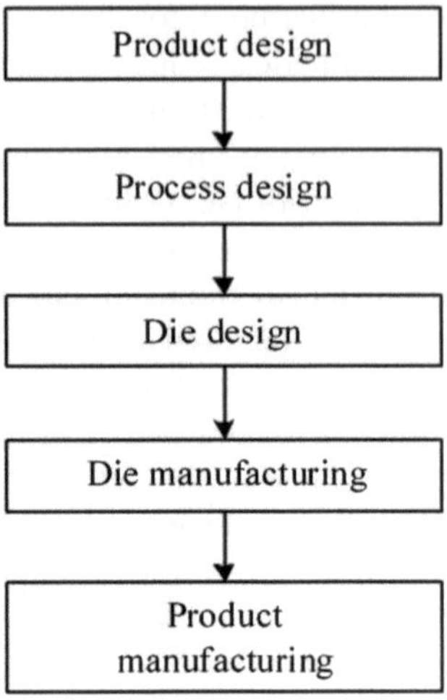

Fig. 1.2 Production process of stampings

Table 1.2 The basic contents and requirements of stamping forming

Item	Basic contents	Basic requirements
Process design	For a given product drawing, a reasonable process scheme (including the property and amount of processes, their sequence, their location mode and combination mode) should be made, according to its production batch, stamping equipment types and specifications, die manufacturing capacity and worker's level of technology. The process scheme should be based on the analysis of the stamping process of part drawing and the process calculation. At last, the stamping process card is obtained	(1) The raw materials should have high utilization rate, that is to say, low material consumption (2) According to the specific production conditions, the process scheme should be technologically advanced and feasible, and economically reasonable (3) Process combination mode and process sequence should comply with the stamping deformation law to ensure qualified workpieces will be produced through stamping (4) The scheme should have low process number and high production efficiency (5) The technological procedure should be made convenient for production organization and management
Die design	According to the stamping technological procedure, the overall structures of corresponding dies are calculated and then designed with the consideration of blank orientation, unloading, waste elimination, stamping tool manufacturing and maintenance convenience, safe and	(1) The structure and size parameters of stamping tools should guarantee the shape, size, and accuracy of the products by stamping should be in line with drawing's requirements (2) Simple structure, reasonable processing accuracy, manufacturing

(continued)

Table 1.2 (continued)

Item	Basic contents	Basic requirements
	reliable operation. After the calculation and design of structures, the assembly drawing and nonstandard part drawing should be plotted to ensure the implement of stamping process	and maintenance convenience, and low cost are required in the design process (3) The die should be firm and durable and satisfy the requirements of mass production (4) It is easy, safe, and reliable to operate the die with low labor intensity. (5) The preparation period of production should be shortened
Die manufacture	The die should be manufactured according to the demand on die structure, die material, size and geometrical accuracy, work characteristics, service life etc. During the process of die manufacturing, the characteristics of existing equipment, the machining method and assembly method should be considered to select an optimal processing scheme, and make out the reasonable die processing technological procedure	(1) To ensure the quality of products and the performance of die, the die should be made with high accuracy (2) The die should be manufactured with a long service life (3) The manufacturing cycle should be short (4) Low manufacturing cost is required

simultaneously guarantee product quality. Excessive pursuit of production efficiency, die accuracy, and its service life will inevitably lead to the increase of cost. However, only considering the reduction of cost and a shorter manufacturing period and ignoring die accuracy, its life will inevitably lead to a drop in quality.

1.3 Materials for Cold Stamping and Its Formability

1.3.1 Requirements on Materials for Cold Stamping

Materials used in stamping should not only meet the technical requirements for product design, but also meet the requirements of the stamping process and subsequent processing requirements (such as cutting, welding, electroplating, etc.). The basic requirements on materials in stamping process are concluded below:

(1) Formability

In order to improve the deformation and stamping parts quality, materials should have good formability. The formability of the material is closely related to its mechanical properties. Therefore, the materials should have good plasticity, small yield strength ratio, high elastic modulus, large normal anisotropic coefficient, and

Table 1.3 Specific requirements of different stamping procedures on the sheet performances

Procedure	Performance requirement
Blanking	Sufficient plasticity, and no cracking in punching; The hardness of materials should be lower than that of punch die
Bending	Sufficient plasticity, low yield limit, and high elastic modulus
Drawing	Good plasticity, low yield limit, large normal anisotropic coefficient, small yield strength ratio σ_s/σ_b, and small plane orientation

small plane orientation coefficient. Specific requirements of different stamping procedures on the sheet performances are shown in Table 1.3.

(2) Thickness tolerance

The thickness tolerance should comply with the national standard. A specific die clearance applies to the material with a specific thickness. Large thickness tolerance not only can directly affect the quality of the product, but also may cause the damage of die and punch.

(3) Surface quality

The surface of materials should be bright, smooth, no stratification, no mechanical damage, and no rusty spot, oxide skin and other attachments. Materials with good surface quality do not easily tend to crack and scratch the die surface during stamping; stamping parts with good surface quality can be manufactured.

1.3.2 The Formability of Materials

1. Concept of stamping formability

Just as other processing methods, stamping processing method is based on the material performance. The materials used in stamping processing must have a good stamping formability.

The stamping formability of the material is defined as the capacity of adapting itself to different stamping processing methods. Materials with good formability refer to its handling ease, high limited deformation and total limited deformation, high productivity, high-quality products, long stamping tool service life, etc. This shows that stamping formability is a comprehensive concept, which involves many factors. From the perspective of its contents, there are only two factors involved, forming limit and forming quality.

(1) Forming limit

The maximum deformation degree of the material achieved during the forming process is defined as forming limit. Coefficients of deformation limit are used to represent different forming limits in different forming process. Most stamping forming is conducted in a state of plane stress which ignores stress in the thickness direction, so it is not difficult to analyze the following phenomenon. In the interior of the deformation blank, excessive tensile stress will make part of it severely thinned, even cracked and scraped, and excessive pressure stress will make part of it lose stability and then wrinkle. Therefore, in order to improve the forming limit, the blank material requires better plasticity and stronger resistance to tension and compression.

In the stamping process, the deformation formed by tensile stress with maximum absolute value on the deformation region of the blank is called tensile deformation (such as bulging, flared, hole flanging, etc.). The deformation formed by compressive stress with maximum absolute value on the deformation region of the blank is called compression deformation (such as drawing, necking, etc.). Coefficients of deformation limit of tensile deformation mainly depend on material plasticity, while coefficients of deformation limit of compression deformation is usually affected by the carrying capacity of blank force transmission zone, and sometimes affected by instability and wrinkle of deformation zone or force transmission zone.

(2) Forming quality

The main quality indicators of the stamping parts are dimensional accuracy, thickness variation, surface quality, and the physical and mechanical properties of materials after forming. There are many different factors which influence the quality of stamping parts in different stamping processes.

The plastic deformation always accompanies with the elastic deformation in the deformation process of materials. Due to the elastic recovery resulting from loading, the size and shape of stamping parts deviate from the stamping tool, which affects their size and form accuracy. Therefore, it is very important to grasp the springback law to control the springback.

After pressing stamping, the thickness of blanks generally becomes thinner or thicker. The thickness thinning directly affects the strength and the use of stamping parts. Therefore, the maximum thinning should be defined if requests are made regarding the strength of stampings.

After plastic deformation, materials suffer from work hardening and residual stress caused by nonuniform deformation. The residual stress on materials causes changes in the size and shape of workpieces, and leads to the cracking of workpieces under severe conditions. All of these circumstances should be considered in the course of making stamping process.

The factors affecting the surface quality of workpieces are the surface state and grain size of raw materials, the condition of the stamping tool sticking to material and the abrasion of the stamping parts' surface by the stamping tool. The surface state of raw materials directly influences the surface quality of workpieces. When stretched, the steel plate with coarse grains forms the so-called "orange peel" (with rough surface). Stamping materials that are easily sticking to stamping tool will scratch the stamping parts and reduce stamping die life. In addition, the uneven die clearance and rough surfaces will also scratch the stamping parts.

2. Testing methods of sheet formability

The formability of sheet is measured by different test methods. These tests can be roughly divided into indirect and direct test. Indirect test methods include tensile test, shear test, hardness test, metallographic test, etc. In the indirect tests, the stress state and deformation characteristics of test samples are somewhat different from those of the actual stamping, so the results of these tests can only indirectly reflect the stamping performance of sheet metals. However, these tests can be conducted on general-purpose test equipments, so they are usually adopted to measure the formability. Direct test methods include repeated bending test, bulging performance tests and deep drawing performance test, etc. In such tests, the stress state and deformation characteristics of test samples are basically the same as those of actual stampings, so these direct tests can provide direct and reliable identification for the formability of certain types of stamping. But in these tests specialized test equipment or tooling are required. The following part is the introduction of tensile test, which is known as the most frequently used indirect test.

Tensile specimens which are cut from different positions and directions of the sheet are made according to the standard shown in Fig. 1.3. The specimens will be stretched by using a universal testing machine. According to the test results or by using the automatic recording devices, the stress and strain curve (or stretch curve) shown in Fig. 1.4 can be obtained.

The mechanical properties of sheet metal can be measured by the tensile test. The formability of sheet metal has a very close relationship with its mechanical properties, so its formability can be reflected by these properties from different perspectives. In general, the higher its strength is, the greater the force producing the same amount of deformation is. The higher its plasticity is, the greater the amount of limit deformation is. The higher its stiffness is, the greater the ability of resistance to instability and wrinkle is. Some essential mechanical properties are illustrated as follows:

(1) Total elongation δ and uniform elongation δ_b

δ represents the total elongation of the destroyed specimen in tensile test, called elongation for short. δ_b represents the uniform elongation when local concentration of deformation begins to generate in the tensile test (first appearing necking). δ_b is

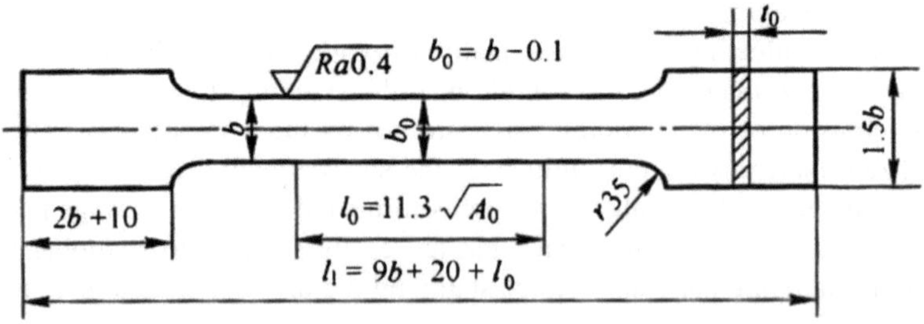

Fig. 1.3 Standard tensile specimen

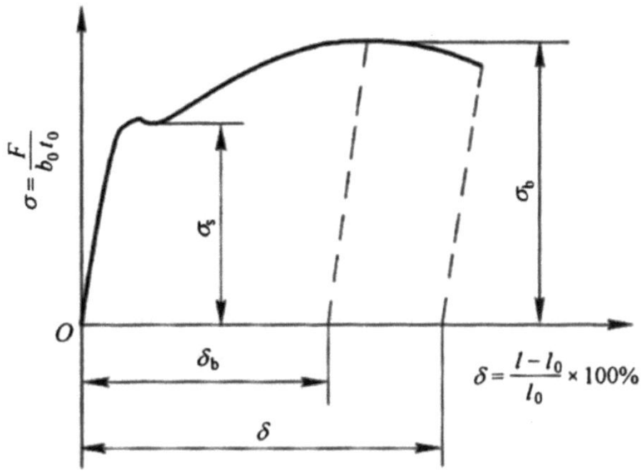

Fig. 1.4 Stress and strain curve

an indicator of the capacity of producing uniform or stable deformation. In general, sheet metal forming is carried out within the scope of uniform deformation, so δ_b affects the sheet metal forming more directly. In the process of tensile deformation, such as hole flanging, bulging, and other processes, the larger δ_b is, the greater the limit deformation is.

(2) Yield ratio (σ_s/σ_b)

σ_s/σ_b, called yield ratio, is the ratio of material yield limit to ultimate strength ratio. Small yield ratio means a big difference between σ_b and σ_s. The material with this yield ratio can withstand a large plastic deformation without rupture, which is beneficial to stamping forming.

In the process of the compression deformation, such as deep drawing, small yield ratio and low material yield point descend the tangential stress of the deformation zone, the trend of sheet metal to wrinkling and instability, and the binder force to prevent binder from wrinkling and frictional force. Thus the total deformation force and the force transmission zone loads are decreased. The higher the ultimate strength is, the greater the bearing capacity of the power transmission area is. In a word, the small yield ratio is in favor of improving the forming limit.

In the process of the tensile deformation, such as bulging, the small yield ratio which means a great difference between the tensile force in plastic forming and that in blank rupture makes the stability of plastic deformation ascend and the probabilities of blank cracking during drawing and accompanying waste products descend.

(3) Elastic modulus E

Elastic modulus is the stiffness index of the material. The greater the elastic modulus is, the stronger the ability of resisting compression and instability in forming is, and the smaller the elastic recovery is after unloading. This is helpful to improve the dimensional accuracy of parts.

(4) Hardening exponent n

Hardening exponent n is the hardening strength of the material in cold plastic deformation. The material with higher n value has the greater hardening effect, which is beneficial to tensile deformation. When the n value increases, the local material deformation will increase the deformation resistance. This compensates for the weakening of the carrying capacity caused by the decrease of cross-sectional area, which thereby prevents the local concentration of deformation from further developing, and makes contributions to expand the uniform deformation zone and make deformation uniform and increase deformation limit. It is proved that [9] the value of material hardening exponent n is the necking strain ε_j. So the higher the hardening exponent n is, the stronger the deformation capacity is.

(5) Thickness direction coefficient r

Thickness direction coefficient r is the ratio of width strain to thickness strain in the specimen sheet uniaxial tension (also known as the plastic strain ratio),

$$r = \frac{\varepsilon_b}{\varepsilon_t} = \frac{\ln \frac{b}{b_0}}{\ln \frac{t}{t_0}}$$

where b_0, b, t_0, and t are, respectively, the width and thickness of the specimen before and after deformation. In general, r value is calculated by the results measured under the elongation rate 20 % of the sample.

The value of r reflects the comparison of the difficulty level of the deformation between the plane direction and thickness direction. When the value of r is greater, the deformation in the plane direction occurs more easily, and the deformation in the thickness direction occurs more hardly. This is of great benefit to stretch forming. For example, for stretch forming of curved surface part, the deformation of the middle part of the sheet in the thickness direction is more difficult under the tensile stress. That is to say the degree of thinning is small. While the compression in the direction perpendicular to tensile stress is easier inside the plate plane, the middle part of the sheet is less likely to wrinkle. It is conductive to stretch smoothly and improve the quality of the workpiece. Similarly, with a large r of the sheet for cylindrical part drawing, under the function of the tensile stress the cylinder wall is not easy to thin, not easy to be broken down. The tangential compression of the flange area is easy, wrinkling trend reduces, binder force decreases. This inversely reduces the tensile stress in the cylinder wall, and increases deformation of drawing limit in the cylindrical pieces.

All the sheet metals applied in the stamping process are rolled materials. Due to the influence of fibrous tissue, there is a significant difference in the performance of its vertical and horizontal properties. The value of r is different in different directions, so the mean value $\bar{r}$ is often used.

$$\overline{r} = \frac{r_0 + r_{90} + 2r_{45}}{4}$$

where r_0, r_{90}, and r_{45} represent the thickness direction coefficients of vertical, horizontal, and 45° respectively.

(6) Plane orientation

The anisotropy, presented in the mechanical and physical properties of the sheet after rolling is called plane orientation. The more obvious the anisotropy is, the greater influence will perform on the formability. As for bending, when the bending line of the curved pieces is perpendicular to the fiber direction of the sheet, the allowed deformation limit is greater. But when the bending line is parallel to the fiber direction, the allowed deformation limit is smaller. For another instance of cylinder drawing parts, the mouth of the drawing parts may become earrings for the anisotropy. The more obvious the anisotropy is, the greater the height of the earrings is.

Plane orientation mainly performs as the mechanical property's distinction in different directions of the plate. In the indexes which express the mechanical properties of the sheet, the thickness directivity coefficient has an obvious impact on the stamping performance. The plane orientation is generally measured by the mean difference of $\boldsymbol{r}$ in specific directions, which is given as

$$\Delta r = \frac{r_0 + r_{90} - 2r_{45}}{2}$$

As the plane orientation is disadvantageous to the stamping deformation and quality of finish products, it is often required to reduce the value of Δr in the production.

1.4 Summary

Based on the introduction of a basic concept in sheet metal stamping forming, this chapter analyzes the forming thinning in traditional cold stamping technology and the difficulty for high strength steel to form in room temperature, and then points out the necessity and feasibility for high strength steel to form into auto parts by adopting hot stamping technology. This chapter introduces the basic knowledge on forming process of sheet metal stamping, including stamping process procedure, stamping tools, stamping press, production process of stampings, etc. Subsequently, the requirements of material property for stamping process are discussed, including the formability, thickness tolerance, and surface quality of sheet metal. Finally, the evaluation index for stamping performance and the test method for corresponding index are introduced, too, which lay the foundation for the subsequent hot stamping technology.

References

1. Hongyuan Y (2013) Research on hot forming process improvement of automobile high strength steel and its application. Dalian University of Technology, p 72 (in Chinese)
2. Hu P, Ying L, Guo R et al (2012) Development and independent R&D of hot stamping automobile high strength steel. Forging Meltalforming (in Chinese)
3. Kezhi W, Chen S (2006) Achievements of metal plastic working in ancient China. J Plast Eng 6:114–125 (in Chinese)
4. Liang Y (2013) Research and application on key process experiment of high strength steel for hot forming. Dalian University of Technology, p 179 (in Chinese)
5. Ma N (2011) Research on hot forming of high strength steel. Dalian University of Technology, p 224 (in Chinese)
6. Qi J (2005) Analyses of the development dynamic of sheet metal forming industry in China. Machinist Metal Form 1:85–87 (in Chinese)
7. Society TCMESSE (2008) Forging workshop equipment. In: Forging manual, volume 3, 3rd edn. China Machine Press, Beijing
8. Xianbin Z, Yan Z (2005) Development of Chinese stamping industry. China Metalforming Equipment Manuf Technol 1:10–16 (In Chinese)
9. Xuewei S, Ling Y, Sun J et al (1995) A method of determining strain-hardening exponents. J Mech Strength 4:27–28 (In Chinese)
10. Yongzheng S, Qi J (2009) The development of the metal forming industry in China 1. China Metalforming Equip Manuf Technol 2:13–19 (in Chinese)
11. Yumin Z (2003) Five stages in the development of stamping technology in developed countries in 20th century. Mach Electr New Prod Guide 3:48–50 (In Chinese)

Chapter 2
Hot Stamping Technology and the Main Equipment

2.1 The Hot Stamping Technology of High Strength Steel

2.1.1 Brief Introduction of Hot Stamping Technology

In recent years, automotive lightweight has led to the increasing demand of high strength steel (HSS) auto parts. Lots of correlative companies and research institutes in America, Europe, Japan, and China have invested large amount of energy into the research of advanced automotive HSS forming technology. Though the advantage of HSS is high strength, it still has some disadvantages such as the poor forming performance, the uncontrollable springback, and easy to crack during forming process. Therefore, it is difficult to realize the manufacture of complex auto parts by adopting traditional cold stamping technology, where the hot stamping technology is needed.

HSS hot stamping technology is a new manufacturing technology developing in recent years, which combines the traditional hot forging and cold stamping technology. It is a mode of production that integrates the stamping of steel under the condition of high temperature and forming and quenching in dies. It is also known as Hot Forming, Hot Stamping, Hot Press, Press Hardening, or Die Quenching [51] abroad. As early as in the middle of the last century, the HSS hot stamping technology research had appeared in aviation and nuclear industry, such as the United States NASA. Norrbottens Jernverk exploited the hot stamping technology in 1973, which was suitable for automobile parts manufacturing. Volvo Car Company did research on the applicability of hot stamping parts in one type of its cars, and since 2000, more hot stamping parts were used in automobile body manufacturing. The total quantity had reached more than 100 million in 2007. It was from 1990s that the world's largest steelmaker, Arcelor Mittal company, began to make a study of hot stamping technology industrialization and developed the famous Al–Si coating hot stamping steel USIBOR1500 and retrofit 1500p series [21, 39, 42]. At present,

P. Hu et al., *Hot Stamping Advanced Manufacturing Technology of Lightweight Car Body*, DOI 10.1007/978-981-10-2401-6_2

the HSS hot stamping technology has become a hot topic which many car manufacturers pay attention to around the world [45].

As shown in Fig. 2.1, hot stamping technology of HSS can be used to manufacture car structural components whose tensile strength is up to 1500 Mpa, such as the front bumper, tail bumper, A column, B column, C column, roof frame, floor frame, door panel, door anti-beam, and so on [35, 57]. The new China automotive lightweight technology innovation union work conference, held in Shanghai in May 2013, pointed out that China's auto market maintained steady growth in 2012, and the average monthly production and sales were totally more than 1.5 million vehicles. Annual cumulative production and sales were more than 19.2718 million vehicles. Hot stamping parts, as a necessary choice of advanced bodywork, have become a specification. New models, especially SUV and b-class more luxury cars, gradually raised in the market share, which has huge demand for hot stamping parts, and the market demand of hot stamping production equipment and production lines also arise at the historic moment.

According to the related information of EuroCarbody from 2008 to 2012 [55], represented by Audi, international auto giants had used hot stamping parts as a mainstream configuration of lightweight body structure and security design, which were applied widely in the automobile body design and manufacturing. The use ratio of hot stamping parts around the world auto enterprises had greatly increased from 4–15 to 10–30 %: the number of hot stamping parts used in advanced cars in Land Rover and Audi had reached or was close to 25, and the proportion accounted for more than 20 % of the total number of body in white; FIAT plans to use hot stamping parts in the subsequent models, and the proportion is more than 16 %; there are about 6–10 % usage rate of hot stamping parts in Ford and Daimler-Chrysler's. There are totally 165 body parts on the first five-star collision protection utility vehicle, Ford wind star, while nearly 100 of the parts are made with hot stamping of HSS and the proportion is more than 60 %; the usage ratio of hot stamping of boron steel in VOLVO cars has gradually risen from 7 % of XC90

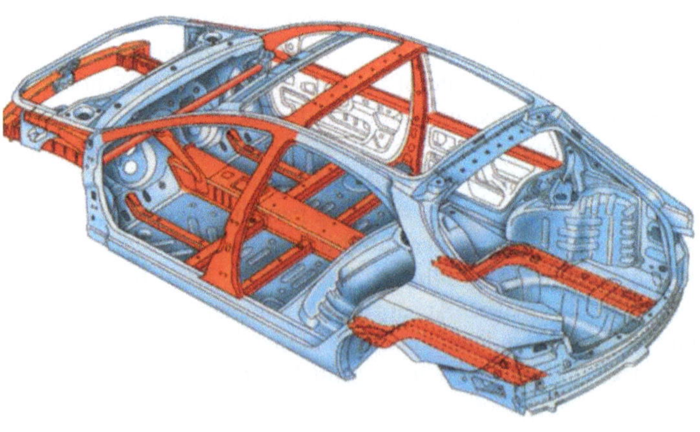

Fig. 2.1 Applications of hot stamping parts in auto BIW

models to 17 % of the S60 series, which is expected to reach 45 % of the total body in white in the future, as shown in Fig. 2.2.

Along with the integration of Chinese automobile manufactures and the international auto industries, the application of hot stamping products will be gradually improved among domestic automobile manufacturers [13, 50]. Hot stamping products are used widely in body structure, which will enhance the safety of the car and greatly promote lightweight prospects of Chinese independent automobile brands. In order to improve their product's competitive advantage within the industry both at domestic and overseas market and strengthen the lightweight and safety performance of the models, many auto private enterprises, such as Great wall, Geely, Changan, Chery, etc., are actively seeking superior cost-effective hot stamping parts suppliers and opportunities, so that it will be the trend of future development to apply hot stamping auto parts widely in automobile body manufacturing.

2.1.2 Hot Stamping Process

Hot stamping process is the necessary means to realize the product forming and strength increasing of hot stamping steel, and it is the indispensable premise for HSS of hot stamping to acquire high strength performance. Its typical technological process is: first, heat the sheet to a specific temperature range for austenization and austenitize it completely, and then transfer it into the water cooling hot stamping dies for stamping and quenching, and finally realize the microstructure's phase transformation and the increase of strength. According to the complexity of

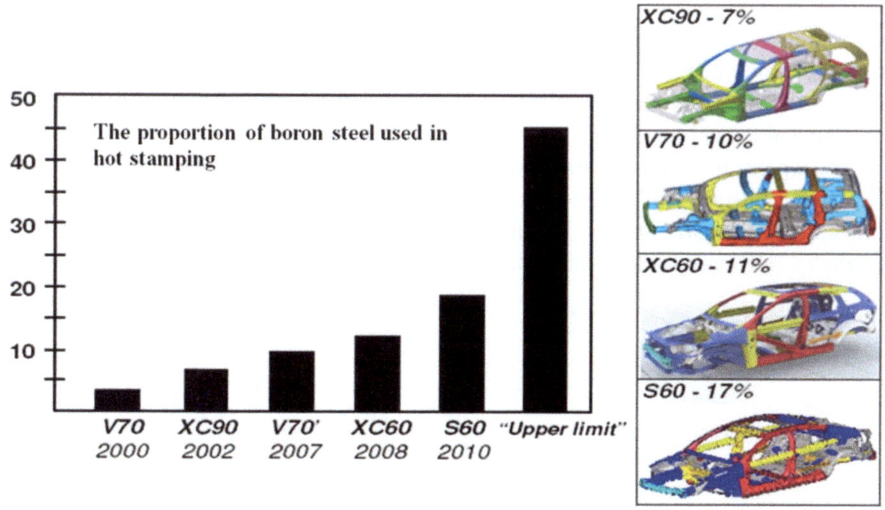

Fig. 2.2 The proportion of hot stamping boron steel applied in VOLVO vehicles

processing and the forming process of actual parts, hot stamping process can be divided into directly hot stamping and indirectly hot stamping process.

1. Direct hot stamping process

In the direct hot stamping process, a blank is heated up in a furnace, transferred to the press and subsequently formed and quenched in the closed dies [6, 10, 19, 29]. As shown in Fig. 2.3, the steel plate is first uncoiled and cut according to the shape of the product, and then the blank is transferred to a continuos furnace, in which it is heated and fully austenitized. Thus, the product is formed and quenched after the blank is transferred to the hot stamping dies with cooling system. Afterwards, the product is trimmed by laser and finished through other follow-up processes.

The advantages of direct hot stamping process are as follows:

(1) The blank is formed and hardened in one mold, which saves the cost of preforming and accelerates the pace of production.
(2) The blank is flat, which not only saves heating area and energy, but also can be heated by a variety of heating methods, for example, induction heating.

The disadvantages of hot stamping process are as follows: it cannot be used for forming automobile parts with complex shapes, and it needs the laser cutting equipment. In addition, the design of cooling system of molds is more complex.

For automotive body structure, the parts with simple shape, not necessary for deep drawing, can be manufactured by the direct hot stamping, such as the inside and outside pieces of b-pillar, the inner plate of side panels, the inner plate of the threshold, the central pillar of front bezel and door beam, and so on (Fig. 2.4) [19, 40].

Figure 2.5 shows the hot stamping tools for an automobile front fender center pillar and the forming process. The shape of this center pillar is simple and the drawing depth is small, so it can be manufactured by the direct hot stamping process. Put the original blank into furnace, heat to 950 °C and fully austenitize for 5 min, then quickly transferred it into the water cooling hot stamping tools (Fig. 3.3) to form it and quench it. The finally obtained hot stamping parts are shown in Fig. 2. 6. In order to verify the feasibility of forming process, the microstructure and the mechanical properties for hot stamping parts were tested.

Six typical test samples in different regions of the center pillar are selected for hardness measurement and metallographic observation, as shown in Fig. 2.7. From the hardness results shown in Table 2.1, it can be found that all the samples' hardness (HR) are higher than 47, far more than that of the original steel, which

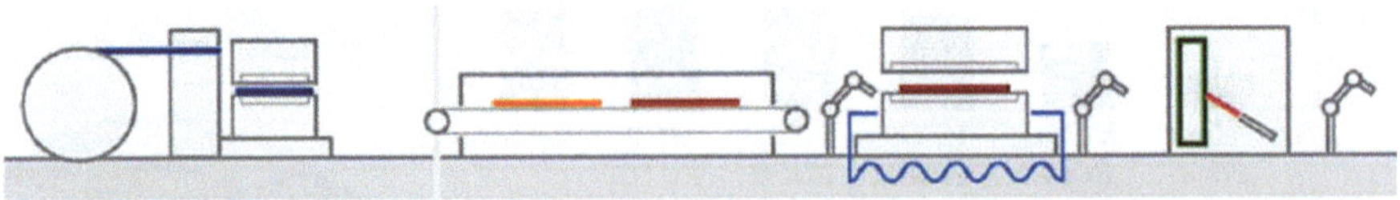

Fig. 2.3 Sketch map of direct hot stamping process

Fig. 2.4 Parts with the direct method of hot stamping

Fig. 2.5 Hot stamping die of brace dase panel, CTR and its process

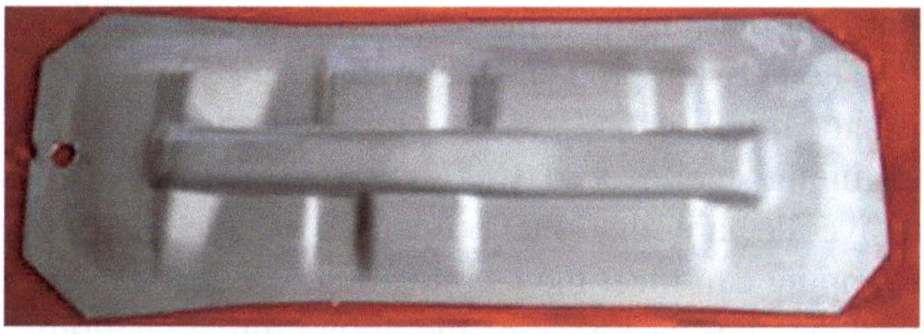

Fig. 2.6 Brace dase panel, CTR with one-step method of hot stamping

indicates a microstructure transformation in the steel part during hot stamping process. The metallographic results illustrated in Fig. 2.8 shows that the microstructure of original steel is mainly mixture of ferritic and pearlitic, together

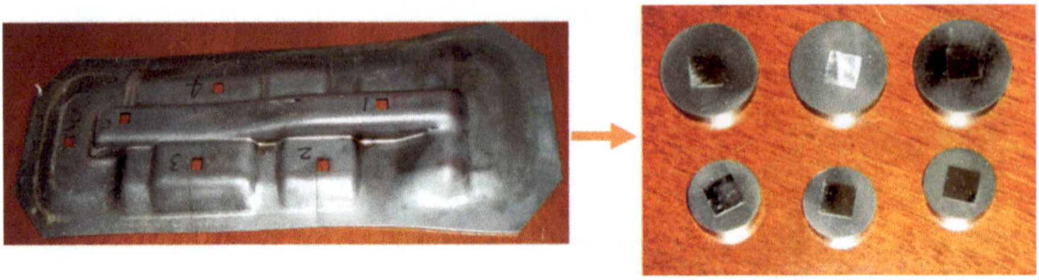

Fig. 2.7 Testing samples for microstructure and hardness of the hot stamping part

Table 2.1 Distribution of rockwell hardness (HRC)

Sample 1	Sample 2	Sample 3	Sample 4	Sample 5	Sample 6
47	48.2	49.8	49.5	50	48.7

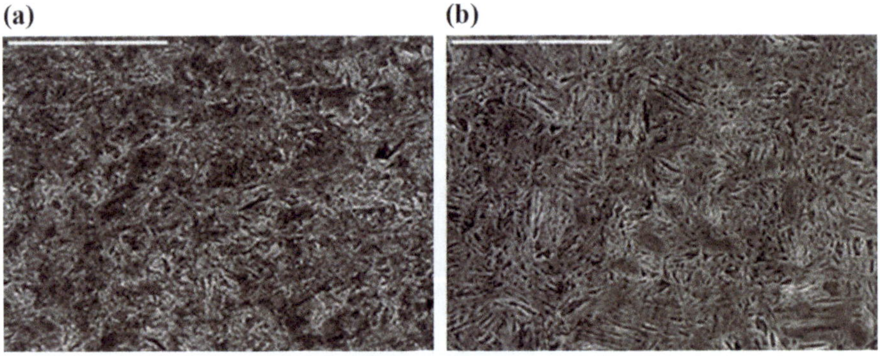

Fig. 2.8 Microstructure of the material before hot stamping and after hot stamping. **a** Microstructure of the raw material. **b** Microstructure of the hot forming product

with a bit of carbide. While the microstructure of steel after hot stamping is the uniform martensite, which has a martensite content of more than 95 %.

The hot stamping steel's engineering stress–strain curve obtained by tensile test is shown in Fig. 2.9, from which we know that the yield and tensile strength of hot stamping steel exceed 1000 and 1600 MPa, respectively. All the results illustrated above declare that the properties of the central pillar produced by direct hot stamping can meet the technical requirements of hot stamping [26, 41]. And it also declares the feasibility of the above direct stamping process.

2. Indirect hot stamping process

In the indirect hot stamping process, the sheet metal is preformed by cold forming before it is heated to austenite temperature in a furnace. After being held for a period of time for full austenization, it will be transferred into the hot stamping tools with cooling system to be stamped and quenched at the same time [29]. The indirect

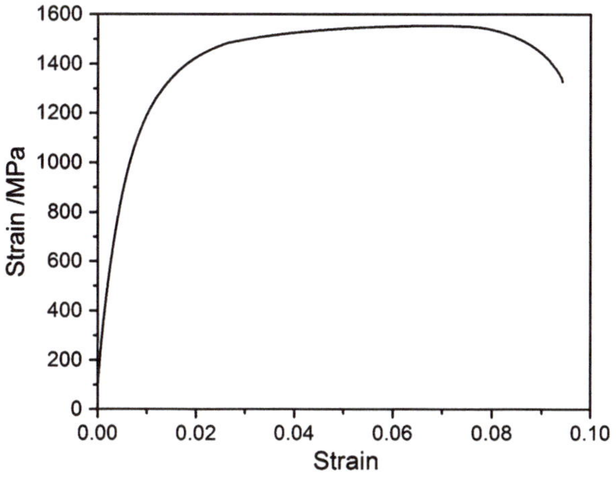

Fig. 2.9 Engineering stress–strain curve of the front fender center pillar

hot stamping process is also named as "multi-step" hot stamping as shown in Fig. 2.10. Compared to the cold stamping steel, though the flowability of the heated steel increases, the tensile strength would decrease accordingly, making it easy to crack for complex part during forming process. The indirect hot stamping process is the exact method invented to solve the problem and form complex auto parts that may not be formed by the direct hot stamping process.

Compared to the direct hot stamping, the HSS sheet is uncoiled and cut according to the shape of the product, then preformed by the traditional process such as cold forming process, flanging, punching and cutting-edge, and so on. After that, the preformed semi-finished product is transferred to the continuous furnace to be heated and insulated before it is sent to be formed and quenched. Afterwards the product will be trimmed by laser processing or other necessary follow-up processes according to the characteristics of the components, or directly output finished product.

For automotive body structure, the parts that have complex shape or need deep drawing or punching, trimming or other complex technology must be manufactured by the indirect hot stamping process, such as the inside and outside pieces of B-pillar, the inner plate of side panels, the central pillar of front fender and door beam, etc. (Fig. 2.11).

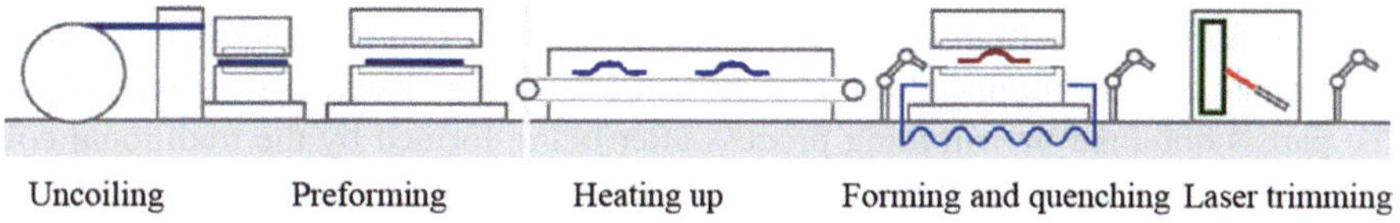

Fig. 2.10 Sketch map of hot stamping process with indirect method of hot stamping

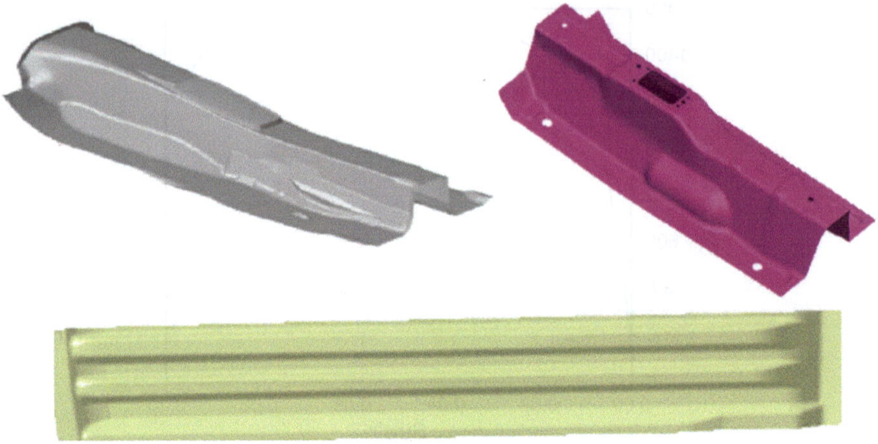

Fig. 2.11 Parts produced by indirect hot stamping technique

The advantages of indirect hot stamping process are as follows:

(1) The parts with complex shapes and almost all of the current stamped carrying parts can be formed by indirect hot stamping process.
(2) After the preforming of the blank, it is unnecessary to worry about the forming performance of the blank at high temperature in subsequent hot stamping process, which can ensure the martensite microstructure of the blank followed by complete quenching.
(3) The blank can be processed by trimming, flanging, punching and other processing after being preformed so that it will be easier for processing after it is quenched. For example, the blank that is quenched must be trimmed by laser cutting equipment, which greatly increases the cost.

The enforced beam shown in Fig. 2.11 is applied to the hot stamping experiment. First, the direct hot stamping process is used and the formed part is shown in Fig. 2.12. It is obvious to find that the cracks happened at both ends of the beam. This is due to the fact that the enhanced beam has three U-shaped deep drawing areas, which greatly increase the difficulty of forming. Aimed at this kind of beam, the indirect hot stamping process is developed based on the numerical simulation [11] and two sets of dies are used to form the enhanced beam. The preforming die is shown in Fig. 2.13a and the dies for quenching, the productive process and the final formed product are illustrated in Fig. 2.13b.

First, according to the shape of the part, the blank size is got in software by inverse forming algorithm before the blank is uncoiled and cut, and then the blank is sent into the preforming dies, as shown in Fig. 2.13a. The required geometry of the part is obtained by trimming process after being formed by the traditional cold

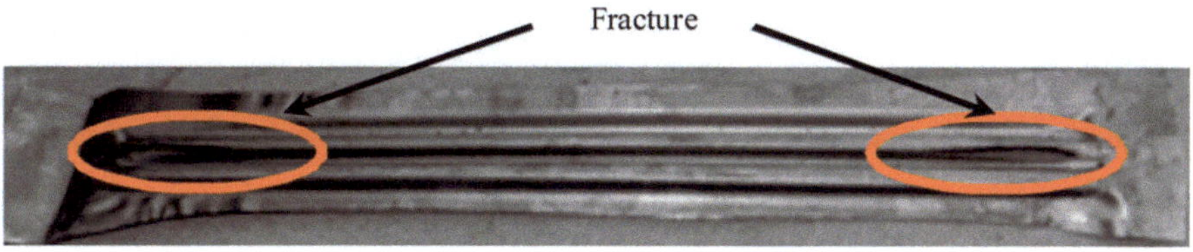

Fig. 2.12 Reinforced beam with direct method of hot stamping

Fig. 2.13 Reinforced beam with direct method of hot stamping. **a** Preforming mold. **b** Final forming and quenching

forming. Afterwards, the preformed part is put into the furnance and fully austenitized at the temperature of 950 °C for 5 min before it is quickly sent into the dies with cooling system, as shown in Fig. 2.13b, for forming and quenching. The obtained hot stamping part is shown in Fig. 2.14. In order to verify the feasibility of forming process, microstructure and mechanical properties of some parts are tested. The results of tensile test for the samples cut from the enhanced beam are shown in Fig. 2.15. The stress–strain curve shows that the yield and tensile strength of the hot stamping part is more than 1000 and 1600 MPa, respectively. The above experimental results show that the enhanced door beam formed by indirect hot stamping meets the technical requirements of hot stamping [40, 41], which also proves the validity of the indirect hot stamping process and the design of hot stamping tools.

Fig. 2.14 Reinforced beam with multi-step method of hot stamping

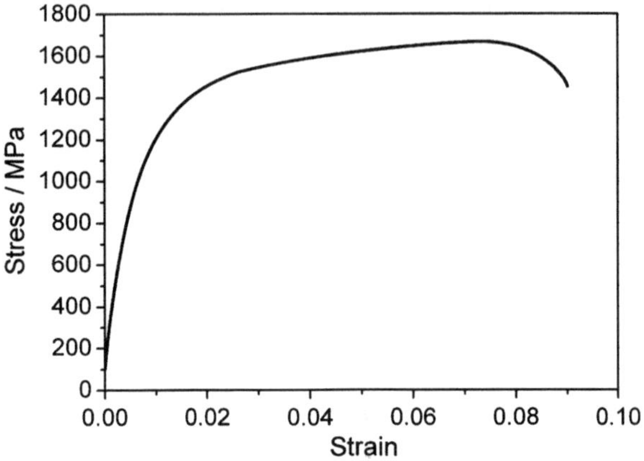

Fig. 2.15 Engineering stress–strain curve of the reinforced beam

2.1.3 Finite Element Simulation Analysis of Hot Stamping Technology

In the late 1980s, with the rapid development of computer technology and the maturity of finite element method, numerical simulation technology of sheet metal forming ushered in a vigorous development, which was driven by strong application demand in the world of automotive industry; and this technology is still in the ascendant.

There are three important signs of progress during this period

(1) The establishment of three-dimensional nonlinear shell theory made it possible to analyze the forming process of some complex parts, such as car body covers, the algorithm for contact and friction problems, took geometric nonlinearity into consideration;
(2) To promote the research of sheet metal forming simulation and investigate the reliability of numerical analysis algorithms, international research organizations have designed a series of standard questions, such as OSU (Ohio State University) standard questions, VDI (Verein Deutscher Ingenieure) standard questions, NUMISHEET numerical simulation of sheet metal forming standard questions. These questions aim to assess finite element softwares' performance of forecasting splitting, wrinkling, buckling and springback from different angles. The examination questions are related with simple parts and complex panels;

(3) A number of element finite element software has been developed and has taken certain standard questions previously mentioned. Some software, such as DYNAFORM, AUTOFORM, and PAMSTAMP, has already been widely applied in automotive industry. These software can solve high geometric nonlinearity, material nonlinearity, contact and friction problems occurred in forming process, so as to make predictions for wrinkling and fracture.

Great progress has been made in sheet metal forming simulation in recent years. Many international companies, especially automobile manufacturers, have established relevant sheet metal forming simulation systems to provide strong support for the stamping mold design, process design and mold test in terms of failure analysis and quality improvement. The well-known automotive companies of United States and Germany have taken stamping formability analysis as a necessary step in the development process, while automobile companies of Japan conducted formability analysis selectively.

However, in hot stamping, due to the effect of temperature, the material's flow stress and forming characteristics in the forming process will change constantly, making the finite element simulation analysis of the forming process much more complicated. Actually, hot stamping process of HSS sheet is a complex process with the interaction of the temperature field, stress field, and phase transformation field. First of all, most thermal and mechanical properties of sheet metal sheet such as the thermal conductivity, specific heat, elastic modulus, the flow rule, etc., depend on its temperature. Second, the thermal stress will produce due to the uneven temperature distribution at internal and external surface of sheet metal during cooling process. And large deformation in forming process can produce plastic work, most of which converted into heat energy and affect the temperature field distribution in turn. Under the effect of stress, the parameters such as initial temperature, finish temperature, and the rate of phase transformation are changed producing the so-called phase transformation plasticity at the same time. The microstructure changes in the process of phase transformation and influences the thermal and mechanical properties and formability. Volume expansion happens when the austenite changes into martensite or bainite, but uneven temperature distribution would lead to the difference in amount of phase transformation and different expansion rate, which can also produce phase transformation stress. During the quenching process, latent heat is released when the microstructure changes from austenite to martensite or bainite, which will influence the temperature field distribution. The start temperature of phase transformation and the final phase transformation amount is decided by both sheet metal temperature and cooling rate, while strain rate may also affect the thermodynamic properties and phase transformation field. Thus, the key to the finite element simulation analysis of hot stamping is the above field coupling analysis.

At present, although the generic software for sheet forming, such as Autoform, Dynaform and Pamstamp, have attempted to add hot stamping module, its commercial background makes them pay more attention to the calculation efficiency and the simplification degree in multi-field coupling hot stamping simulation model is relatively bigger, which cannot meet the demand of comprehensive simulation study for multi-field coupling of hot stamping.

Research team of Professor Ping Hu from Dalian university of technology has taken more than 10 years to develop the KMAS (King-Mesh Analysis System) software system, which is a piece of high and new technology software with fully independent intellectual property rights. The KMAS was commercially available at the end of 2003 and successively provided technical support for China FAW, GM and other famous automobile and software enterprises, and became one of the global best partners of the UGS in United States. With the development of hot stamping technology, the application range of KMAS software system becomes much wider while the function to solve problem becomes stronger, too.

Based on the established constitutive equation coupling thermal, mechanical and phase transformation of HSS hot stamping, editors of this book, together with the other members in hot stamping team, have considered the interface heat transfer between blank sheet and dies. By using the heat transfer theory, the general shell temperature field finite element analysis theory and three-dimensional tetrahedral finite element theory, they developed the algorithm program, perfecting the pre-processing module and the post-processing module of HF/KMAS. By analyzing the change law of temperature field for shell and 3D tetrahedron units, the numerical simulation can obtain reliable temperature field, which can be used for further analyzing the coupling relationship of thermal, stress and phase transformation in hot stamping process to implement the simulation analysis of key variables as thermal, mechanical and phase transition of hot stamping in KMAS.

2.1.4 The Research Status of Hot Stamping

The material required by hot stamping is a special boron steel with high strength, which can be divided into coating hot stamping sheet steel and non-coating hot stamping sheet steel. The world's largest steelmaker Arcelor Group has developed the hot stamping sheet Usibor 1500 [1, 2, 5, 43] and then made it into mass production. Besides, this company also possesses the production patent of Al–Si coated hot stamping steel, whose characteristic is that the material organization is a uniform mixture of ferrite and pearlite after rolling forming, with a yield strength of 280–400 MPa and tensile strength over 450 MPa. After hot stamping, the organization transforms into uniform martensite with a yield strength of 1200 MPa and tensile strength 1600 MPa (3–4 times larger than ordinary steel's strength). Sweden's SSAB has developed and then produced hot-rolled quenched boron steel of Domex series, including 20MnB5, 27MnCrB5, 30MnB5, 33MnCrB5, 38MnB5, etc. The thickness, width and length of these plates are 1.80–12.0 mm, plate width 800–1600 mm and plate length 1500–13000 mm, respectively. After hot-rolling, its yield strength reaches 400 MPa, and its tensile strength reaches 600 MPa. After heating, the tensile strength of water quenching reaches 1480–2050 MPa, while the tensile strength of oil quenching reaches 1360–1845 MPa. In addition, Japan's

Nippon Steel and Kobe Steel, South Korea's Pohang Iron and Steel and other companies are able to produce hot stamping micro-alloy steel in batch production [38]. At present, all these major steel mills are making great efforts to develop high strength hot stamping steel with better performance.

China's Baosteel Group has developed and produced two types of hot stamping boron steel, which are cold rolled B1500HS and hot-rolled BR1500HS, and realized mass supply. It is China's largest supplier of hot stamping steel at present [52]. In addition, Anshan Iron and Steel, Tonghua Steel, and other steel companies have also been developing smelting and rolling process of hot stamping steel that can supply mass production.

Hot stamping technology of HSS was first used in the automotive industry in Europe. The application prospects and the complexity of research in hot stamping technology attracted the attention of scholars in the related fields. Thus, the hot stamping seminar is held during the international famous conference on Metal Forming after 2008, such as IDDRG2009, NUMIFORM2010, and METALFORMING2010 [41, 46]. Currently, three major research echelons [25] in the hot stamping technology of HSS have been formed according to the research time sequence. The first echelon is the European team represented by Germany and Sweden. The second is China, Japan and South Korea while the third is the United States and other countries.

Classified by research groups and research institutes, the main representatives are Lulea University of Technology in Sweden, University of Nuremberg in Germany, Technical University of MUNCHEN in Germany, University of Padua in Italy, Yokohama National University and Toyohashi University of technology in Japan. In addition, the Iran Arak University has also systematically studied the hot stamping principle and process for HSS plate [4, 20, 36, 37, 42, 44].

In China, there are a number of universities and research institutes, such as Dalian University of Technology, Jilin University, Tongji University, Shandong University, Shanghai Jiaotong University, Harbin Institute of Technology, Iron and Steel Research Institute, Baosteel Research Institute and Chongqing Automobile Research Institute, which have paid much attention to the study and development of hot stamping technology. According to their own actual situation and the relevant project from national ministries or enterprises, the above units are actively involved in correlational study of constitutive relation of hot stamping material, rolling experiment, stamping process, the simulation analysis, analysis of manufacturability, etc. [3, 7, 23, 24, 56, 57].

As the developer of domestic first hot stamping batch production line [13, 28, 45] with fully independent intellectual property rights, the research team of professor Hu Ping in Dalian University of Technology has conducted systematic basic experiment and simulation research on hot stamping technology about the thermodynamics constitutive relation of material, heat transfer analysis, process improvement, tool optimization design and manufacturing process [3, 8, 9, 17, 22, 53], and successfully established the constitutive relation coupling the thermal, stress and phase transformation [11, 27, 30–34, 49] for HSS hot stamping. In addition, they have also successfully developed the hot stamping simulation CAE module (KMAS/HF) (King-Mesh Analysis System/Hot Forming) [12, 14], which can accurately simulate the formability of steel plate at elevated temperature, simulate and analyze the features of temperature field, stress field and phase transformation field in hot stamping process and be used for optimization design and manufacture of water cooling tools. They further investigated the basis about manufacturing of hot stamping hardness gradient composite material and parts and realized the tool design and the analysis of manufacturing process for hardness gradient hot stamping [18, 47, 49]. They put forward the method and main technical points suitable for the design of hot stamping water cooling tools [15, 54], and have successfully manufactured several tool sets, which have been applied to the products such as door beam, bumper, and the inner plate of B-pillar [16, 48]. Years of technological research and study have forged their leading position in the integration of industry-university-research-application for the hot stamping technology in China.

2.2 Hot Stamping Production Lines and the Key Equipments

Similar with the cold stamping parts, the acquisition of continuous and large quantity of hot stamping parts also need automatic production lines. However, the mechanical properties of boron steel sheet and the total hot stamping process are closely related to the temperature, resulting in the different requirements for the equipment on production line.

As shown in Fig. 2.16, hot stamping production line is comprised of such devices as anti-oxidation continuous ring heating furnace, fast transmitting device, high-speed hydraulic press, water cooling hot stamping die, laser cutting and shot blasting, etc. With the design and optimization for production line system based on virtual reality technique, the above-mentioned key equipment can be integrated into automatic production line with central control.

Due to technical difficulties and technology monopoly and blockade from foreign enterprise, China did not have the ability to develop the hot stamping Continuous

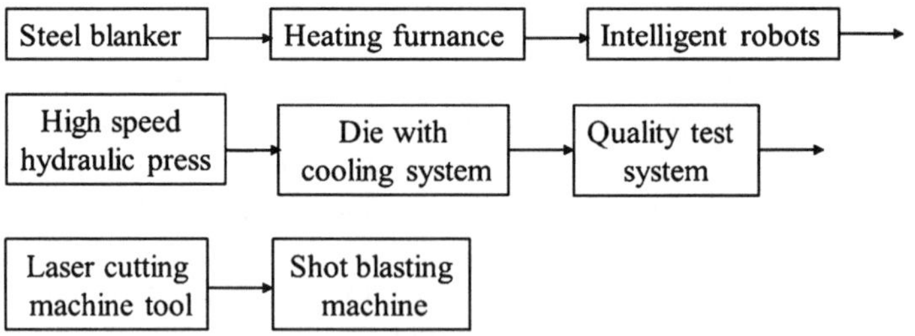

Fig. 2.16 Sketch of production line for hot stamping

ring heating furnace for long-term until JiLin VAFT auto parts co., LTD cooperated with Dalian University of Technology. After several years of continuous technological research, they developed the energy-saving heating system with completely independent intellectual property rights in china. Different from the tunnel furnace adopted by foreign production line, this furnace covers an area of only a quarter of the tunnel furnace and can save almost 30 % of energy from consideration of reducing cooling energy consumption and power-on and power-off consumption. Controlled by the servo device, it is able to ensure the furnace operating stably for a long term, and sheet metal can be heated to realize austenization in furnace stably. China's first hot stamping production line in VAFT with completely independent property right is shown in Fig. 2.17. It is mainly composed of high-speed hydraulic press, industrial Continuous ring heating furnace, water cooling tools, transport agencies, high-speed manipulator and automation control system, etc. This section will make a brief introduction of the key equipment and technology in hot stamping based on the independent developed hot stamping production line [6, 10, 19, 29].

2.2.1 Continuous Ring Heating Furnace

The continuous ring heating furnace used in the hot stamping should ensure that the blank is heated up to the setting temperature for a complete austenitization. At the same time, the high-temperature oxidation and decarburization should be avoided for those blanks without anti-oxidation coating. The continuous ring heating furnace has its unique core technology in comparison with other heating furnaces, and its brief descriptions are as follows:

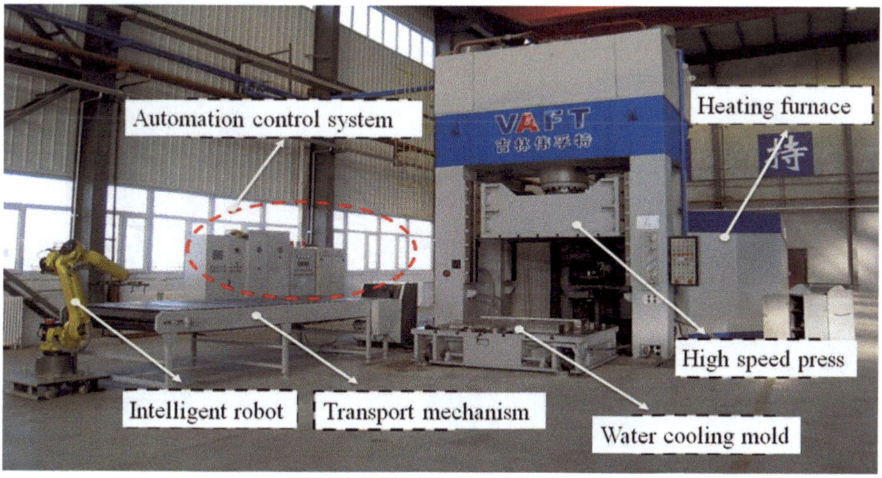

Fig. 2.17 The first domestic hot stamping automatic production line

(1) Most of the automotive HSS blank is sheet. The sheet should be heated up to the specified temperature and then hold heat in order to be fully austenitized. Therefore, the temperature distribution in the heating furnace is required to be homogenous;
(2) For the steel sheet without protective coating from oxidation, its oxidation and decarburization is very serious at elevated temperatures. Therefore, the furnace is needed to take strict measures of gas protection;
(3) All the parts of the heating furnace run under the conditions of long term and continuous operation at a high temperature (over 800 °C), so their expansion property, high temperature fatigue, failure rate and maintainability must be taken into consideration;
(4) Hot stamping production line needs to satisfy the setting cycle time, so it needs the equipment for loading or unloading which can meet the requirements for the high-temperature work and corresponding automatic control system;
(5) One of the purposes of hot stamping technique is to save energy and protect environment, thus it also requires energy-efficient heating furnace.

Figure 2.18 shows a self-developed energy-efficient continuous ring heating furnace, namely ring rotary hearth heating furnace [14]. The core techniques of ring rotary hearth heating furnace are the deformation of the moving spindle with resistance to high temperature, moving bias control, thermal fatigue durability, etc. By using infrared temperature measurement technology and thermocouple temperature measurement technology, the temperature distribution inside the furnace under the no-load and full load condition is monitored, which is used for further optimal design to achieve the temperature control and monitoring system inside the

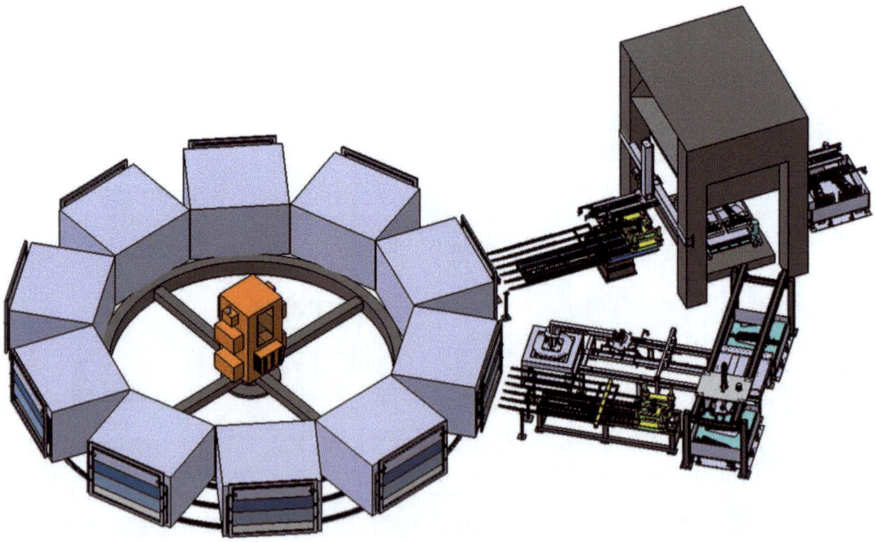

Fig. 2.18 Continuous ring heating furnace for hot stamping

heating furnace. In order to ensure the reliability of furnace, appropriate materials are required to use to produce the core spindle and blade of the converter. Furthermore, the high-temperature resistant property and thermal fatigue property of main heat resistant furnace parts should be tested and validated. Besides, the reliability and precision of the core spindle's continuous operating should also be tested and validated. In order to prevent sheet from oxidation and decarburization at elevated temperature, this continuous ring heating furnace is equipped with anti-oxidation gas-filled device of controlled flow and pressure.

2.2.2 *High-Temperature Resistant Robot Arm and Automatic Transfer Device for Loading and Unloading*

During the forming process, the blank is heated to be austenitized in a heating furnace at a temperature ranging from 800 to 1000 °C. And then high-temperature manipulator is used to remove the blank from the furnace and put it into die for forming and quenching, as is shown in Fig. 2.19. In this process, the wireless temperature testing sensor is placed on the robot arm in order to make online detection of the temperature changes of sheet transfer. Then the rotating speed of the robot arm is controlled according to the temperature information collected by the sensors. The information about temperature is transferred to the total console through the data lines, which can ensure the sheet temperature would not decline out of the prescribed limits in the transfer process of high-temperature sheet. The main structure of the jaw is clamp-like attached with reticulation to avoid excessive force and the drop of the sheet. The high-temperature position sensors are placed on the arm of robot to find accurate position in order to guarantee the product quality.

Fig. 2.19 Intelligent robot arm with high temperature resistance

The design and manufacture, sensing technology, automatic control technology and reliability of high-speed intelligent transfer device (including the high-temperature resistant tooling) are the key techniques which are needed to ensure that high-temperature steel sheet can be put into shaping dies rapidly and accurately, and that the stamping parts can be removed from the dies after quenching, in the hot stamping production line. As shown in Fig. 2.20, after automatic feeding and austenitic transformation in the heating furnace, the sheet is transferred to the specified position rapidly and accurately by the automatic unloading device to ensure the accurate subsequent operations of the robot arm.

High-temperature resistant gripper is designed and manufactured for automotive structural parts of different sizes and shapes. Cooperating with the high-speed intelligent transfer device, the gripper completes the grab and delivery of hot sheet under the precondition of process requirements. In order to achieve the reliable mass production, the control system of high-speed intelligent transfer device and the temperature monitoring device attached on transfer device are set up in the central control system. This temperature monitoring device and die temperature monitoring device are combined together to form a temperature collection system in the production line. Thus, the temperature parameters in production are recorded, which is helpful to optimize the process, predict the property of the steel sheet and die fatigue, etc., according to the actual production data.

2.2.3 *Key Technologies for Design and Manufacture of Hot Stamping Dies*

After the steel plate is heated in the furnace, its forming into expected shape and strengthening simultaneously must rely on the dies with cooling system. The key technologies for design and manufacture of hot stamping dies are

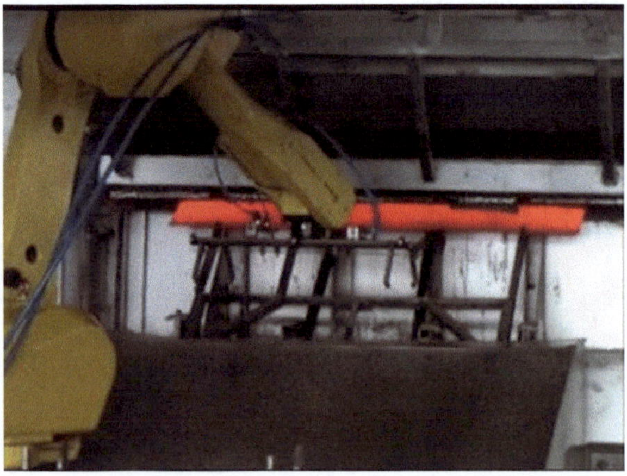

Fig. 2.20 Feeding by automatic loading/unloading device after transfer process

(1) After the steel plate is heated, its ductility will increase, while its strength will decrease. In order to avoid cracking and wrinkling of sheet, the die clearance and its accuracy of manufacture should be arranged reasonably;
(2) In order to meet the martensitic transformation and the uniformity of microstructure distribution, the cooling ducts' design should be optimized to ensure that the sheet can be cooled uniformly and formed at a specified cooling rate;
(3) Hot stamping dies work in an environment where the temperatures are continuously changing between the high temperature and low. In this process the dies withstand expansion and contraction. Thus, the materials for dies and processing technologies need to be taken into consideration for a proper service life.

According to different performance requirements of the products, the layout of the dies with different pipes is shown in Fig. 2.21. Figure 2.21a shows a layout of uniform cooling pipes, in which the hot stamping parts can be obtained with uniform martensitic microstructure [40]. Figure 2.21b shows a layout with nonuniform cooling pipes, in which the microstructure of the parts obtained is segmented, continuous and mixed [26]. Figure 2.21c shows a layout of temperature variable district cooling tools with both cooling pipes and heating pipes, in which the hot

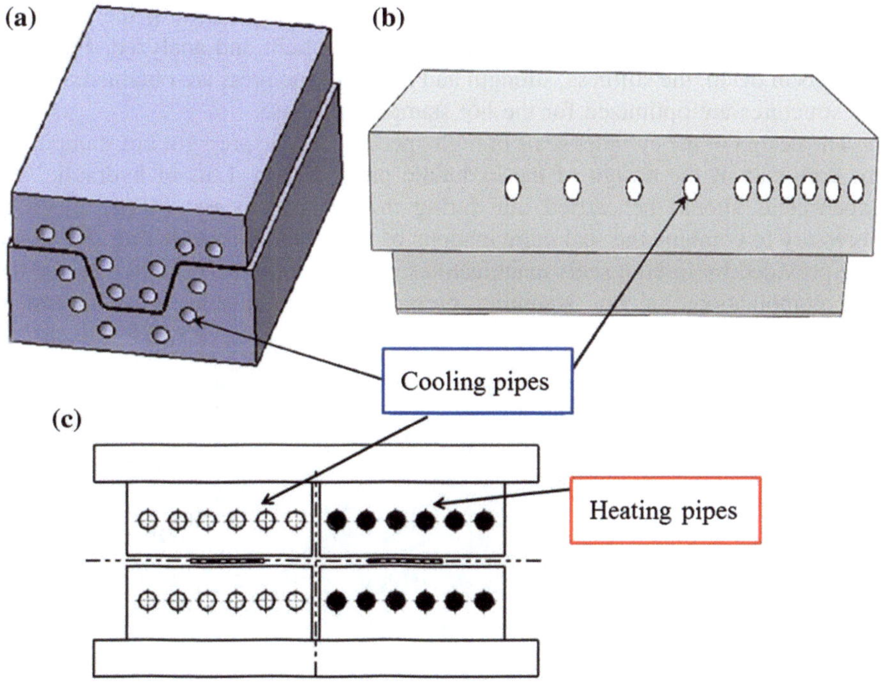

Fig. 2.21 Sketch map of die cooling pipes. **a** Layout of uniform cooling pipes. **b** Layout of non-uniform cooling pipes. **c** Layout of tools with both cooling pipes and heating pipes

stamping parts with feature of changing hybrid organization and continuous gradient hardness can be obtained.

Figure 2.22 shows a picture of die for experiments. This water cooling die made from special material possesses a cooling system, in which the pressure and the flow can be servo-controlled [41]. As the hot stamping die, especially its surface, experiences frequent hot and cold alternation during the stamping process, it is easy to result in fatigue failure in the process of mass production. Therefore, hot stamping die materials with high-temperature resistance, high wear resistance and high fatigue resistance should be developed on the basis of existing heat resistant die materials.

2.2.4 High-Speed Hydraulic Press for Hot Stamping

Hot stamping press should possess the functions of quick die closing, stamping and pressure holding. At present, the presses for cold stamping cannot be applied to hot stamping because the conventional hydraulic press has a low speed of die closing and the mechanical press does not have the function of pressure holding. Thus, it is necessary to tailor a press for hot stamping [1].

The hydraulic press for hot stamping is shown in Fig. 2.23. On the basis of the accuracy requirement of forming parts, the finite element models of the hydraulic press with the coupled dies and stamping parts are built and analyzed. From the analysis in detail, the stiffness, strength and modal of the press are obtained. Finally the structures are optimized for the hot stamping purpose.

The design of the cylinder seals of high-speed hydraulic press for hot stamping is the core part of the design of the hydraulic press sealing. Lots of hydraulic seal experiments should be carried out during the design process. Meanwhile it is necessary to combine the seal combinations of Lancaster closure-Y ring seal-guide ring provided by foreign seals manufactures, with the purpose of searching for the seal combinations of hot stamping press which is suitable for independent

Fig. 2.22 Water cooling hot stamping die for experiments

Fig. 2.23 800t hydraulic press for hot stamping

development. Only by this way, can we ensure the sealing effectiveness and stable pressure of hydraulic press under high-speed operation in order to meet the accuracy requirements of forming stampings.

2.2.5 Central Intelligence Control Automatic Integrated System

Due to the high-temperature heating in hot stamping production line, sheet metal in red state is extremely easy to deformation and many factors like high-temperature condition determine that the loading device should use special catcher, which should connect to the mechanical arm to realize the up and down movement of sheet material. The design of catcher should only consider the high-temperature state of blank, but also ensure the smooth and steady during transfer process and decrease the local cooling in the contact area of red blank as far as possible. To minimize the oxide on the surface of the steel plate, feeding time should be as short as possible. By using the mechanical arm equipped with complex motion structure computer, together with AC variable frequency control system and hydraulic buffer with composite positioning with electric braking, it is able to realize high speed, precise, stable positioning and synchronous control of take time. In addition, the loading device should also be equipped with temperature monitoring device, such as infrared temperature measurement, inspection, etc. Based on the data obtained from actual production process, the function of process optimization, performance test, fatigue performance prediction and process fault self-locking can be realized, together with the closed loop control system.

Based on TOP–DOWN system design and idea of industrial integration, to build global flexible, efficient, stable, reliable and integrated controllable hot stamping production line is the key aspect to realize equipment nationalization. Integrated system should first complete the top-level design, then decompose layer by layer to guarantee each functional unit meets the unified requirement of system accuracy and reliability, realizing the central controlled industrial network online integration system.

The control system should adopt highly reliable and stable field bus industrial network technology to form three-step control system as enterprise, technology and equipment level, while the external-related equipment are controlled and interlocked by the bus control protocol. In solving the common key technical problems in the major functional units, equipment such as heating furnace, high-speed press and high-speed conveyer are preset according to the take time of production to realize synchronous linkage and system integration, and ensure the reliability and stability of the local hot stamping production line. The central control system is shown in Fig. 2.24.

2.2.6 Subsequent Shot Blasting, Trimming, and Punching Equipment

HSS for hot stamping can be divided into two kinds: one with protective coating and the other without protective coating. Boron steel with protective coating can avoid high-temperature oxidation effectively during the hot stamping process. While oxide scale formation occurs in the process to the boron steel without protective layer. In order to ensure the surface quality of components, shot blasting is needed to remove the scale generated by high temperature from hot stamping components. Figure 2.25 shows the shot blasting equipment and the work site. Of course, shot blasting should not reduce the dimensional accuracy of components.

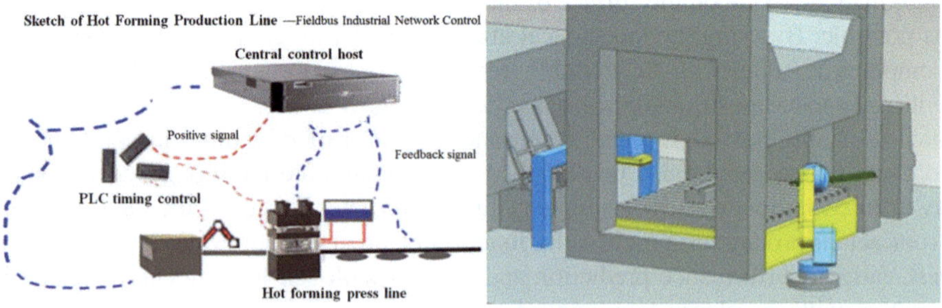

Fig. 2.24 The schematic diagram of domestic hot stamping production line with central control network

Fig. 2.25 Shot blasting equipment

Hot stamping or reinforced beam which is made by steel without protective coating of anti-oxidation and anti-decarburization after hot stamping is shown in Fig. 2.26. It can be seen from the figure that there is scale on the sample surface. Hot stamping door reinforced beam product after laser cutting and shot blasting which meets the requirements of size and surface treatment is shown in Fig. 2.27.

In terms of cold stamping, dies are used for batch processing of trimming and piercing. But for hot stamping of steel plates, it is hard to use the dies to trim or pierce due to the high strength of parts. The approach of laser cutting is often used to achieve it. The so-called laser cutting is to use the released energy generated by the laser beam falling on the workpiece surface to make the workpiece melt and evaporate, so as to achieve the purpose of cutting and engraving. The laser cutting has the advantages of high precision, fast cutting, few cutting pattern restrictions, automatic layout and material-saving, even cuts, low processing costs, etc. It may gradually improve or replace conventional cutting equipment. It should be noted

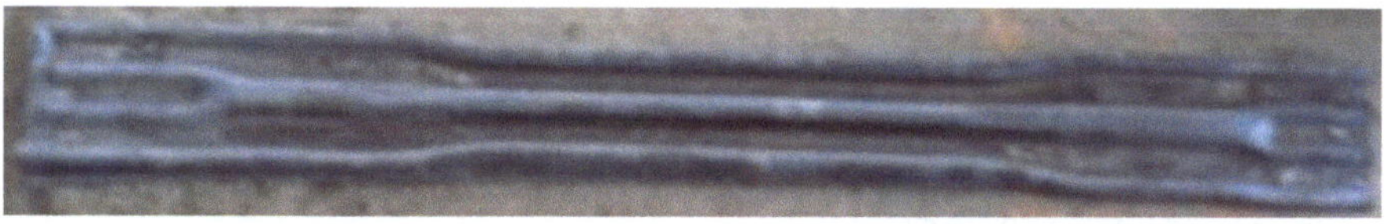

Fig. 2.26 Door reinforced beam by hot stamping (without protective coating)

Fig. 2.27 Door reinforced beam after shot blasting

that with the method of laser cutting, process parameters should be better controlled to avoid parts annealing under local high temperature which will influence the properties of the final parts.

2.3 Summary

This chapter mainly gives a brief instruction of the application of HSS hot stamping technology, the technological process, research status, the automatic production line with main equipment and the key technologies to systematically describe the basic content of hot stamping advanced manufacturing technology, so that readers are familiar with the research content in the following chapters. In addition, the finite element simulation analysis of hot stamping is particularly introduced. Characteristics of thermal-stress-phase coupling in hot stamping process and the role of multi-field coupling thermal forming simulation and thermal mechanical material parameters in hot stamping research are emphasized. This chapter also focuses on the hot stamping production line and the main equipment and elaborates the special requirements of hot stamping equipment comparing to traditional equipment in order to provide important guidance for the readers engaged in equipment development.

References

1. Arcelormittal (2008) Usibor 1500 and hot-stamping. In: AP&T Press Hardening seminar, AP&T Press Hardening proceedings. Dearborn MI, USA %\2014-09-19 08:49:00
2. Borsetto, F., Ghiotti et al (2009) Investigation of the high strength Steel Al-Si coating during hot stamping operations. In: Key engineering materials, vol 410–411, 289–296 %\ 2014-2009-2019 2020:2036:2000
3. Chang Y, Meng Z-H, Ying L et al (2011) Influence of hot press forming techniques on properties of vehicle high strength steels. J Iron Steel Res Int 18:59–63
4. Geiger M, Merklein M, Lechler J (2008) Determination of tribological conditions within hot stamping. Prod Eng Res Devel 2:269–276
5. Goedicke S, Sepeur S, Frenzer et al (2008) Wet chemical coating materials for hot sheet forming—anti scaling and corrosion protection. In: 1st International conference on hot sheet metal forming of high-performance steel. Kassel, Germany %\ 2014-09-19 20:37:00
6. Guo W, Hu P, Zhai S et al (2010) The high efficiency and energy saving rotary hearth furnace for warm forming, p 3 (in Chinese)
7. Guo Y, Ma M, Zhang Y et al (2013) Numerical simulation and experiment of hot stamping for front bumper of automobile. Forging Stamping Technol 3:46–50 (in Chinese)
8. He B, Ying L, Hu P et al (2012) Design of water cooling tool for hot stamping process of high strength steel. China Metalforming Equip Manuf Technol 6:62–65 (in Chinese)
9. Hou H, Ying L, Wu X et al (2012) The influence of heating temperature on 22MnB5 microstructure and austenite grain. China Metalforming Equip Manuf Technol 6:89–91 (in Chinese)
10. Hu P, Guo W, Ma N et al (2010) High temperature intelligent robot for warm forming, p 4 (in Chinese)

11. Hu P, Ma N (2011) Development of hot forming technology for ultra high strength steel and its mechanical problem. Adv Mech 3:310–334 (in Chinese)
12. Hu P, Shi DY, Ying L et al (2013) Coupled of thermal-mechanical-transformation numerical simulation on hot stamping with static explicit algorithm, pp 394–405
13. Hu P, Ying L, Guo R et al (2012) Development and independent R&D of hot stamping automobile high strength steel. Forging Meltalforming (in Chinese)
14. Hu P, Ying L, Jiang DX et al (2013) The Investigation of 3D temperature field numerical simulation in hot forming of high strength steel, pp 104–113
15. Hu P, Ying L, Zhao X et al (2012) Digital prototyping developing of hot stamping tool based on CAx. In: The advanced manufacturing technology top BBS and the 11th manufacturing automation and information technology seminar. Xi'an
16. Hu P, Yu H, Ying L et al (2013) Development of hot forming high strength components based on side impact. Automobile Technol 1:57–61 (in Chinese)
17. Jiang D, Wu W, Hu P et al (2012) Thermo-mechanical-martensitic transformation numerical simulation of high strength steel in hot forming. J Mech Eng 12:18–23 (in Chinese)
18. Kan DB, Liu LZ, Hu P et al (2011) Numerical prediction of microstructure and mechanical properties during the hot stamping process. In: 8th International conference and workshop on numerical simulation of 3d sheet metal forming processes, vol 1383, pp 602–609
19. Karbasian H, Tekkaya AE (2010) A review on hot stamping. J Mater Process Technol 210:2103–2118
20. Kolleck R, Veit R, Merklein M et al (2009) Investigation on induction heating for hot stamping of boron alloyed steels. CIRP Ann Manuf Technol 58:275–278
21. Lechler J, Merklein M (2008) Hot stamping of ultra high strength steels as a key technology for lightweight construction. In: Materials science and technology (MS&T) 2008. Steel Product Metallurgy and Applications, Pittsburgh, Pennsylvania
22. Liao Z, Ying L, Hu P et al (2013) Basic study on heat transfer of high strength sheet during hot stamping. Hot Work Technol 5:1–2 (in Chinese)
23. Lin J, Wang L, Tian H et al (2009) Research on hot forming behavior of ultrahigh strength steel. J Plast Eng 2:180–183 (in Chinese)
24. Liu H, Xing Z, Lei C (2012) Hot formation quality of high strength steel BR1500HS for hot stamping without cooling system. Trans Nonferrous Metals Soc China 2:542–547 (in Chinese)
25. MaN (2011) Research on hot forming of high strength steel. Dalian University of Technology, p 224 (in Chinese)
26. Ma N, Hu P (2010) Research on formability and application of hot forming technology. pp 26–32 %\ 2014-2009-2019 2008:2039:2000
27. Ma N, Hu P, Guo W (2010) Experiments and analysis of relations among heat, stress and transformation of boron steel for hot forming. Trans Mater Heat Treat 11:33–36 (in Chinese)
28. Ma N, Hu P, Guo W (2009) Technology and equipment of hot forming for ultra high strength steel automobile & parts technology 45:28–30 (in Chinese)
29. Ma N, Hu P, Guo W et al (2010) The online fast detection control method and device for warm forming, p 4 (in Chinese)
30. Ma N, Hu P, Wu W et al (2011) Constitutive theory and experiment analysis of hot forming for high strength steel. Chin J Theor Appl Mech 2:346–354 (in Chinese)
31. Ma N, Hu P, Yan K et al (2010) Research on boron steel for hot forming and its application. Chin J Mech Eng 14:68–72 (in Chinese)
32. Ma N, Shen G, Zhang Z et al (2011) Material performance of hot-forming high strength steel and its application in vehicle body. Chin J Mech Eng 8:60–65 (In Chinese)
33. Ma N, Wu W, Shen G et al (2011) Study of hot forming for high strength steel: numerical simulation-static explicit algorithm. Chin J Comput Mech 3:371–376 (In Chinese)
34. Ma N, Zhang Z, Hu P et al (2011) Microstructure and mechanical behavior of new type multi-layer metallic composite material in hot forming. J Mater Eng 5:88–92 (in Chinese)
35. Mayyas A, Qattawi A, Omar M et al (2012) Design for sustainability in automotive industry: a comprehensive review. Renew Sustain Energy Rev 16:1845–1862

36. Merklein M, Lechler J, Stoehr T (2008) Characterization of tribological and thermal properties of metallic coatings for hot stamping boron manganese steels. In: Proceedings of the 7th international conference. Chalkidiki, Greece
37. Merklein M, Lechler J, Stoehr T (2009) Investigations on the thermal behavior of ultra high strength boron manganese steels within hot stamping. IntJ Mater Form 2:259–262
38. Mori K, Ito D (2009) Prevention of oxidation in hot stamping of quenchable steel sheet by oxidation preventive oil. CIRP Ann Manuf Technol 58:267–270
39. Mori K, Saito S, Maki S (2008) Warm and hot punching of ultra high strength steel sheet. CIRP Ann Manuf Technol 57:321–324
40. N. M, P. H, W G et al. (2010) Coupled constitutive relation and numerical simulation of hot forming, pp 937–943 %\ 2014-2009-2019 2008:2037:2000
41. N. M, P. H, W. G et al. (2009) Feasible methods applied to the design and manufacturing process of hot firming. In: IDDRG2009 conference, Golden. CO USA %\ 2014-09-19 08:49:00
42. Nader M (2007) Hot stamping of ultra high strength steels. In: Rheinisch-Westfälischen Technischen Hochschule Aachen
43. Naderi M, Durrenberger L, Molinari A et al (2008) Constitutive relationships for 22MnB5 boron steel deformed isothermally at high temperatures. Mater Sci Eng A 478:130–139
44. Naderi M, Saeed-Akbari A, Bleck W (2008) The effects of non-isothermal deformation on martensitic transformation in 22MnB5 steel. Mater Sci Eng A 487:445–455
45. Ning M, Hu P, Zhai S et al (2009) Hot forming technology of high strength steel and its engineering implementation. Automobile Technol Mater 12:12–14 (in Chinese)
46. Olle P, Behrens et al (2008) Numerical modeling of phase transformation in hot stamping and deep drawing. In: The 9th international conference on technology of plasticity. CIRP %\ 2014-09-19 20:41:00
47. Shen G, Li X, Ying L et al (2013) Bending behavior based on hardness gradient for hot forming high strength steel. Trans Chin Soc Agric Mach 2:261–266 (in Chinese)
48. Shi D, Ying L, Hu P et al (2013) The hot forming experiment and thermal-mechanical coupled simulation of B-pillar for vehicle. China Metalforming Equip Manuf Technol 1:82–85 (in Chinese)
49. Sun H, Hu P, Ma N et al (2010) Application of hot forming high strength steel parts on car body in side impact. Chin J Mech Eng 23:252–256
50. Toros S, Ozturk F, Kacar I (2008) Review of warm forming of aluminum-magnesium alloys. J Mater Process Technol 207:1–12
51. Xu W, Ai J, Luo A et al (2009) Introduction of sheet metal hot-forming. J Plast Eng 04:39–43 (in Chinese)
52. Xu W, Guan S, Ai J et al (2009) Steel hot stamping key equipment and core technology. World Iron Steel 2:30–33 (In Chinese)
53. Ying L, Chang Y, Hu P et al (2011) Influence of low tempering temperature on fracture toughness of ultra high strength boron steel for hot forming, pp 160–165
54. Ying L, Li Y, Hu P et al (2012) High strength steel hot stamping digital prototype engineering. In: The 3rd China and Germany advanced manufacturing technology seminar (CDFK2012). Shang Hai
55. Ying S, Zhang Y, Cao G (2012) Automobile lightweight and the application of high strength steel plate. Automobile Technol Mater 10:11–23 (in Chinese)
56. Zhu L, Gu Z, Lu Y et al (2013) Mechanism of hot stamping hardening for ultra-high strength steel. J Jilin Univ (Eng Technol Edn) 2:376–379 (in Chinese)
57. Zhuang B, Dan Z, Jiang C (2010) Hot stamping technology and its application in car body. Metal Process 21:62–64 (in Chinese)

Chapter 3
Performance of Hot Stamping High Strength Steel (HSS) Technology

Hot stamping process should be optimized systematically, to produce premium quality vehicle components. Three major areas need to be addressed: (1) Understanding the theory and law of phase transformation in hot stamping process, capable of analyzing the phase transformation with the help of CCT (Continuous Cooling Transformation) curves. (2) Grasping producing process and specifying the influence of different factors, such as heating temperature, soaking time, cooling rate, etc. (3) Identifying the material performance indicators of the hot stamped steel, including strength, hardness, springback, uniformity, toughness, etc. and understanding the influencing factors of each performance indicators and their influence to the final hot stamping products. In this chapter, on the basis of fundamental theory and law of phase transformation, major factors in hot stamping process will be optimized, and optimal design has been conducted in order to optimize the producing process [14, 21, 42, 57, 58].

3.1 Process and Principle of Hot Stamping HSS

3.1.1 Hot Stamping Plate

Major steel suppliers all over the world have come up with a series of light-weight projects involving ultra-light body (ULSAB), ultra-light cover (ULSAC), and lightweight concept car project (USSAB-AVC) [24], which are mainly proposed to develop lower cost materials with more competitive potentials. In recent years, the demand of HSS with yield strength over 1000 MPa has increased dramatically in automobile industry. Figure 3.1 shows the differences in mechanical properties between common steel and hot stamping HSS where HSS is categorized as Ultra-High Strength Steel, with apparent advantages compared to other types of steel.

P. Hu et al., *Hot Stamping Advanced Manufacturing Technology of Lightweight Car Body*, DOI 10.1007/978-981-10-2401-6_3

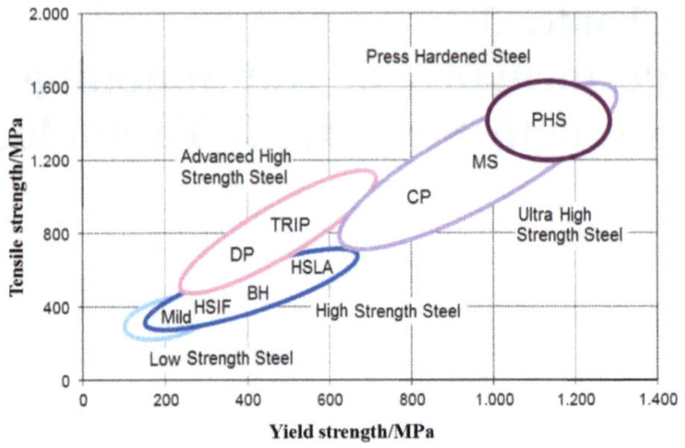

Fig. 3.1 Classification map of mechanical property for automotive steel

Hot stamping steel can be generally classified in two types: with the protection of antioxidation coat and the other without the protection coat. Currently, in market, one of the widely used hot stamping steel is USIBOR1500 and its improved heir with Al–Si coat from Arcelor in Europe. Thickness of this product varies from 0.7 to 2.4 mm. Other companies and researchers [12, 18, 28, 40], however, restricted by intellectual property law, have to target on steel with galvanized coat, zinc-alloy coat, zinc–10Ni coat, etc. To this day, some new techniques aiming at antioxidation in hot stamping have been put in application in the production process.

Besides the coated steel, steel without protection coat is also used in large quantity. SSAB, a Swedish company, has developed and manufactured hot rolling quenchable steel named as Domex including 22MnB5, 24MnB5, 26MnB5, 27MnCrB5, 30MnB5, 33MnCrB5, 38MnB5, which belongs to the same type with BTR series from BENTLER, Bo 02–05 from Lucoil, Boron steel series from RAEX®B, etc. US SivoriSpartak and Daimler Chrysler also jointly developed hot stamping boron steel; Nippon Steel, Kobe Steel, Pu-seok Korean mills are also actively developing hot stamping steel, respectively; domestic Bao steel, Wuhan Iron and Steel, Shougang, Anshan Iron and Steel, Tonghua Iron and Steel, and other steel companies are also actively engaged in steel development about hot stamping steel. Products like cold-rolled Baosteel B1500HS, hot-rolled BR1500HS, Wuhan Iron and Steel WHT1300HF, Angang AC1500HS represented by hot stamping steel have gradually entered the market.

The hot stamping steel, developed by whichever company mentioned above, can be enhanced by quenching process. All these steels can also be specified by their composition, respectively. For example, Mn–B series in China; Mn–Mo–B series in North American and Europe; Mn–Cr–B series (highly quenchable); Mn–Cr system (part of the martensitic hot stamping steel), Mn–W–Ti–B system, (such as South Korea POSCO company developed high bake-hardening fine-grained hot stamping steel), etc. [6, 7]. Most steel can be strengthened by adding alloy like Mn, Cr, B, Ti, Mo on the basis of C–Mn and its tensile strength could be improved from 500–600 to 1500 MPa or even 2000 MPa. Table 3.1 shows typical chemical composition of hot stamping steel.

Table 3.1 Main chemical composition of common hot forming steel (%)

Code	C	Mn	P	S	Si	Al	B	Cr
22MnB5	0.19–0.25	1.10–1.40	0.025	0.015	0.015	0.08	0.0008–0.005	0.30
30MnB5	0.27–0.33	1.15–1.45	0.025	0.010	0.10–0.40	0.015–0.08	0.001–0.005	0.01–0.30
USIBOR1500	0.2210	1.2110	0.019	0.003	0.2580	0.0360	0.0037	0.1910
Docol Boron02	0.2–0.25	1.0–1.3	0.0019	0.01	0.20–0.35	–	0.0050	0.14–0.26

3.1.2 Hot Stamping Technology and Mechanism

The hot stamping process is one kind of brand new heat-treatment process that combines the traditional hot forging with cold stamping. Figure 3.2 shows the changes of temperature and microstructure for high strength steel during hot stamping process. The boron alloy steel with microstructure of mixture of ferrite and pearlite at room temperature, can be fully austenitized after heating in furnace and heat preservation, then phase transformed into martensite by stamping and quenching in the mould with water cooling system, finally end up with a martensite phase steel with tensile strength as high as 1500 MPa.

Hot stamping technology is a new plate forming technology, completely different from traditional cold stamping technology. There are a number of process parameters, and the process is complexed. The key technical aspects, include feeding, heating, preserving heat, transferring, stamping, pressure quenching, shot blasting, trimming, piercing, liquoring, rust protection, and so on, which are illustrated in Fig. 3.3. Focusing three key process: heating, forming and quenching, temperature-related process parameters, include heating temperature (*T*), holding time (*t*), cooling rate (*V*). In order to achieve the transformation from austenite to

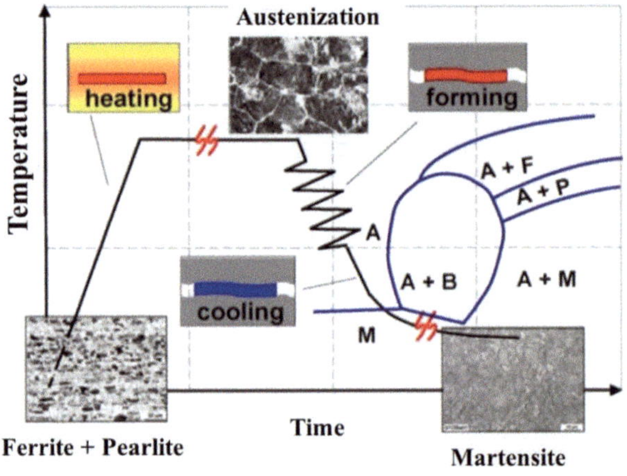

Fig. 3.2 Diagram of temperature history for plate in hot forming process

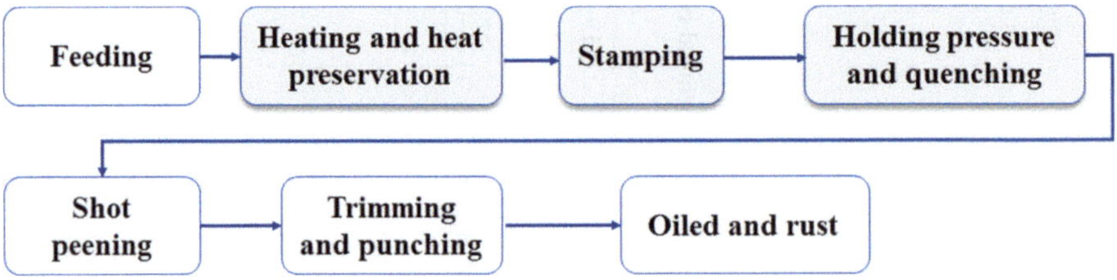

Fig. 3.3 The key procedure of hot forming process

martensite, and to ensure the mechanical properties of the products, these parameters are of great importance.

During the heating stage, heating temperature and holding time is the major process parameters. The heating temperature must be kept above the recrystallization temperature, then the microstructure can achieve austenitizing, but the heating temperature should not be too high either, because high temperature will lead to the burning of sheet metal surface and grain growth, then the components quality and performance will be affected. Holding time will affect the uniformity of austenitizing, sheet metal should be holding a period of time when it is heated to the specified temperature in order to promote austenitizing processes and realize fully austenitizing. But a long time of heat preservation can also lead to grain growth, then deterioration parts and mechanical performance, as well as increase the production cycle and reduce the production efficiency. The sheet metal after high-temperature austenitizing would be sent to the water cooling mold by the manipulator, and it will be cooled in the air during the process. If the transportation time is too long, the sheet metal of high-temperature oxidation would aggravate, and the sheet metal may transfer to bainite and ferrite. Therefore, on the premise of meeting the production rhythm, ensuring the plate into the mold temperature above 600, and making the hot forming plate will achieve a superior performance [26].

In the cooling stage, forming components cooling quenching touch with the surface of mold, then transform from austenite to martensite and the organization become strengthened. But this kind of phase change process is associated with the speed of cooling quenching. Austenite into martensite transformation is realized only when the cooling rate passes a certain critical value. On the contrary, if the cooling speed is too low, forming would appear bainite and other organizations which affect the strength of the forming components. The continuous cooling transformation curve of the hot forming (CCT curve) is illustrated in Fig. 3.4. Research shows that

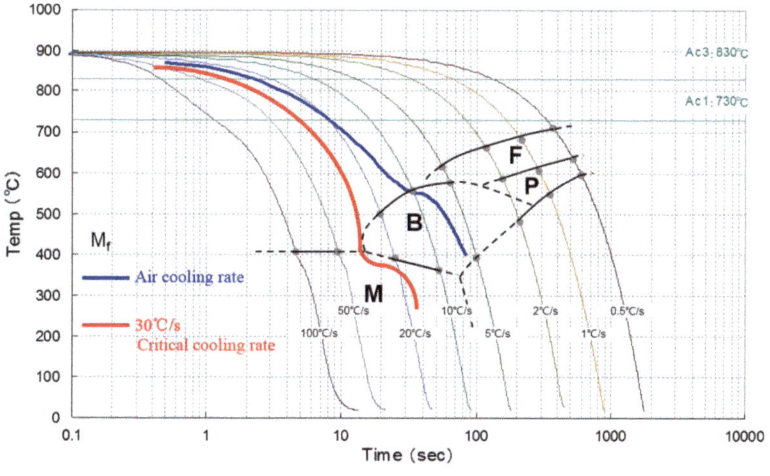

Fig. 3.4 The continuous cooling transformation curve of the hot forming

during hot stamping process, the austenite to martensite transformation minimum cooling speed (or is called the critical cooling rate) is 30 °C/s for 22MnB5 [25, 29]. As a result, the cooling speed of die for forming must be greater than this value. Therefore, it is critical to increase the cycle speed and the cycle pressure of cooling medium, and take away heat on the surface of the mold timely, and keep the same cooling effect and the thermal stress distribution to the shape. Meanwhile, the cooling medium should be kept in a certain temperature range. However, excessive cooling speed will result in the crack of the forming parts.

In order to apply hot forming technology into manufacture, it is necessary to thoroughly and systematically study the factors above that influence law of macro-micro material mechanical performance of the steel plate, so as to achieve complete and stable hot stamping process, and optimize the process parameters.

3.2 Research in the basic technology of hot stamping high strength steel

This section will focus on the three critical parameters related to temperature in hot stamping process, and study the macro-mechanical properties of hot stamping steel under different techniques. The macroscopic mechanical properties including yield strength, tensile strength, hardness, elongation rate, microstructure, including form of organization, grain size, are all taken as the performance indicator, aiming to obtain the optimal range of process parameters suitable for hot stamping production.

3.2.1 Mechanical Properties of Hot Stamping Steel

The similar steel products, like hot stamping steel with tensile strength reaching 1500 MPa, or quench hardening steel with manganese and boron components, but carbon less than 0.2 %, have been widely used throughout the world. In Europe, The quench hardening steel, commonly known as 22MnB5, named in accordance with the EN10027 standard system. "22" indicates the carbon component in percentage; "Mn" stands for the manganese element; "B" stands for boron element; "5" indicates the composition of Mn and B (1.25 % represents Mn composition), and thus, the composition of 22MnB5 is 0.22%C–1.25%Mn–B. Therefore, the boron steel mentioned above is also known as the manganese boron steel. Typical hot stamping 22MnB5 steel plate can be divided into two types: cold rolling and hot rolling. The advantages of the cold-rolled steel sheet are that the plate thickness accuracy is higher than that of hot-rolled steel, and the gap between die and sheet can be kept minimum. This chapter studies the three kinds of hard steel, including cold rolling, hot rolling high strength, hot stamping 22MnB5 high strength steel, coded as HFBG (cold rolling), HFTG (cold rolling), and HFSG (hot rolling), respectively, and determine the content of their composition and quality with the use of CENTER XRF-1800 LAB scanning X ray fluorescence spectrometer. Table 3.2 shows the results.

The table illustrates that all the three kinds of steel plate contain a large amount of alloy elements which can help to improve the stability of the 22MnB5 steel over cold austenite, such as Mn, Cr, Mo, Cu, Ni, Si, and nonmetal elements C, B, P, N. The boron content of 22MnB5 less than 0.004 % is an important factor to improve hot stamping material strength. Because of the similar size of the radius of the boron atom and the carbon atom, approximate 0.077–0.082 nm, the boron atom can displace the internal carbon atoms, and the partial poly grain boundary can delay the nucleation of ferrite, pearlite, bainite, and finally lead to solid solution strengthening. Boron element also can promote the homogenization of the microstructure in stamping.

The matrix of 22MnB5 hot stamping steel is mainly ferrite (Ferrite) and pearlite (Pearlite) mixed microstructure before quenching. After rapid quenching, the microstructure changed into dislocation type low carbon martensite structure (Martensite), three typical hot stamping steel 22MnB5 before and after the metallographic structure, as is shown in Fig. 3.5.

Tensile test was used to test the strength of 22MnB5 plate. The whole test procedure was based on and guided by "the mechanical properties of 2975-1998

Table 3.2 Chemical composition for experimental hot forming steels in this paper (%)

22MnB5	C	Mn	Cr	Si	B	P	Al	Ti	Ni	S
HFBG	0.240	1.29	0.165	0.24	0.0037	0.016	0.040	0.029	0.019	0.006
HFTG	0.225	1.24	0.163	0.24	0.0034	0.016	0.023	0.026	0.015	0.003
HFSG	0.230	1.26	0.210	0.18	0.035	0.013	0.050	0.025	0.010	0.002

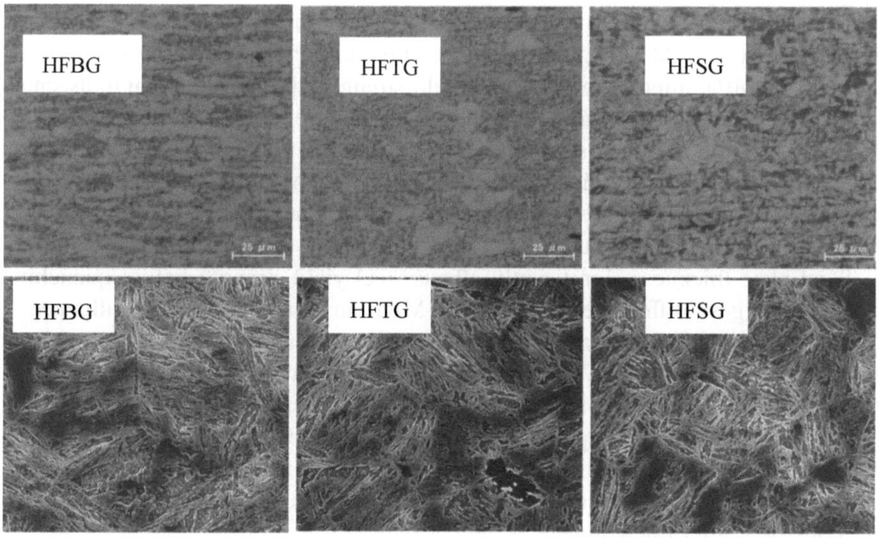

Fig. 3.5 Metallographic structures of three typical hot forming steel 22MnB5 steel at RT&AQ

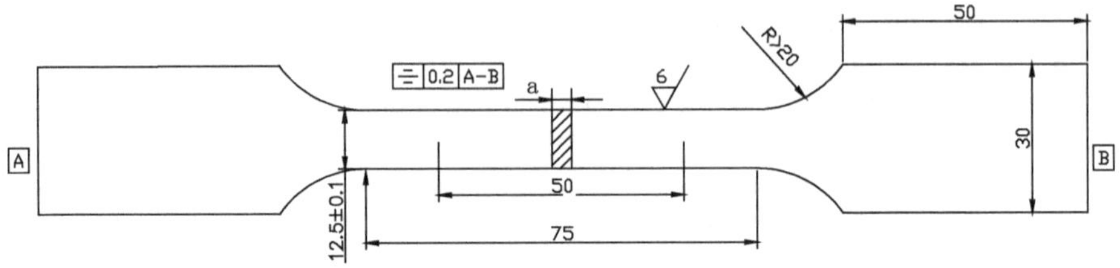

Fig. 3.6 The dimension of axial tensile test specimen (Unit: mm)

GB/T steel and steel products sampling position and sample preparation," "228-2002 GB/T metallic materials tensile test method" and "5028-2008 GB/T metal thin sheet and thin strip tensile strain hardening index (n value)," and the shape and size of the sample (Fig. 3.6) were defined. Tensile process used WDW3100 microcomputer control electronic universal testing machine, which is based on the GB/T16825 test, to reach Level 1 accuracy. The typical engineering stress–strain curves of the 2.0 mm thick steel plate are obtained by the uniaxial tension test of the three types of steel plate before and after quenching (Fig. 3.7).

Table 3.3 shows the mechanical properties of the hot stamping materials with different thickness before and after quenching. The ratio of the yield limit and the tensile strength of the material σ_s/σ_b is called the yield ratio, and the smaller yield strength ratio, the better the tensile strength. Because smaller σ_s value means better plastic flow, yield ratio $\sigma_s/\sigma_b \leq 0.65$, and elongation $\delta \geq 28\,\%$ are generally considered to show better elastic performance. The mechanical property parameters of hot stamping steel plate before and after quenching are compared. The result shows that there is no obvious yield phenomenon after quenching of 22MnB5 steel plate. The yield strength and tensile strength in quenching after hot stamping samples increased significantly compared to the before stamping samples. Taken HFBG as an example, the yield strength and tensile strength before quenching were 297.75 and 496.90 MPa; after quenching, the yield strength and tensile strength ascended to 1028.12 and 1554.21 MPa, which is an increase of 245.30 and 212.78 %, respectively. And the yield strength ratio is within the range of 0.6–0.7. The break elongation rate δ and strain strengthening coefficient n of samples have been reduced by 79.4 and 79.4 %, respectively, showing that hot stamping process leads to a substantial decline in its toughness.

Steel plate thickness significantly affects steel yield strength before quenching, with the biggest difference value 45.93 MPa. In comparison, the influence of thickness on the yield strength of samples after quenching diminishes, with the biggest difference value 27.62 MPa. This shows that hot stamping process can effectively shield effects of different performances caused by boron steel with different thickness and composition. Under the precondition of meeting the harden ability, thickness effect on the mechanical properties of the hot stamping of boron steel after quenching effect is not obvious. The mechanical properties of steel plate are consistent after quenching.

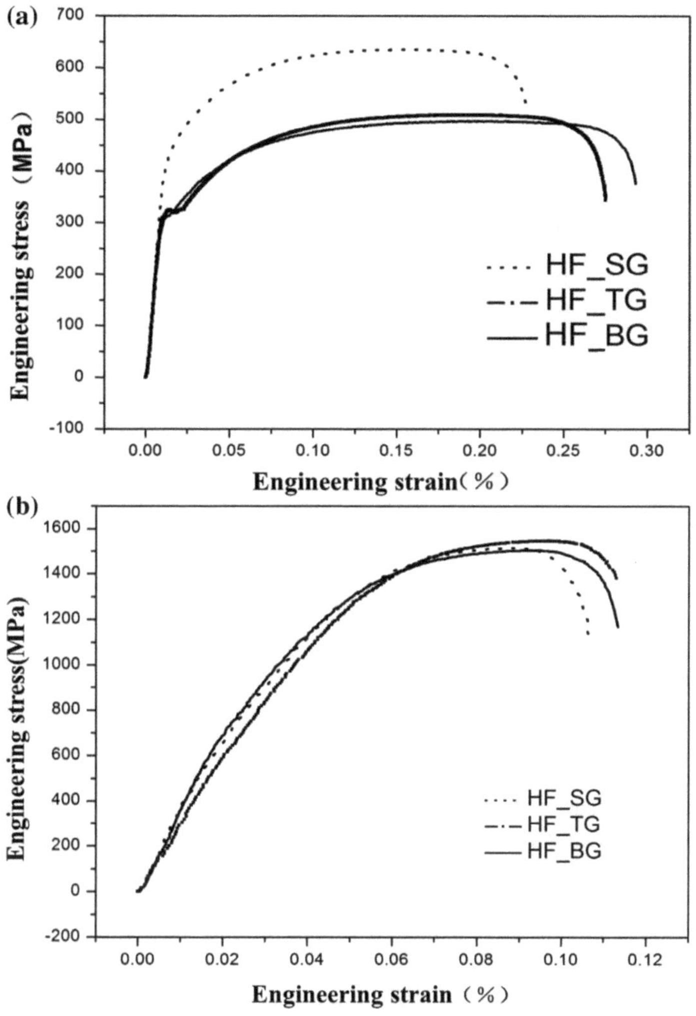

Fig. 3.7 Engineering stress-strain curve of hot forming blank **a** Before quenching, **b** After quenching

3.2.2 *Effect of Heating Temperature on the Mechanical Properties of Hot Stamping Steel*

Heating temperature is one of the critical parameters in the process of hot stamping process, which should be kept above the recrystallization temperature to ensure sheet austenitizing, and to avoid sheet metal surface burnt and grain overgrowth, caused by long-time heating, which will affect the product quality and performance

Table 3.3 Mechanical property comparison of 22MnB5 before/after hot forming

State	Thickness	Quenching state	Yield strength σ_s (MPa)	Tensile strength σ_b (MPa)	Yield strength ratio σ_s/σ_b	Elongation rate δ	Hardening coefficient n
22MnB5	1.0	Before quenching	343.68	499.80	0.68	0.30	0.22
		After quenching	1010.11	1522.23	0.66	0.05	0.11
	1.2	Before quenching	321.60	503.19	0.64	0.31	0.20
		After quenching	1007.64	1554.78	0.65	0.06	0.11
	1.6	Before quenching	297.75	496.90	0.60	0.34	0.20
		After quenching	1028.12	1554.21	0.66	0.07	0.11
	2.0	Before quenching	318.37	496.26	0.63	0.32	0.23
		After quenching	1000.50	1548.36	0.65	0.08	0.10

after quenching[15, 34]. At the same time, grain production affected by austenitizing temperature also plays an important role in martensite initial transformation start temperature (point M_s). Naderi et al. [30] have claimed that delicate and smooth grain will also cause rise of the martensite transformation temperature; coarse grain will lead to the reduction of the martensitic transformation temperature; thereby, the final material properties are influenced. This section focuses on the influence of different heating temperature on 22MnB5 experiment, and the hot stamping material can be determined through testing micro-austenitic grain size and mechanical strength of sheet metal, then determining the scope of the reasonable heating temperature.

Pilot scheme determines the 860, 880, 900, 920, 950 °C as five groups of different temperature testing setting, holding time 5 min, after water quenching. HFSG hot stamping steel plate, 2.0 mm in thickness, is chosen as testing material. Then the sample after quenching was cut into unidirectional tensile, as is shown in Fig. 3.6 by thread cutting machine for the purpose of mechanical properties testing.

At the same time, hardness testing and metallographic of the samples after quenching were carried out. According to GBT6394-2002 “metal measuring method of the average grain size” microscopic austenitic grain size of hot stamping steel after quenching was measured. Metallographic texture of hot stamping steel microstructure after quenching at different temperature settings are shown in Fig. 3.8. When the temperature is lower than 900 °C, there are still some ferrite and austenite in the steel matrix, and the martensite in the plate is obviously increased when the temperature is higher than 900 °C. It can inferred that the martensite slab beam increases with the increase of heating temperature process of hot stamping steel plate.

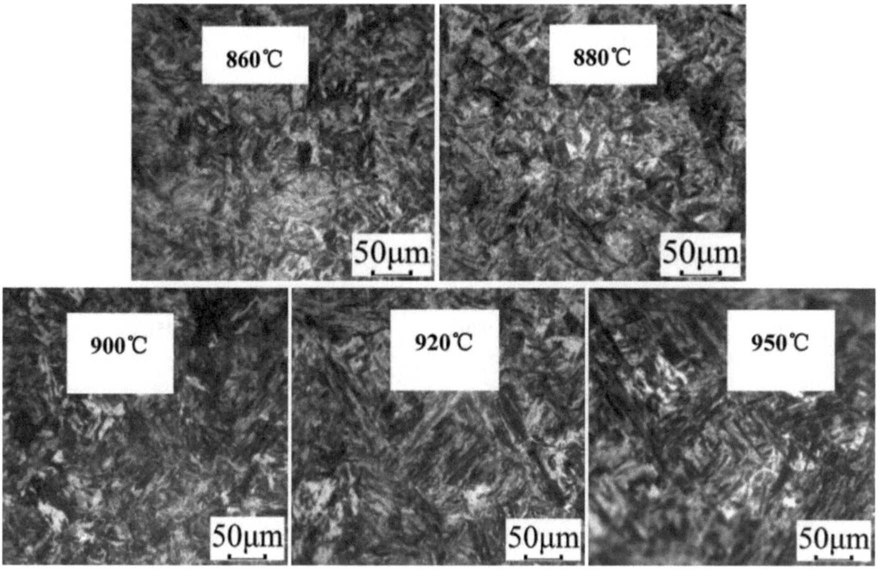

Fig. 3.8 Metallographic structure of hot forming blanks at different heating temperature after quenching (HFSG)

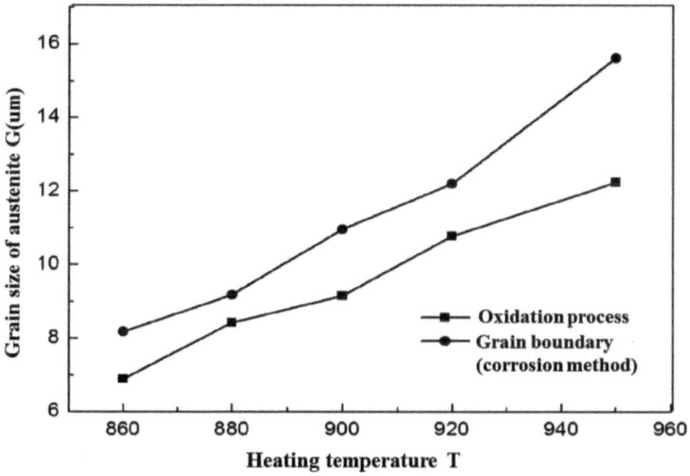

Fig. 3.9 Curve of austenitic grain size at different heating temperature (HFSG)

Figure 3.9 shows the 2.0 mm * HFSG austenite grain size curves and the size of the austenite grain size measured by the method of oxidation and grain boundary attack. With the increase of temperature, austenite grain size shows linear increasing trend. When the temperature rises to 900 °C, DEGC after grain size distribution is more uniform. When the temperature increases to 940 °C, DEGC

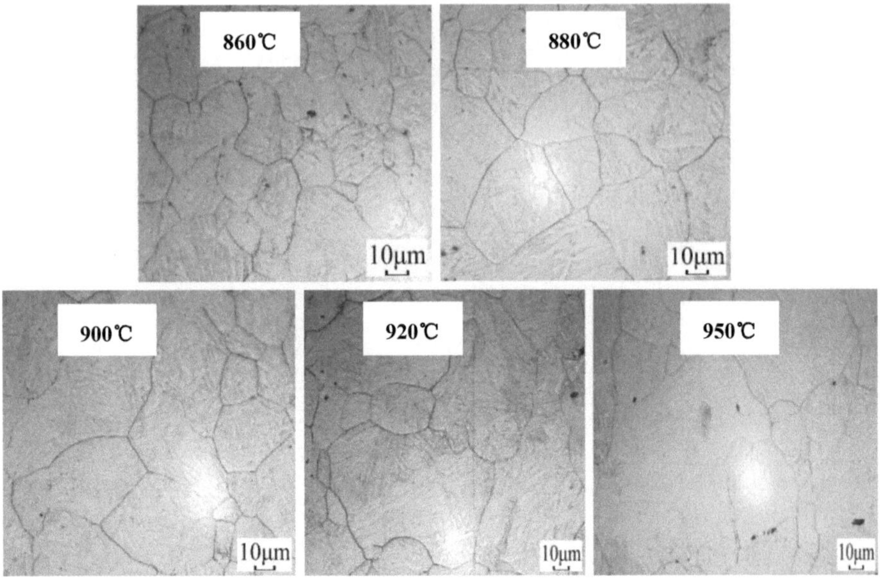

Fig. 3.10 Morphology of austenitic grain boundary at different heating temperature (HFSG)

austenite grain size tends to increase, with the grain size increasing to 15.62 μm. The grain size is smaller measured by the oxidation method compared to boundary attack method, which is because the grain boundary oxidation hinders the growth of austenite grain [55].

Figure 3.10 is HFSG hot stamping steel plate morphology of the austenite grain boundary under different heating temperature settings. The tendency of austenite grain growth is very obvious at 950 °C. Coarse grains microscopic mechanism caused by high temperature mainly lies in the atom diffusion. The higher the heating temperature, the more intensely atoms diffuse. The austenitic grain size will increase rapidly, but stabilize after reaching a certain temperature. In the actual production process, special attention should be paid to the heating temperature control factors, and avoid properties of final sheet affected by the coarse austenite grain size in hot stamping products [2]. Figure 3.11 shows the curve changes of mechanical properties of 22MnB5 under different heating temperature settings. It can be seen from the graph that, with the increase of heating temperature, there was a maximum tensile strength at 900 °C, and with the temperature increase, the tensile strength declines. Hardness measurement of specimen is tested under 900 °C, with the hardness value of about 550HV. To conclude, the results show that the plate material, under the heating temperature of 900–920 °C, can obtain premium mechanical properties.

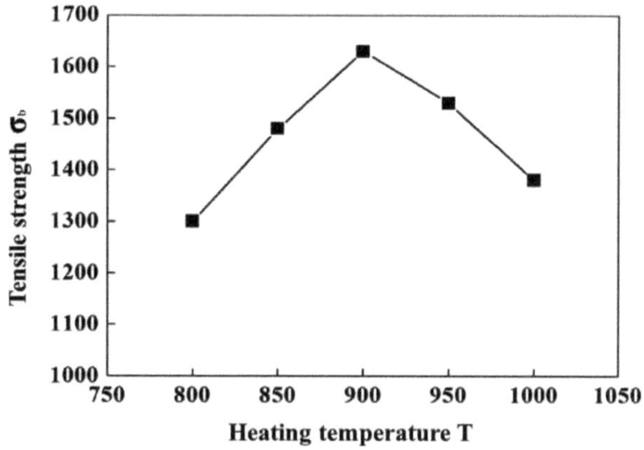

Fig. 3.11 Curve of mechanical property at property at different heating temperature (HFSG)

3.2.3 Effect of Holding Time on the Mechanical Properties of Hot Stamping Steel Plate

After the austenitic state, the sheet material needs constant temperature control treatment to ensure the uniformity of the microstructure and properties, so as to ensure the quality of the steel plate. Heat preservation time will directly affect the hot stamping process. If heat preservation time is too short, constant temperature of the soaking effect cannot be realized. If holding time is too long, single beat time will increase. Holding time refers to the internal temperature of the fingerboard material after reaching a specified heating temperature in the furnace to the constant temperature heating time. The internal temperature of steel plate heat through the time may differ due to the unequal heating equipment, the heating power, heating the sheet thickness. Therefore, the thermal insulation time should be defined as the holding time after reaching the specified heating temperature in the furnace heating time. The length of thermal insulation time will decide whether uniform austenite can be obtained after quenching.

Foreign scholars Turetta et al. [38] drew 22MnB5 continuous isothermal transformation CCT curve, and studied the evolution law of austenite phase under different circumstances. They claimed that austenitizing time of heat preservation under 900 °C within 5 min is most effective, based on austenitic grain size variation law of the sample after austenite with time. To further define the most optimum heat preservation time, this section will determine the effects of duration of heat preservation over the mechanical properties of hot stamping sheet materials, through the study of the austenitic grain size under different heat preservation time.

In the experimental process, the HFSG boron steel blank with the thickness of 2.0 mm was heated in the furnace, with the sheet size of 255 × 140 mm. The furnace temperature was set at 900 °C and thermocouples were installed to record the temperature changes. When the temperature of the sheet metal reached 900 °C, time of heating preservation was recorded. Samples were taken for a test after 0–15 min, quenching with water and the austenitic grain size of samples with different

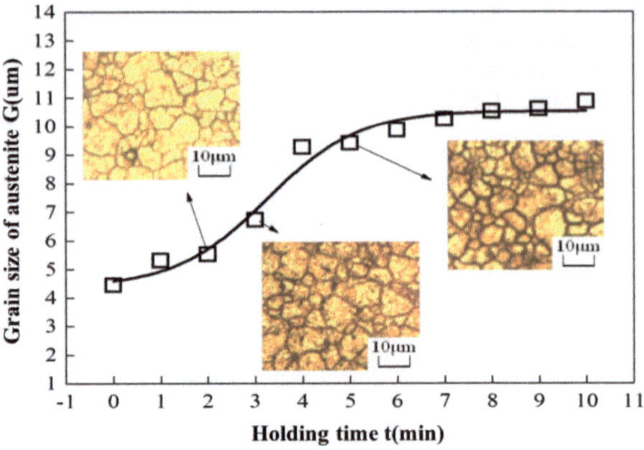

Fig. 3.12 Curve of the austenitic grain size for different holding time (HFSG)

heat preservation time were measured, and mechanical strength and hardness were measured. Figure 3.12 shows the changes of the austenitic grain sizes of the hot stamping HFSG boron steel under different heat preservation time.

Austenitic grain size of the hot stamping steel sheet significantly influences the microstructure and mechanical properties of the product after cooling. Figure 3.12 shows that under the heating temperature of 900 °C, the austenitic grain size increases gradually with the extension of heat preservation time. When the heat preservation time is 3–6 min, the austenite grain size of materials increased gradually. When it reached 6–7 min, grain growth trend slowed down. To ensure uniform fine austenite grain size, and avoid the coarse grains, caused by long heating time, the heat preservation time of 22MnB5 controlled within 3–6 min is most appropriate.

Mechanical strength and elongation pattern of 22MnB5 steel under different heat preservation time was shown in Fig. 3.13. Between the holding time 1–3 min, tensile strength, hardness, and break elongation of 22MnB5 increased gradually along with the extension of heat preservation time, reaching maximum of 1626 MPa, 49.3 HRC and 8.56 %, respectively, when the holding time was 3 min. When the holding time was between 3 and 10 min, mechanical properties of the material quenching with the mold reduced gradually with the extension of heating preservation time. When the heating preservation time was 10 min, the tensile strength of HFSG was 1549 MPa, a decrease of 77 MPa compared with the holding time of 3 min, intensity decrease of about 5 %. It can be inferred that, the material mechanical strength sensitivity on heat preservation time is low. The results shown in Figs. 3.12 and 3.13 show that, when the heat preservation time is within 1–2 min, cementite within the organization fails to melt completely due to the short austenitizing process time. When the holding time reaches 3 min, there exists almost no cementite between the martensite lath formation by quenching, and carbon in austenite achieves uniformity. When the holding time reaches 4 min or more, the austenitic grain size increases gradually along with the extension of heat preservation time. The formation area of the martensite after quenching also increase gradually, causing the corresponding tensile strength and hardness decrease gradually, and elongation decrease gradually with the extension of

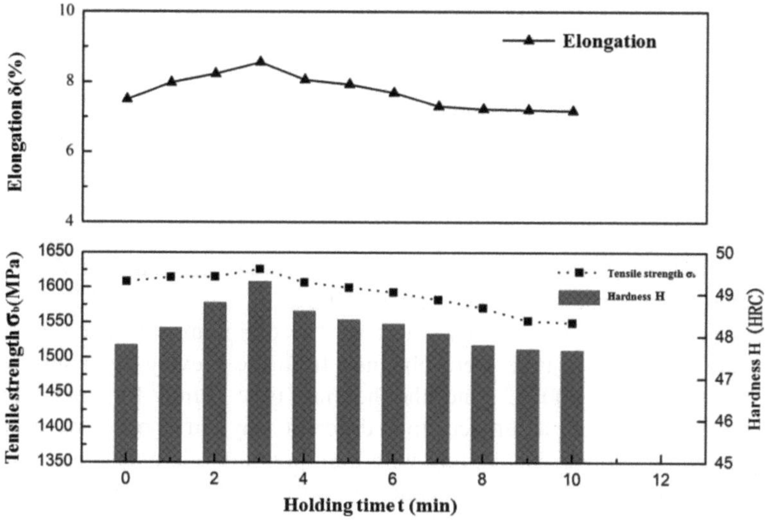

Fig. 3.13 Curve of mechanical properties under different holding time (HFSG)

heat when the preservation time is over 3 min. The results are caused by the increase of austenitic grain size and the decrease of the hardness of the formation martensite after quenching, and their corresponding plasticity and toughness deteriorate consequently.

Above all, when the mechanical strength is over 1500 MPa, considering the actual hot stamping production efficiency, strength, and elongation index, heat preservation time between 3 and 5 min in hot stamping process is most optimum and effective.

3.2.4 *Effects of Cooling Rate on the Mechanical Behavior of Hot Stamping Steel Plate*

Cooling rate is one of the important hot stamping process that influencing the temperature field [56]. It is critical that hot stamping material is shaped in mold under high temperature and form martensite structure in an appropriate quenching rate. Studies have shown that, the minimum critical cooling rate to realize the transformation from austenite to martensite for high strength plate 22MnB5 is $V_k = 30$ °C/s [27]. The transformation from austenite to martensite can be realized when the cooling rate is $V > V_k$, and the cooling speed range can ensure the supercooled austenite may not decompose and can directly obtain martensitic (including residual austenite) organization. When $V < V_k$, part or all of the bainite transformation and pearlite transformation occur. Organizational change processes with the cooling rate changes are shown in Fig. 3.4, and the greater the cooling rate of the steel plate, the more easily it hardens. But the interior of the plates is also prone to huge quenching stress, causing deformation and cracking. It is critical to treat the process in a scientific and reasonable manner, to determine the applicable rate of cooling hot stamping process, to ensure that the needs of the process parameters of hot stamping technology, so as to produce quality and performance of the final products. This section studies hot stamping steel mechanical properties and microstructure under different cooling rates, to determine 22MnB5 sheet in the process of optimum cooling rate range, and to provide practical guidance for the actual hot stamping production process optimization and temperature values simulation studies.

Rectangular hot stamping sheet metal HFSG of the size of 120 × 25 mm, thickness of 2.0 mm, was heated in the test. The optimum parameters of heating temperature and holding time were obtained from the previous two sections, the temperature reaching 900 °C, and the holding time 3 min for soaking. After removing the plate from a furnace, four different ways of cooling were applied, including natural cooling, air-cooling with high pressure, die quenching and water cooling, to achieve 22MnB5 sheet quenching between 10 and 300 °C/s at different cooling rates, wherein the air natural cooling (cooling rate 10–15 °C/s) and water quench (cooling rate >300 °C/s) are the two extremes of the cooling state.

According to characteristics of temperature change curve in the production process, the cooling process can be divided into three phases: the heating and insulation phase, the fast-moving and packing phase, and the quenching phase. The classification is based on the prototypical characteristics of cooling curves in steel HFTG and HFSG, as is shown in Fig. 3.14. Because during the hot stamping 22MnB5 martensite transformation in the vicinity of 400 °C (circled in black in Fig. 3.14), actual cooling rate of the sheet is defined as from the beginning of the cooling phase (Transfer phase and Cooling curve turning point) to the end at 300 °C

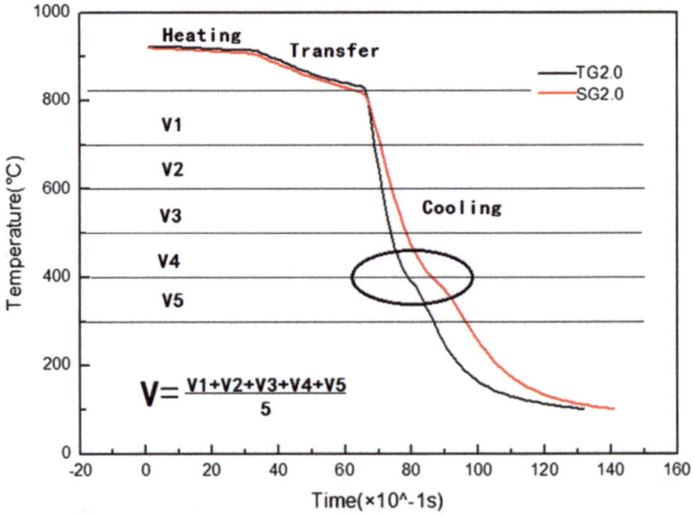

Fig. 3.14 Curve of the temperature characteristic for 22MnB5 in the process

temperature change rate. In order to improve the accuracy of the cooling rate, the whole cooling process is divided into five sections, with figures averaged to calculate and assess the value of the corresponding cooling rate.

Uniaxial tensile test and hardness test were taken on samples of different cooling rates, and material mechanical strength, elongation, and hardness of hot stamping 22MnB5 with cooling rate variation were obtained and were shown in the Fig. 3.15. The endpoints (A-Q) and (W-Q) represent the properties of the materials after natural cooling and water quenching. Air quenching sheet cooling rate is 9.27 °C/s, hardness 19.2HRC, a tensile strength of 709 MPa, elongation of 19.9 %. The material at this state is similar to the annealing process, heated and cooled naturally material, basically failing to achieve strength and increased performance, with the performance close to the basic mechanical properties of the original material. With the increase of the cooling rate, tensile strength, and hardness in hot stamping sheet metal showed a gradual increasing trend, but the elongation rate was just the opposite. When the quenching rate reached >300 °C/s (when water quenching), the material hardness reached 52.2HRC, tensile strength of 1650 MPa, elongation of 7.6 %. When cooling rate was within 10–175 °C/s range, the tensile strength increased from 709 to 1598.34 MPa, and elongation rate decreased from 19.9 to 10.68 %. When the cooling rate was within 175.7–310 °C/s, the tensile strength increased from 1598.34 to 1617.24 MPa, with an increase of 18.9 MPa, a ratio of about 1.18 %. At the same time, the elongation decreased from 10.6 to 7.23 %, a drop ratio of 31.8 %. The cooling rate needs to be over 80 °C/s to achieve the ultimate tensile strength of hot stamping material, with tensile strength >1500 MPa.

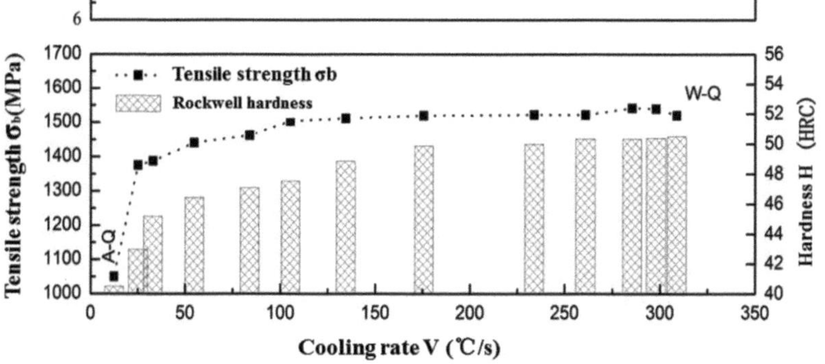

Fig. 3.15 Curve of mechanical property of hot forming blank at different cooling rates

Under different cooling rate of hot stamping material microstructure as shown in Fig. 3.16, air quenched organization is mainly pearlite and austenite, hardness is 19.9HRC. When the cooling rate is increased to a critical cooling rate of 30 °C/s, the cooling rate is relatively small, corresponding to the martensite lath width larger and have incomplete residual austenite; With the increase of the cooling rate, the microstructure of the /s is obtained when cooling rate reaches to 105 °C/s the microstructure is almost martensite shown in (d), and only a very small amount of residual austenite is obtained. The microstructure of the material in the (d), (e), (f) is basically the same as that of the martensite, and the corresponding hardness and tensile strength of the material are the largest.

After comprehensive analysis with Figs. 3.15 and 3.16, to meet the critical cooling rate (>30 °C/s), with the quenching cooling rate increases, the content of martensite is increased and its volume is also expanded so that strength of material is improved, but at the same time due to increase of the organization stress and thermal stress will have a great quenching stress, resulting in the elongation of 22MnB5 decline after quenching. Since the transformation temperature Ms increases as the cooling rate decreases and the martensitic transformation finish temperature M_f increases with the increase of the cooling rate [30], the martensitic transformation process will be completed in a very short time, although the organization was small grains, it retains a larger residual stress.

Based on the analysis above, it shows that: to optimize hot stamping process, the cooling rate range should be controlled between 85 and 175 °C/s so that microstructure of material is uniform martensite and its mechanical properties can exceed over 1500 MPa.

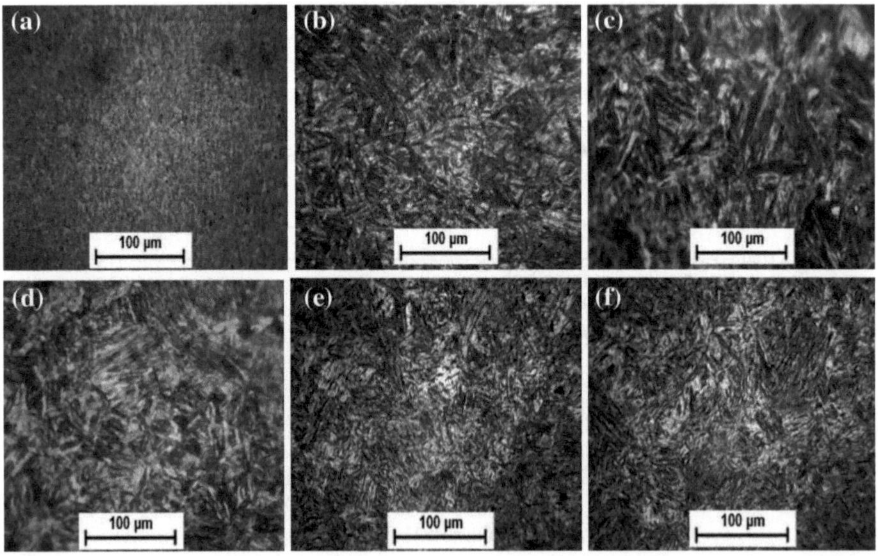

Fig. 3.16 Metallographic structures of quenched 22MnB5 at different cooling rates. **a** =10 °C/s. **b** =30 °C/s. **c** =85 °C/s. **d** =105 °C/s. **e** =175 °C/s. **f** =310 °C/s

3.3 The Study of Hot Stamping Material Toughness Process Experiment

To improve the strength of the material is the core of the materials research for centuries. So far, there are many kinds of methods widely used in all kinds of new materials on the strength of ascension, such as solid solution strengthening, dispersion strengthening, work (strain), and the grain refinement strengthening. These methods are used in introducing all kinds of defects in the material (point, line, surface, and body defects, etc.), making it difficult to produce plastic deformation, and improving the strength. With the strengthening of the material, the plasticity or toughness will decrease sharply. The toughness of material will form the so-called inverted relationship. To ensure the material has high strength and toughness properties has become a major scientific problem and the important bottleneck which restricts the development of the material in the field of advanced manufacturing.

High strength steel represented by 22MnB5 is the most wide research and application of hot stamping materials. There has been widespread attention in various fields [1, 19, 46], no matter in the ultra-high strength mechanical properties research, or in the trace elements affecting the strength-toughness mechanism experimental exploration. After hot stamping, the high strength steel 22MnB5, 30MnB5 have relatively low plasticity and high hydrogen-induced fracture sensitivity, so the strength will improve and the plasticity and toughness reduce sharply [17]. On the one hand, by optimizing the composition proportion adding and adjusting the microelement, such as Mn, Cr, B, Ti, Mo to inhibit pearlite, bainite and ferrite it can control fine-grain strengthening and improve elongation; on the other hand, micronize martensite can obtain good strength and toughness, and it also can obtain high strength and good resistance to delayed fracture by multiphase structure control to form effective hydrogen trap. The above two aspects which to improve the properties mainly happen in the production process of steelmaking and organization, but they are unable to be applied universally due to the difficult process control and high production costs. Therefore, it is a real problem demanding prompt study on how to improve the hot stamping products in the actual process of strength-toughness.

3.3.1 Hot Stamping Steel Strength-Toughness Tempering Process

This section is based on Sect. 3.2 in which the impact of mechanical property for 22MnB5 boron steel's before and after quenching and technological parameters on performance, to introduce the tempering technology to improve the high strength hot stamping material toughness, and explore change laws for the 22MnB5 hot stamping material under different tempering system of fracture toughness [44, 48]. Introduce different tempering toughness evaluation index to research the influence law of process parameters on the strength-toughness, then it is concluded tempering process effectively which can improve the tenacity improves strength-toughness inverted relations. It can provide improvement foundation and reference for the actual production process.

The temperature variation of hot stamping is almost the same with that of direct quenching (DQ process) in iron and steel industry. After the DQ process, it can improve the internal material microstructure of the quenching steel, reduce the residual internal stress and optimize material strength-toughness. The process of tempering is simple, and the time of process is short. Meanwhile, this process has advantages in cost, date of delivery, and energy consumption. Therefore, the process is suitable for mass production. Traditional tempering process is divided into mid-low temperature and high-temperature tempering process. Subcritical quenching (Hypoeutectoid steel heating quenching below the temperature of the AC3) can improve toughness, reduce the brittle transition temperature, and eliminate the temper brittleness. Thus, it can improve the toughness of the high strength steel by adopting different tempering processes [5, 6, 41]. In this section, through establishment of hot stamping production process system of quenching and tempering, to analyze the influence of the process parameters on the performance of hot stamping material; using the KAHN tear test and toughness evaluation index to research on the effect rule of the process parameters on the strength-toughness of the hot stamping 22MnB5 boron steel under the system of mid-low temperature tempering. And then, it is concluded that there is an effective tempering process route.

Hot stamping tempering process test [43] uses two sets of category four kinds of hot stamping boron steels of different thickness, boron steel type 1 is as follows: HFTG_1.0 and HFTG_2.0 mm; boron steel type 2 is as follows: HFBG_1.2, and HFBG_1.6 mm. The chemical composition of test material is shown in Table 2.1. The six different tempering programs were designed based on the tempering feature of the hot stamping boron steel. And then mainly explore the relationship between the fracture toughness at different temperature tempering conditions. The tests use SX-12-10 box-type resistance furnace. The specimen heating temperature is 900–920 °C. Hold time for 3 min. Take out the specimen from the furnace, then quench in the mold with water cooling system and tempering. The tempering system is as follows: 100, 150, 200, 250, 300, 350°C, and holding time for 40 min.

Since the sheet thickness of the hot stamping products are less than 3 mm, traditional Charpy impact test is not suitable for assessing toughness of sheet steel [4]. In order to show hot stamping material fracture toughness, the introduction of tear strength and crack nucleation work per unit area[32] is as toughness evaluation by KAHN test. The method is mainly used for evaluating the fracture toughness of aluminum-magnesium alloy. This method measures the work done by plus tearing load during the crack propagation process of tearing. The characteristics of KAHN test are as follows: size of the specimen is small, the direction of cracking and tensile is orthogonal, the operation is simple, the fracture is sharp, crack can get easily, it can accurately measure the ability to crack, and it can obtain fracture surface. Figure 3.17 shows the size of KAHN toughness test specimen.

Fig. 3.17 The size of KAHN toughness test specimen (unit: mm)

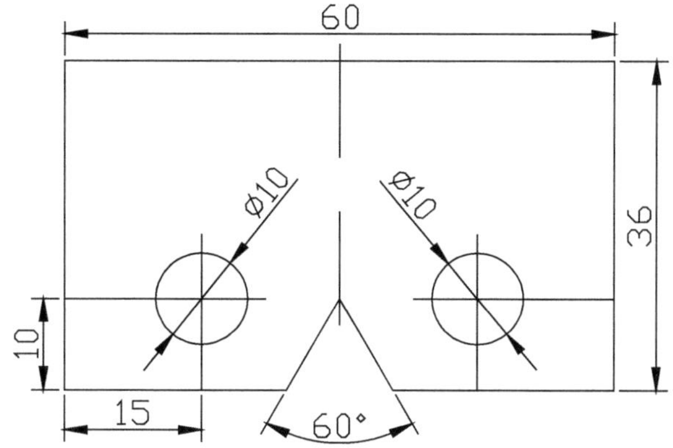

The tensile speed of the test is 1 mm/min. The test force-time and displacement-time curves should be recorded. Figure 3.18 shows load-displacement curve of KAHN toughness test. According to the curve can get hot stamping of boron steel toughness index—TS and crack nucleation work per unit area. In this figure, TS is tear strength, UIE is crack nucleation work per unit area, UPE is crack propagation energy per unit area, P is maximum applied loads, $w1$ is energy before crack propagation, A is effective area of specimen, M is moment of inertia, C is radius of gyration, I is rotational inertia, b is the minimum width of the notch, t is specimen thickness.

According to the two parameters to characterize the fracture toughness of the boron steel, these parameters include crack propagation energy per unit area (UPE) and TS. The TS is given by Eq. 3.1.

$$\mathrm{TS} = \frac{P}{A} + \frac{MC}{I} = \frac{P}{bt} + \frac{3P}{bt} = \frac{4P}{bt}\ (\mathrm{MPa}) \tag{3.1}$$

Fig. 3.18 Load-displacement curve of KAHN toughness test

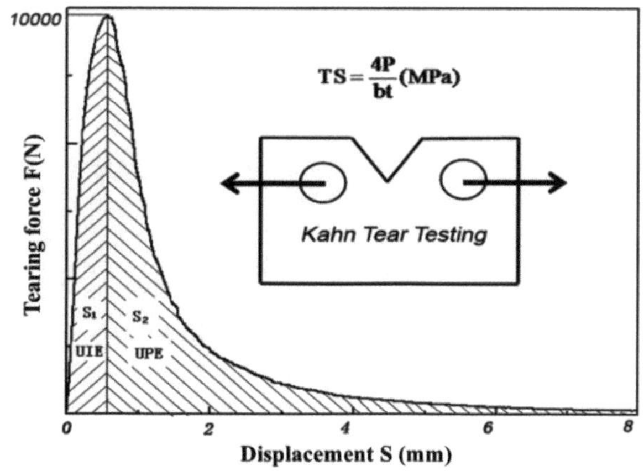

P is maximum applied loads, *b* is the minimum width of the notch, *t* is specimen thickness, *A* is effective area of specimen, *M* is moment of inertia, *C* is radius of gyration, *I* is rotational inertia.

On the basis of the area after crack extension to calculate *S*. The UIE and UPE can calculate by Eqs. 3.2 and 3.3. Both of them can calculate and assess the TS of hot stamping material. The TS and UPE value is obtained according to load-displacement curve. UPE is crack propagation energy per unit area, K_{IC} is plane-strain fracture toughness (stress intensity factor). Relations between the UPE and K_{IC} is $K_{IC}^2 = 3\text{UPE} + 400$ (in imperial units). There is a linear relation between K_{IC}^2 and UPE, the fracture toughness of steel can be represented by the symbiosis between UPE and TS.

$$\text{UIE} = \frac{s_1}{A} = \frac{s_1}{bt}\ (\text{N/m}) \tag{3.2}$$

$$\text{UPE} = \frac{s_2}{A} = \frac{s_2}{bt}\ (\text{N/m}) \tag{3.3}$$

According to the Orowan fracture theory [23, 39], it can be known that the notch of specimen is due to the growth TS and yield strength. The maximum axial stress is located at the junction of the plastic region and the elastic region because of the plastic deformation that occurs in the notched root of KAHN sample. When the maximum axial stress reaches the material fracture resistance level, the junction will crack.

The macroscopic characteristics are as follows: it is the origin of the tear that there is a fiber-rich region in a range of depth from the surface on the fracture. With the crack extend radially toward the center of the elastic region, the final fracture-end is located in the center of the broken or biased toward one side. Figure 3.19 shows KAHN toughness test and the samples after torn.

Determination of hot stamping plate fracture toughness after quenching. Figure 3.20 shows curve of tear and tensile strength for original 22MnB5 steel with

Fig. 3.19 KAHN toughness test and the samples after torn

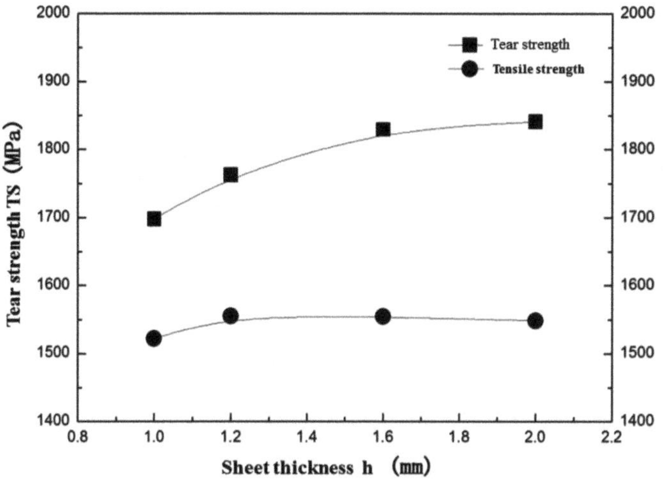

Fig. 3.20 Curve of tear and tensile strength for original 22MnB5 steel with different thicknesses

different thicknesses. It shows tensile TS sample is greater than its equivalent thickness from the figure. With the increase of hot stamped boron steel sheet thickness, TS, and tensile strength of the gap are growing.

The hot stamping samples of different thicknesses were tempered at 40 min, respectively in 100, 150, 200, 250, 300, 350 °C. Then perform the KAHN tear test at room temperature. Figure 3.21 shows the curve of UPE and TS-Tempering temperature for 22MnB5 steel.

From the chart, it can be concluded that UPE and TS have the same trend for the samples of different thicknesses. If the thickness of the sample increased, the sample would store more elastic energy, then the crack propagation energy per unit area and TS would increase. With the tempering temperature increase, UPE and TS will improve in the range of 100–200 °C. The reason is that through improving the tempering temperature in the range of 100–200 °C, the lath martensite will increase and the twin martensite will be eliminated. So, UPE and TS will increase. According to Cottrell air mass theory [11], there are a lot of dislocation exist in the boron steel quenching organization, and they would not disappear at low temperature tempering. In the role of external forces, the steel is not easy to crack because dislocation has certain ability to migrate. The toughness and the strength of the boron steel increase with the temper temperature's rise. In the range of 200–300 °C, UPE and TS will decline with the tempering temperature's increase. Its UPE and TS will reach the peak, while the sample tempering at 200 °C and 40 min insulation. When the tempering temperature is greater than 300 °C, it is not conducive to improving the fracture toughness.

Figure 3.22 shows the relationship between TS, UPE and tempering temperature for the HFBG_1.6 mm. The values of UPE and TS are decreased by 5.8 and 23.7 % than the values at 200 °C. Before tempering (at room temperature), the UPE and TS

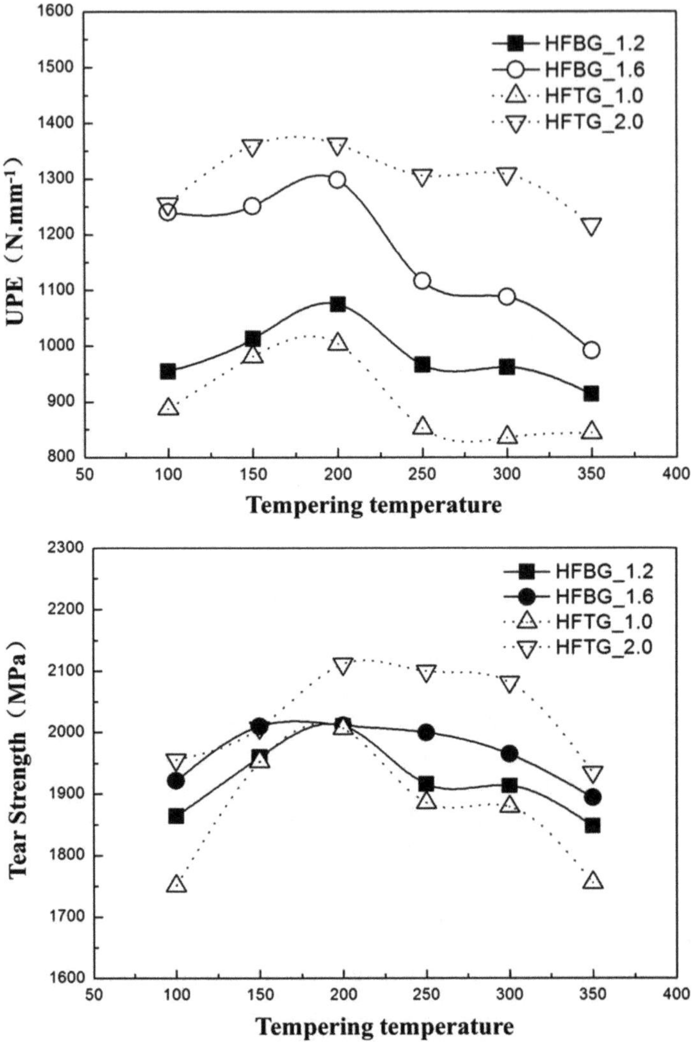

Fig. 3.21 Curve of UPE and TS-tempering temperature for 22MnB5 steel

is 965.7 N/mm and 1829.5 MPa. Tempering process can improve the mechanical properties of high strength hot stamping material. Tempering at 200 °C after quenching can obtain tempering martensite structure with high toughness. It is useful to improve the strength-toughness of the material.

Figure 3.23 shows the relationship between hardness and tempering temperature of hot stamping steel. When tempering temperature is greater than 200 °C, the hardness of materials decreases. So, tempering temperature should not be greater than 200 °C. The mechanisms of improving the strength-toughness by tempering temperature are as follows [9, 35]: with the increase of tempering temperature, the

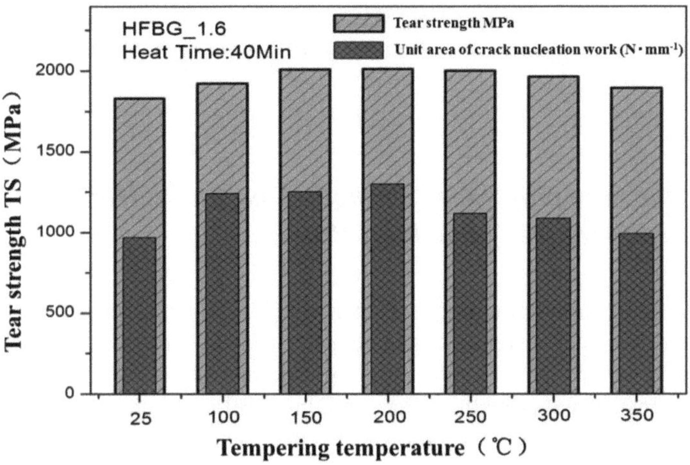

Fig. 3.22 Relationship between TS, UPE, and tempering temperature (HFBG1.6)

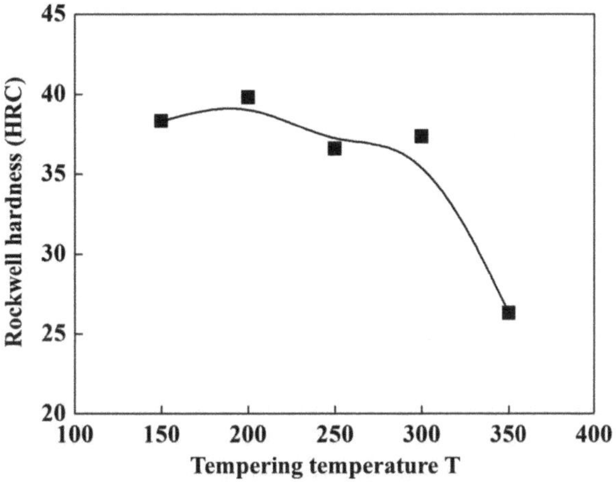

Fig. 3.23 Relationship between hardness and tempering temperature of hot forming steel

carbides precipitated from the martensite will increase. Meanwhile, residual austenite begins to break down, and cementite brittle film is formed on the grain boundaries so that the grain boundary is embrittled. At the same time, external force lead the brittle fracture trend to increase, and dislocation motion becomes difficult [52]. In the meantime, with the increase of tempering temperature, the quenching martensite starts to break down for carburizing size carbides, and the carbon content of the martensite decreases. So, the toughness and strength of the boron steel decrease with the increase of tempering temperature, while the UPE and TS of the samples decline.

In conclusion, hot stamping boron steel after quenching and low tempering at a certain time could improve the strength and fracture toughness. The best low-temperature tempering process for hot stamping steel should be proceeded at

200 °C and 40 min insulation. The UPE and TS is 1298.0 N/mm and 2011.6 MPa, compared to these properties before tempering treatment increased by 34.4 and 10 %, respectively.

3.3.2 Hot Stamping Steel Strength-Toughness High Temperature Quenching Process

The tissue of the hot stamping steel after high-temperature quenching consists of martensite or residual austenite and martensite. Tensile strength exceeds 1500 MPa, elongation is lower than 6 %, and strong plastic product is less than 10 GPa% [8]. In order to ensure the strength of hot stamping steel and improve the toughness of the product after quenching, editor uses the method improving technological parameter to enhance the strength-toughness of the material [7, 47, 51] based on Sect. 3.2. The scholars from Harbin Institute of Technology and Shandong University [13, 20] have studied the conventional hot stamping process, and introduce the response surface method to optimize technological parameters. But their study was based on evaluation of conventional mechanical properties, they did not consider to improve technological parameters based on index of the strength-toughness. In this section, hot stamping boron steel is taken as the object of study, while introducing the index of toughness to research optimal scheme for strength-toughness high-temperature quenching of hot stamping steel using L9 (3^4) orthogonal experiment to analysis the heating temperature, holding time, and temperature starting quenching. Meanwhile, select tensile strength, elongation, TS, and crack nucleation work per unit area to composite score. Then obtain an optimized strength-toughness high-temperature quenching process.

Orthogonal design [3, 33] is a high-efficiency test design to seek the optimal level combination for multifactor. The advantage is that this method can get most possible assessments by as little as possible factors. So, this method has been widely adopted in the study of process test optimization scheme. In this experiment, the orthogonal test table is designed with three factors and four levels. As illustrated in Table 3.4, *T*(*h*) is heating temperature, *t*(*h*) is heat-up time, *T*(*q*) is forming temperature.

According to the above test scheme, uniaxial tensile tests were conducted on the test machine to obtain the tensile strength and elongation. Conduct KAHN tear test to obtain TS and crack nucleation work per unit area. Figure 3.24 shows the influence of quenching process parameters on the test index.

Table 3.4 Process factors for quenching and tempering experiment in high temperature

Level of factor *T*(*h*) (°C)	A *t*(*h*) (min)	B *T*(*q*) (°C)	C	D
1	890	1	650	1
2	920	3	700	2
3	950	5	750	3

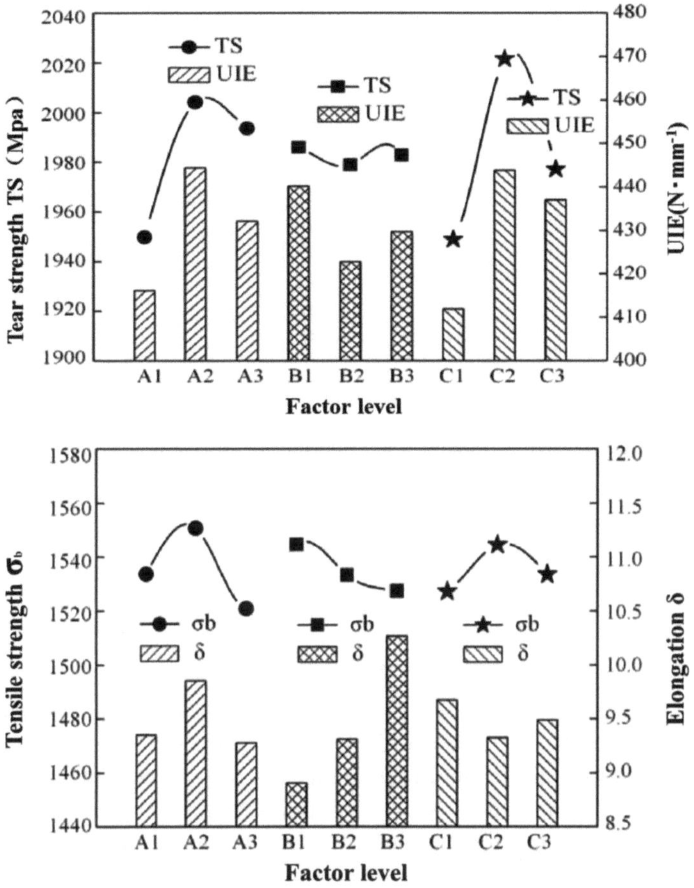

Fig. 3.24 The influence of quenching process parameters on the test index

Orthogonal design can greatly reduce the number of tests. But the drawback is that this method can only analyze isolated sites. When the test indicators are analyzed solely, the optimum conditions are inconsistent. So, the editor furthers the optimization of high-temperature quenching process parameters by comprehensive scoring method [16], taking the multi-objective problem into a single-objective indicators to study the strength and toughness of the process.

Comprehensive scoring method studies the importance of corresponding indicators in the entire process by setting weight. Then determine the optimal process plan of single target. The calculation method is as follows: Composite score = Tensile strength score + Elongation score + tear strength score + crack nucleation work per unit area score. Figure 3.25 shows intuitionistic analysis diagram of weight allocation.

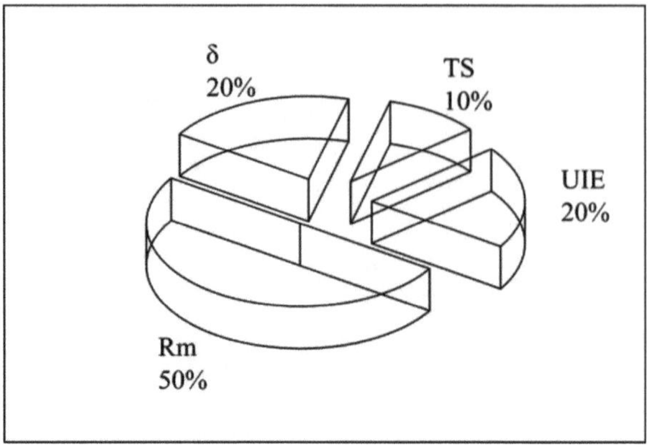

Fig. 3.25 Intuitionistic analysis diagram of weight allocation

Equations (3.4)–(3.6) represent the calculating formula of weighted comprehensive score,

$$(y_i)'_k = (100/R_k)\left[(y_i)_k - (y_{\min})_k\right] \tag{3.4}$$

$$R_k = (y_{\max})_k - (y_{\min})_k \tag{3.5}$$

$$y_i = \sum w_k \times (y_i)'_k \tag{3.6}$$

While, $(y_i)'_k$ (i = 1, 2,… 9; k = 1, 2, 3, 4) represents the score of No i test item k index, R_k represents the range of test indicators, $(y_i)_k$ represents the experiment value of No i test item k index, $(y_{\min})_k$ represents minimum value of item k index, $(y_{\max})_k$ represents the maximum value of item k index, y_i represents the comprehensive score of No i test, w_k represents weight coefficient.

Table 3.5 shows orthogonal test results of Quenching-Tempering process in high temperature. In the table, the column A represents heating temperature of

Table 3.5 Orthogonal test results of quenching-tempering process in high temperature

No.	Factors				Test parameters			RST	
	A	B	C	D	σ_b (MPa)		δ (%)	TS/MPa	UIE (N/mm)
1	890	1	650	1	1540.72	8.505	1900.50	400.00	0.842
2	890	3	700	2	1548.00	8.690	1984.00	413.50	0.866
3	890	5	750	3	1512.20	10.865	1965.00	435.00	0.932
4	920	1	700	3	1557.60	9.845	2067.00	480.50	1.000
5	920	3	750	1	1552.78	9.230	1976.00	436.00	0.910
6	920	5	650	2	1541.87	10.495	1969.68	416.76	0.876
7	950	1	750	2	1535.45	8.370	1990.50	440.05	0.929
8	950	3	650	3	1499.07	10.020	1976.71	418.98	0.906
9	950	5	700	1	1528.12	9.445	2013.72	437.47	0.928

austenitizing, the column B represents holding time, the column C represents the level of factor of starting quenching temperature and th column D is blank column. σ_b is tensile strength, δ is elongation, TS is tear strength, UIE is crack nucleation work per unit area. The RST = (UIE/450)/(Rm/1500), and it represents the relative size between the toughness and strength. Figure 3.26 shows the index score intuitionistic analysis diagram of comprehensive value image for quenching-tempering process in high temperature.

Conclusions can be obtained from Table 3.6 and Fig. 3.26 as follows: Factor A (heating temperature) and Factor C (temperature starting to quenching) have a significant effect on the test results, the optimal levels are A2 and C2, the effect of Factor B is lesser. So, the optimal case is A2B1C2, namely, heating temperature is 920 °C, holding time is 1 min, the temperature of staring quenching is 650–700 °C, and it will have the highest composite score. In a word, the hot stamping high strength steel with these process parameters have excellent strength-toughness.

Figure 3.27 shows SEM morphology of KAHN torn fracture under different heating treatments. When heating temperature is 900 °C, holding time is 3 min, the temperature of staring quenching is 800 °C. The fracture morphology is mixed of cleavage and quasi-cleavage, which presents the fracture characteristics of low

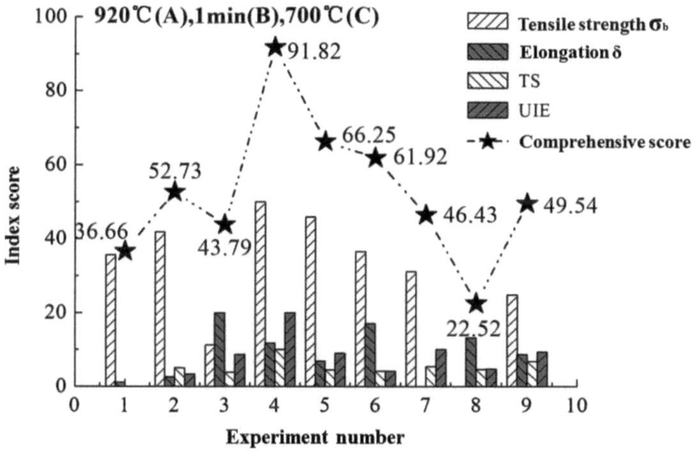

Fig. 3.26 The index score intuitionistic analysis diagram of comprehensive value image for quenching-tempering process in high temperature

Table 3.6 The ANOVA analysis of process test for quenching-tempering process in high temperature

Source of variation	Degree of freedom	Sum of square	*F* ratio	*F* critical value
Factor A	2	2006.002	156.389	19.000
Factor B	2	187.978	14.655	19.000
Factor C	2	888.205	69.245	19.000
Error	2	12.830		

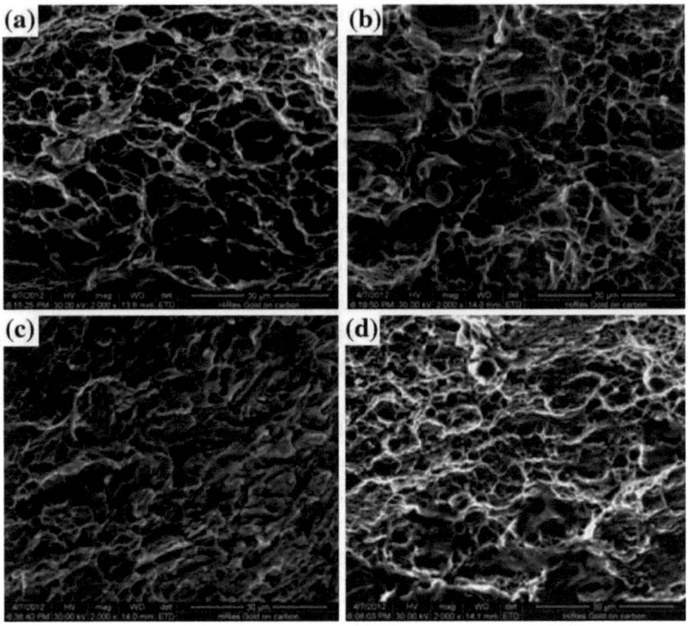

Fig. 3.27 SEM morphology of KAHN torn fracture under different heating treatment. **a** 900 °C, 3 min, 800 °C. **b** 890 °C, 1 min, 650 °C. **c** 950 °C, 1 min, 750 °C. **d** 920 °C, 1 min, 700 °C

energy absorption process (Fig. 3.27a). When heating temperature is 890 °C, holding time is 1 min, the temperature of starting quenching is 650 °C, the fracture morphology also is mixed of cleavage and quasi-cleavage, and the ductile fracture trend of the material enhances (Fig. 3.27b). When heating temperature is 950 °C, holding time is 1 min, the temperature of starting quenching is 750 °C, the tear fracture morphology is quasi-cleavage. The morphology is between cleavage and dimple fracture (Fig. 3.27c). When heating temperature is 920 °C, holding time is 1 min, the temperature of starting quenching is 700 °C, the tear fracture morphology is mixed of quasi-cleavage and dimple fracture, which presents the fracture characteristics of high-energy absorption process (Fig. 3.27d). It can be concluded from the above phenomenon that the variation trend of fracture morphology is in accordance with the value of UIE. Using the optimized process scheme, the fracture has the characteristics of ductile fracture and it becomes plastic deformation before fracture, which can prevent the fracture effectively.

3.4 Tailored Properties of Hot Stamping Part

Compared with the cold formed parts, the hardness and strength of hot stamping parts have been greatly improved so that with the usage of these parts vehicles can become lighter but as safe as usual. However due to the super high hardness and

Fig. 3.28 Application of the ideal strength-hardness gradient A-pillar in BIW

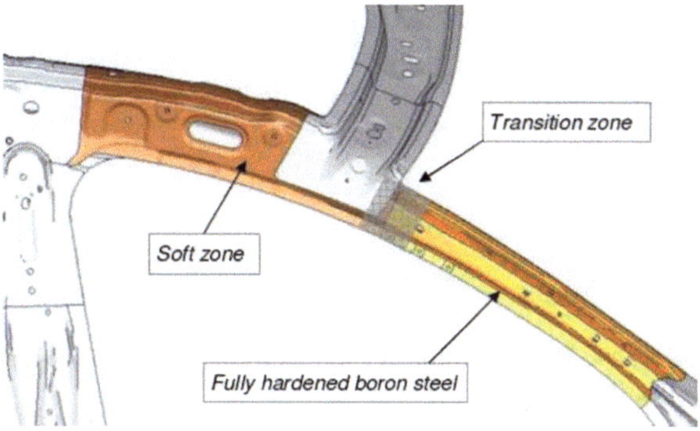

strength of hot stamping part, defects exist like poor formability and energy absorption, etc. Given this problem, component with high strength, better formability and good energy absorption in a certain part are in great need [22] and one of the solutions is TWB(tailored welded blank). In addition, such component with tailored properties can also be obtained in hot stamping by controlling the cooling rate when quenching the blanks. This technology makes it possible to achieve continuous change of property on one part without the use of welding, riveting, bonding or any other procedures therefore on one hand the producing process becomes more efficient and economic, on the other hand it is much easier to design component because of the simplified technic. One typical hot stamping A-pillar inner blank is presented in Fig. 3.28.

Through the control of producing process, components with one part hard and the other soft can be acquired and we call such property "Gradient Strength."

3.4.1 Forming Mechanism of Hot Stamping Gradient Strength Steel

When producing hot stamping blank with gradient strength, the key process lies in the control of the cooling rate. According to CCT curve shown in Fig. 3.4, when austenite is quenched at different cooling rates, different phase, and microscopic structure will be generated so that gradient strength boron steel can be acquired.

From Fig. 3.4, when cooling rates vary from 100 to 0.5 °C/s the microscopic structure of boron steel will change and result in different blanks with different Vickers hardness. If cooling rate is above 30 °C/s, microstructure will transform into martensite; If the rates keep growing the steel will become harder but with poor ductility for the huge internal stress caused in quenching and the state of microstructure is unstable either. While, when the cooling rate remains at 40 °C/s martensite will be kept uniformly in lath shape (shown in Fig. 3.29) so that steel can become harder and ductile as well. However, as cooling rate is lower than

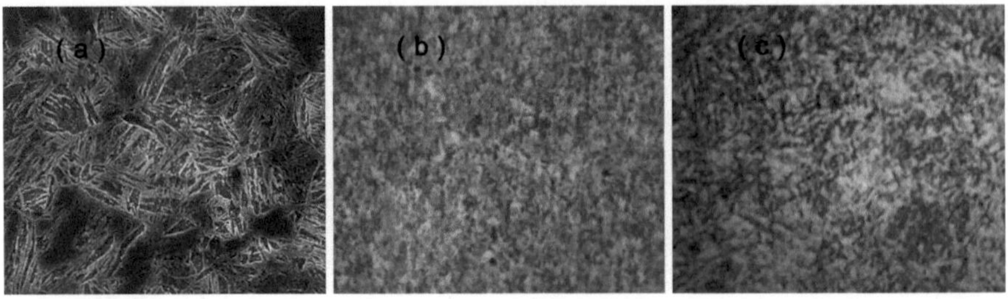

Fig. 3.29 The microstructure of 22MnB5 after different phase transformation **a** Martensite, **b** Pearlite, **c** Bainite

30 °C/s, martensite transformation is incomplete and, on the other hand, residual austenite increases and it will weaken the strength of material.

When the cooling rate is between 6 and 30 °C/s, as the rate raises, upper bainite, lower bainite and mixture of lower bainite, Martensite will be, generated respectively. Especially, when cooling rate is at 10 °C/s the bainite will transform into granular bainite and then decompose into ferrite and carbide known as pearlite (shown in Fig. 3.29). Due to the low strength of pearlite, the final property of material is unsatisfactory. If cooling rate remains between 10 and 25 °C/s, bainite is in low transformation temperature and will transform into black pin-like shaped low bainite which to the contrary has higher strength and ductility. When cooling rate varies from 25 to 30 °C/s, the microstructure may transform into lower bainite or Martensite so that the property of material is between those made of lower bainite or Martensite.

In summary, the phase transition is determined by cooling rate and the mechanical property of material lies to its microstructure which means by controlling the cooling rate hot stamping component with gradient strength can be achieved in practical manufacturing.

3.4.2 Experimental Research on District Cooling Process of Gradient Strength Part

Now that the mechanical properties of blank can be determined by quenching process, it is of great importance to confirm the optimal method to control cooling rate. Usually, the most commonly used cooling medium include water, oil, die, air, etc., and through the control of these parameters, cooling rate remains adjustable.

Nowadays, hot stamping technic has drawn great attention among scholars overseas. Svec et al. [37] found that by controlling temperature of die, it is available in the making of gradient strength component. When the die temperature rises to 500 °C, the hardness of the hot zone will be reduced to 240HV, the hardness of the cold zone is 420HV; Feuser et al. [31] and George et al. [10] achieved B-pillar hard in middle and soft on two sides by heating die locally. In order to find the appropriate process for the making of gradient strength part and establish the relation [49] of strength, hardness, cooling rate, and a kind of parted flat die was designed [36]. Experiments were conducted to achieve incomplete quenching under the control of temperature in cold zone and hot zone. 3 factors are studied in this experiment that are die temperature, pressure, and air gap.

(1) parted-cooling test

During quenching with flat die, the major factors are pressure, and die temperature. By controlling die temperature the heat-transfer rate can be changed that the cooling rate on a certain part can be controlled thereafter. Figure 3.30 shows the two-part flat die designed to carry out experiment about gradient strength material.

Hot stamping parted-cooling test was conducted on 40T press machine. The whole equipment contains 3 major parts including: (1) flat parted die; (2) resistance heating rods; (3) temperature controller which can change the temperature from 50 to 500 °C. The die used in the experiment is made of two main parts, upper and lower die. Each part is divided in two parts too, one is cold zone and the other is hot zone. There are 6400 W heating rods in each hot zone and none heating rods in cold zone. Between hot zone and cold zone, there is an adjustable air gap which is mainly used to isolate heat. Thermocouples are mounted between the surface of upper and lower die in order to monitor temperature in time.

According to experiment, the target temperature of hot zone were: 100, 300, 330, 370, 400, 500 °C. Test specimens were rectangular blank with 2 mm thickness made of 22MnB5. During experiment, specimen and die temperature were recorded by thermocouple imported from Japan. First of all, specimen was heated to 900 °C and insulated for 5 min so that material can be fully transformed to austenite. Then, it was transferred quickly into flat die which was already heated to a certain temperature. As specimen's temperature dropped down to 750 °C the optimal forming temperature, stamping started and when quenching process was finished, the specimen was exposed in air and cooled to room temperature. Part of the experiment was shown in Fig. 3.31.

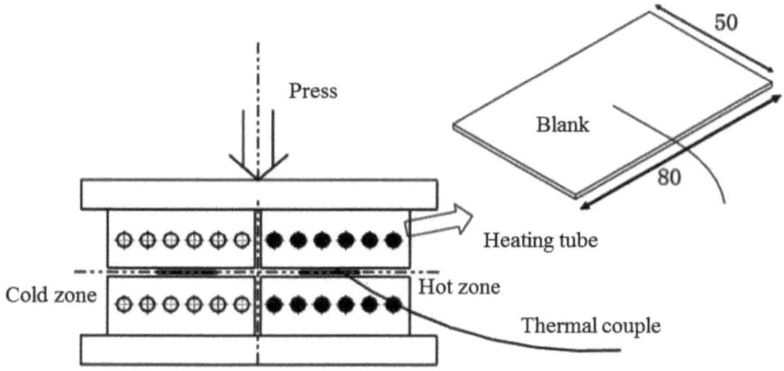

Fig. 3.30 Hot forming divided stage cooling plate tool

Fig. 3.31 Setup of the divided cooling plate experiment

Different cooling curves and cooling analysis are shown in Fig. 3.32, and it can be concluded that cooling rate becomes slower when die temperature is higher. When die temperature is at 300–400 °C, cooling rate is near critical cooling rate 30 °C/s; when temperature stays at 100 °C, cooling rate is about 70.55 °C/s nearly four times larger than 16.88 °C/s at 500 °C; when temperature are 300, 330, 370 °C cooling rate is about 35 °C/s. As temperature of die increases to 400 °C, the cooling rate only remains at 28.6 °C/s.

Figure 3.33 shows the changes of hardness on blank after hot stamping. Hardness decreases from the cold zone to the hot zone and different die temperatures result in different hardness at the beginning. It also reveals that if die temperature increases hardness difference between hot zone and cold zone is larger; when die temperature is no larger than 100 °C, full martensitic structure with hardness about 463HV–488HV will be acquired because cooling rate is high; at 300 °C, the hardness from hot zone to cold zone varies from 343HV to 455HV. Since there is a narrow air gap between the cold zone and the hot zone, blank here is cooled in air at a very low speed nearly 10 °C/s so that hardness in this part is much lower which finally shapes the distribution of hardness like letter "V." However, if die temperature is higher, "V" type phenomenon is less obvious but changes of hardness is much manifested. Take 400 °C curve for example, hardness on blank under cold zone reaches up to 450HV and the microstructure is full martensitic but from hot zone the hardness is only 243HV decreased by 39.3 %; as temperature rises to 500 °C, hardness from hot zone is 243HV, fell as much as 46 %.

In order to study the property of blank with gradient strength, specimens, 100 mm * 20 mm, were cut from blank to do tensile test and small blocks were also cut to measure the hardness. Final test results are shown in Fig. 3.34.

The figure above shows that: specimen (position 1) cut from cold zone has the maximum tensile strength of more than 1400 MPa, which proves that microstructure has fully transformed to fine lath martensite. Specimen (position 5) cut from

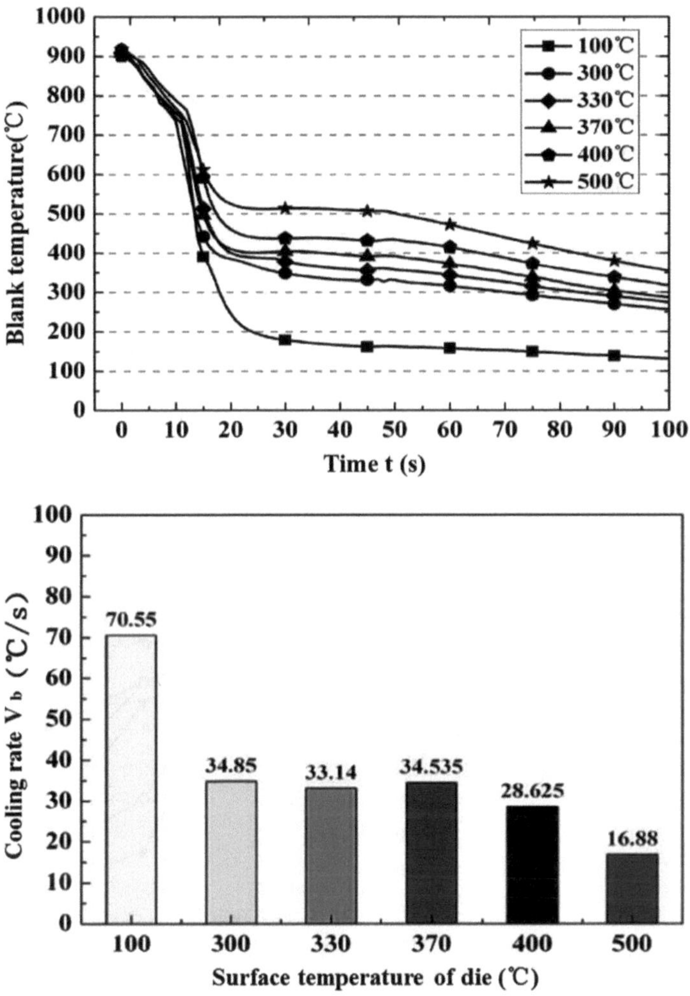

Fig. 3.32 Cooling rates analysis of blanks under different tool surface temperatures

hot zone has revealed strong relation between strength and die temperature for tensile strength reduces from 1460 to 763 MPa as die temperature increases. In addition, at 300–370 °C, tensile strength from position 3, 4, 5 again distributes like "V" letter but as long as temperature reaches to 400 °C or 500 °C, tensile strength is mainly related to die temperature.

(2) The impact of pressure on gradient properties

Controlling the pressure of die can change heat transfer coefficient on the surface between blank, thus changing the cooling rate in quenching. The specific principles are detailed in Chap. 6. Die pressure is a key parameter to realize gradient

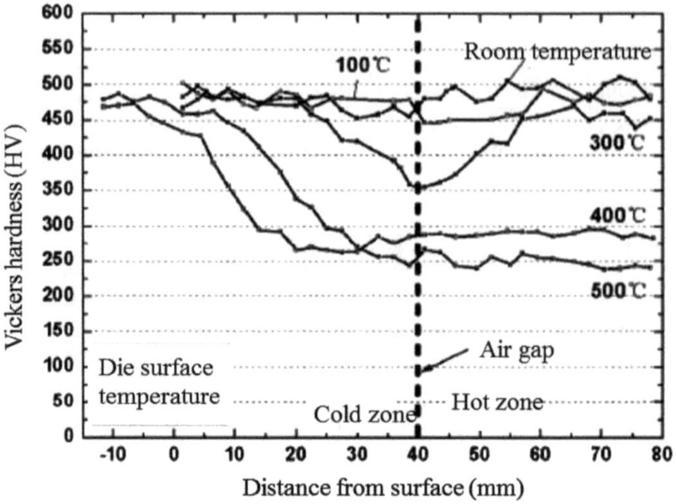

Fig. 3.33 The influence of hardness with different tool temperatures

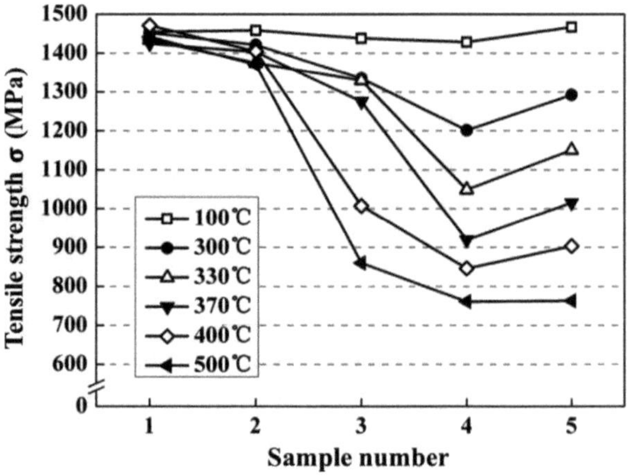

Fig. 3.34 The influence of strength with different tool temperature

properties of blank. To study its influence, three different working conditions, 5, 10, and 15 MPa, were selected in hot stamping experiment. Besides, any other factors like die temperature, air gap are the same with the experiment designed to study influence of die temperature.

Figure 3.35 shows the influence of die pressure on cooling rate. When the die temperature remains fixed, quenching rate decreases as die pressure rises; when die pressure is stable, quenching rate becomes slower when die temperature increases. Therefore, in the case of 100, 15 MPa, blank is quenched completely at the fastest cooling rate; Accordingly, in the case of 50 °C, 5 MPa, the quenching rate is minimal. By increasing the temperature gap between blank and die or increasing die

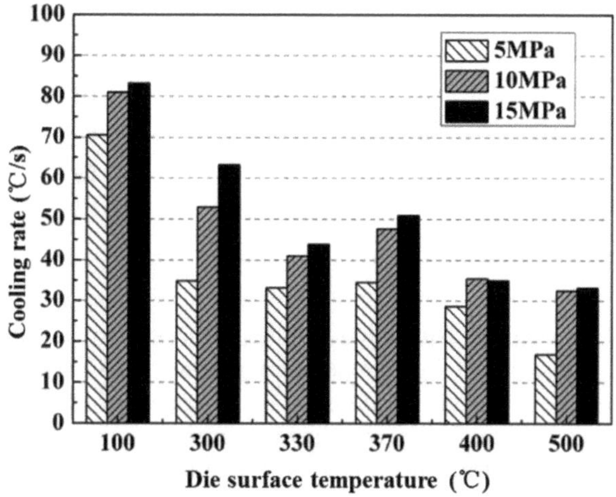

Fig. 3.35 The cooling rate change along with the contact pressures

pressure, both heat transfer rates can be promoted. But when the temperature is relatively higher, the increase of pressure can barely influence heat transfer. For example, as die temperature is over 400 °C or 500 °C, heat transfer rate would notpromoted though pressure increases from 10 to 15 MPa. The main reason might be that once blank is heated, it becomes softer so that it could be easily flatted under high pressure, but if the pressure kept increasing there is not any other room for blank to be further compressed therefore the quenching rate will not increase at last. Hardness scatter corresponding to die temperature and pressure is presented in Fig. 3.36. Table 3.7 shows hardness scatter under different temperatures and pressures.

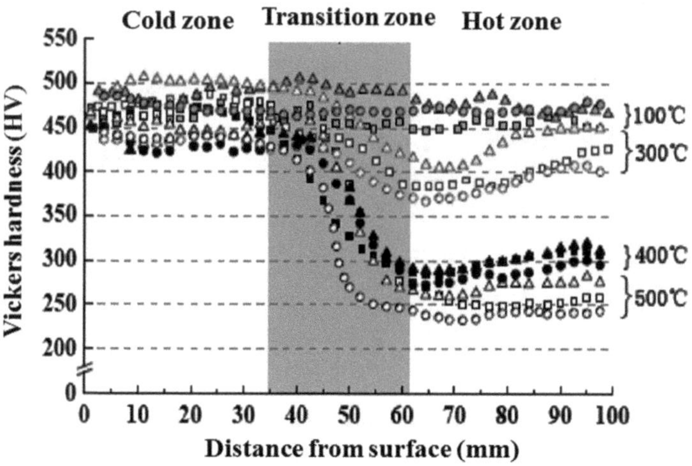

Fig. 3.36 The hardness scatters distribution of contact pressure and temperature

Table 3.7 Hardness scatter under different temperatures and pressures

Temperature in hot zone (°C)	Die pressure (MPa)		
	5	10	15
100	■ (grey)	▲ (grey)	● (grey)
300	□	△	○
400	■	▲	●
500	□	△	○

From Fig. 3.36 it can be concluded that the influence of pressure on hardness is not significant. Only when die temperature reaches to 400 °C or 500 °C, can hardness gradient be achieved. Regions on blank under lower cooling rate, due to the high temperature, cannot be transformed from austenite to martensite sufficiently and by the way structures mixed with bainite and ferrite start to appear so the hardness is lower. Meanwhile, hardness of specimen from cold zone almost exceeds 420HV which indicates that this part has been fully quenched and 90 % microstructure should be martensite; In addition, air gap between hot zone and cold zone plays an important role in insulating the heat wherefore cold zone remains unaffected during the experiments. Through measurement, the transition zone generated by air gap has a width of about 30 mm.

Further analysis of the influence of die pressure is shown in Fig. 3.37. Comparing hardness distribution under 10 and 5 MPa, it can be found that, despite the rising of pressure with cooling rate increasing a little consequently, only a slight reflection on strength gradient can be observed and the rule is not obvious. Therefore, in comparison, the effect of die temperature plays a dominant role in gradient property of material. In summary, component with gradient property requires die temperature at least up to 300 °C and die pressure more than 5 MPa.

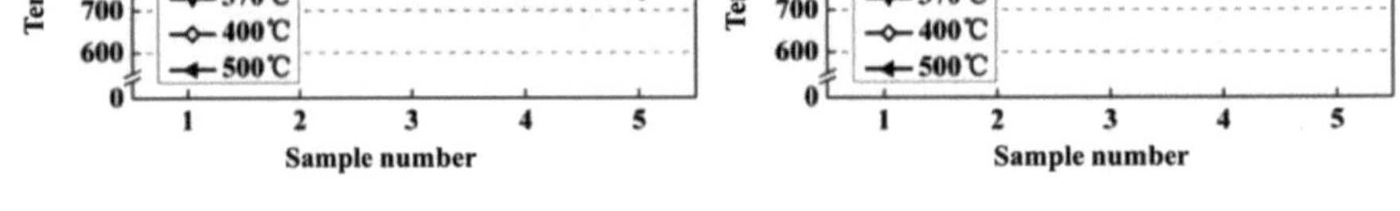

Fig. 3.37 Comparison of tensile strength under different contact pressures **a** 5 MPa, **b** 15 MPa

3.4.3 Exponential Relation Between Strength-Hardness-Cooling Rate of Hot Stamping Steel

From analysis concluded above, there is a certain relationship between strength, hardness, and cooling rate [49, 50]. In order to explore more detailed relationships among these parameters so that this technique can be applied in simulation and manufacture, optimal quenching process is utilized on the basis of data obtained in Sect. 3.4.2 to find the relation between cooling rate and mechanical properties [45]. Besides, the use of dimensional analysis and reverse extrapolation is used to establish exponential model of mechanical strength, hardness, and cooling rate. Next, program this model in our own software named KMAS_HF and comb heat transfer coefficient data related to die pressure, numerical simulation is carried out on typical U-shape part to testify strength, hardness distribution after quenching. The result shows that this program can effectively predict the final properties of hot stamping product and provide important guidance for the formability of hot stamping part with gradient properties as well as development of hot stamping die with such function.

A series of experiments are carried out to study quenching under different cooling conditions based on parted flat die. The whole producer detailed in Sect. 3.4.2 is taken as reference. Material tested is 22MnB5, of which the thickness is 2 mm, tensile strength unquenched is 600 MPa, hardness is 170HV and in each condition at least three specimens are prepared.

To study the effect of cooling rate, the cooling rate is separated in two rates first cooling V_1 and secondary cooling V_2. Researches have shown that, during stage V_1, no ferrite, bainite would be created if cooling rate is faster than critical rate and even if the blank is quickly quenched at MS point it may would not create fully martensite either. At stage V_2 the cooling rate is slower so the hardness of blank will decrease to 400HV and hardness is mainly effected in stage V_2, meanwhile at MS point, if blank is quenched far slower than critical rate (like 20 °C/s) tempering will happen automatically so the microstructure produces a mixture of acicular-and lath-shaped martensite which will decrease blank hardness further. Thus, cooling rate during quenching will influence the hardness of blank at last. That is, separated cooling should be considered when conducting numerical simulation and making prediction.

When the temperature of blank is between 750 and 400 °C, cooling rate during this period is defined as first cooling rate V_1 and if temperature is between 400 and 200 °C cooling rate in this stage is defined as secondary cooling rate V_2. During the experiment, blank was heated to 900 °C and kept about 6 min till the blank was transformed into austenite fully. Then, it was shifted to flat die quickly and cooled down to 750 °C the optimal forming temperature. After that it was quenched to 400 °C [5], the transition point of martensite (MS point), and later shifted to cooling material such as water, air, or die to continue quenching process. Cooling curves from experiments are shown in Fig. 3.38.

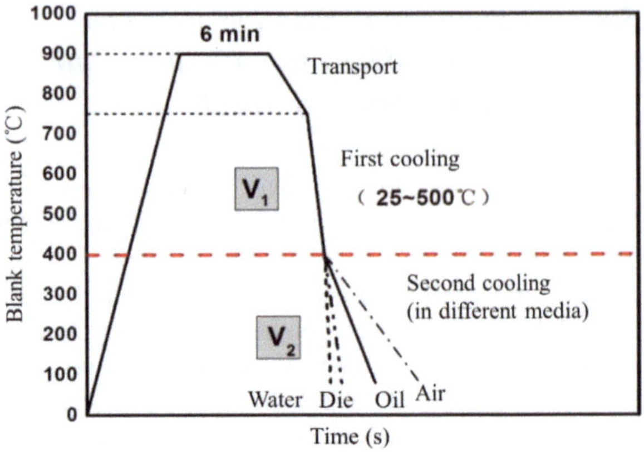

Fig. 3.38 Flowing chart of divided stage cooling test in phase cooling

To change rates of V_1 and V_2, different die temperatures (25, 100, 200, 300, 400 °C) and die pressures (5, 10, 20, 30 MPa) are set to control the cooling rate in stage V_1. During stage V_2, blank is shifted in cooling material so the cooling rate can be changed either. Though the usage of CCT curve can provide prediction in hot stamping, the cooling rate should be stable because this curve cannot guide hot stamping under different cooling rates. In this section, with the help of separated cooling rate V_1, V_2, a new method is obtained to predict the hardness and strength of products. From the result, it shows that hardness will increase by the increase of V_1 and once V_1 is fast enough it would not change further. On the contrary, hardness is barely effected by V_2. The analysis of microstructure reveals that when $V_1 > 60$ °C/s, no ferrite or bainite is generated. Cases are graphed in Fig. 3.39 when V_1 is between 60 and 70 °C/s.

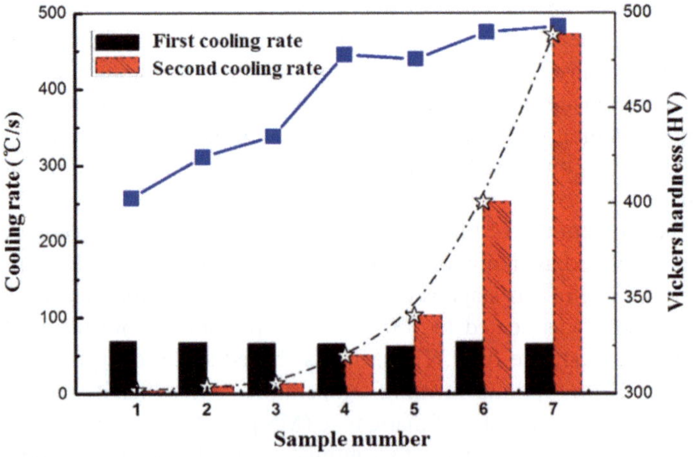

Fig. 3.39 Influence of phase cooling rate V_2-hardness for hot forming steel (V_1 = 60–70 °C/s)

When cooling rate V_2 is fast, 95 % microstructure in blank is lath-shaped martensite and hardness of blank reaches to its peak. With the decline of the cooling rate, the hardness is also reduced, for example hardness decreases to 400HV when cooling rate is 5 °C/s. All of these mean that rapid cooling at Ms point will not necessarily generate single martensite, drop of V_2 will also weaken hardness of blank. In other word, if V_2 is far lower than V_1, only soft structure with 400HV will be obtained according to CCT curve which is a mixture of lath-shaped martensite needle-shaped cementite. After phase transformation, cooling of martensite takes a long time and at the same time the precipitation of needle-shaped cementite will lead to decrease in hardness. This process is called auto-tempering process, smaller V_2 will enhance this process and then cause decrease in hardness.

Data in Fig. 3.40 shows that the strength and hardness of hot stamping steel are not a simple linear relationship, but with growth characteristics in nonlinear relationship. As the hardness increases, the strength also increases and growth rate of strength also increases with the increase of hardness. Therefore, a flexible exponential function model is adopted to describe the relationship as follows:

$$S = a + b * \exp(H/C), \tag{3.7}$$

where S is strength, H is Vickers hardness and a, b, c are constant coefficients which are calculated as 48.9079, 353.8484 and 350.0112, respectively after numerical fitting. Fitting curve is shown in Fig. 3.40. To test the reliability of this equation, determination coefficient R^2 can be used to test this equation, which is defined as follows:

$$R^2 = 1 - \frac{\sum (y_i - \hat{y}_i)^2}{\sum (y_i - \bar{y}_i)^2} \tag{3.8}$$

If R^2 is closer to 1 it means that curve fits better. $R^2 = 0.9662$ is obtained after data is substituted, indicating that it is proper to use flexible exponential model to fit the relationship between strength and hardness.

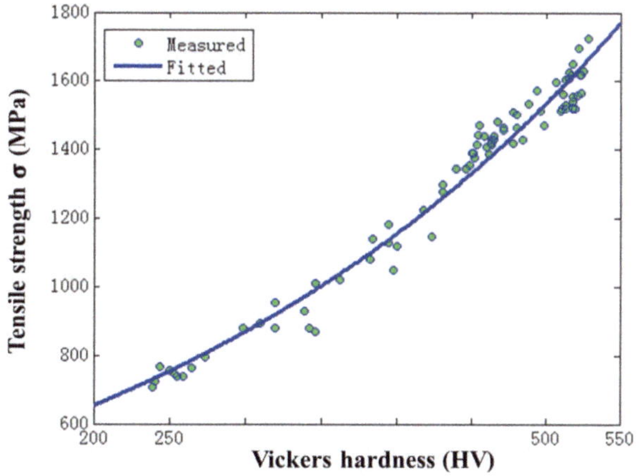

Fig. 3.40 The curve of strength and hardness for hot forming 22MnB5 steel

Dimensional analysis is a method to explore the relationship between physical quantities based on law of physical dimension, and obtain a final function [54]. To obtain a function model between mechanical properties and cooling rates of hot stamping products, inverse method [18] and dimensional analysis [53] are adopted to establish dimensionless function between hardness, strength, and cooling rate. Experimental data are summarized by using inverse method, and meanwhile function model of hardness, strength, and cooling rate is obtained in accordance with the intrinsic relationship between the physical quantities.

During hot stamping, complex changes happen to microstructure in material so do the mechanical properties of material. Mechanical properties of hot stamping products are not only affected by quenching conditions but also strength, hardness, and elongation of material are also closely related [45], for example, cooling rate, mechanical strength, and hardness are bonded closely. First, exponential model of strength, hardness, and cooling rate is established according to dimensional analysis.

The objective function is:

$$H = H(V_1, V_2, V_0, E, \sigma) \tag{3.9}$$

The formula above is generally expressed as:

$$F(V_1, V_2, V_0, E, \sigma, H) = 0, \tag{3.10}$$

where H is the hardness, σ is tensile strength, V_1 is the first cooling rate, V_2 is secondary cooling rate, V_0 is critical cooling rate (V_0 = 30 °C/s), E is Young's modulus. Any physical dimension can be described with three fundamental dimensions, that is mass M, length L, and time T. Through the analysis of physical quantities in objective function the explicit expression can be described on the basis of Buckingham pi theorem in dimensional analysis.

$$\frac{H}{\sigma} = \prod\left(\frac{\sigma}{E}, \frac{V_1}{V_0}, \frac{V_2}{V_0}\right) \tag{3.11}$$

$\prod$ is an unknown function which will be determined by reverse extrapolation.

From Sect. 3.4.2, when $V_2 > 20$ °C/s, hardness and tensile strength is mainly dependent on V_1; when $V_2 \leq 20$ °C/s, a significant temper process occurs which leads to a decrease of material. Therefore, function model can be divided by V_2 so that formula (3.11) can be converted to:

$$\frac{H}{\sigma} = \begin{cases} \Pi_\alpha\left(\frac{\sigma}{E}, \frac{V_1}{V_0}\right) & V_2 > 20\ ^\circ\mathrm{C/s} \\ \Pi_\beta\left(\frac{\sigma}{E}, \frac{V_1}{V_0}, \frac{V_2}{V_0}\right) & V_2 \leq 20\ ^\circ\mathrm{C/s} \end{cases} \tag{3.12}$$

$\prod_\alpha$,$\prod_\beta$ are obtained from experimental data by inverse method. To ensure the continuity and stability of function, nonlinear numerical analysis algorithms are used in MATLAB to fit the exponential model about strength and hardness shown in 3.12, 3.14. Where H_0 is the original steel hardness and other parameters are the same as mentioned above.

$$\frac{H}{H_0}=\begin{cases}\alpha+\beta*\exp\left(\chi*\frac{V_1}{V_0}\right)+\delta*\exp\left(\varepsilon*\frac{V_1}{V_0}\right) & V_2>20\,^{\circ}\mathrm{C/s}\\ \left(\alpha+\beta*\exp\left(\chi*\frac{V_1}{V_0}\right)+\delta*\exp\left(\varepsilon*\frac{V_1}{V_0}\right)\right)*\exp\left(1+a*\left(\frac{3V_2}{2V_0}\right)^b-(1+a)*\left(\frac{3V_2}{2V_0}\right)^c\right) & V_2\leq 20\,^{\circ}\mathrm{C/s}\end{cases}\tag{3.13}$$

$$\frac{H}{\sigma}=\begin{cases}\left(\rho+\theta*\exp\left(\varsigma*\frac{V_1}{V_0}\right)\right)*\left(\frac{\sigma}{E}\right)^{\xi} & V_2>20\,^{\circ}\mathrm{C/s}\\ \left(\psi+\varphi*\exp\left(\tau*\frac{V_1}{V_0}\right)*\exp\left(\phi*\frac{V_2}{V_0}\right)\right)*\left(\frac{\sigma}{E}\right)^{\vartheta} & V_2\leq 20\,^{\circ}\mathrm{C/s}\end{cases}\tag{3.14}$$

For 22MnB5 hot stamping boron alloy steel, constants are shown in Table 3.8.

Based on this exponential model, mechanical properties of blank during hot stamping, such as strength and hardness, can be predicted numerically according to cooling curve. First, numerical approximation is carried out in the way based on quartic polynomial function in formula 3.13 and polynomial coefficients will be set as initial value in prediction program. Prediction process is applied in hot stamping module in commercial CAE software KMAS/HF which is independently developed. Prediction diagram is shown in Fig. 3.41.

To verify the accuracy of exponential model, U-shaped mold is used to do numerical simulation and experiment. Results of simulation and test are compared and analyzed. Three-dimensional finite element model used in simulation is shown in Fig. 3.41. In simulation coupled with thermal and mechanical condition BT shell elements are used, which gives consideration to the effects of temperature changes on heat capacity and thermal conductivity. Besides, contact heat-transfer coefficient is the data mentioned in Chap. 6 measured with pressure changes.

Three-dimensional curve of V_1, V_2 and HV is shown in Fig. 3.42. When $V_2 \geq 20$ °C/s, first cooling rate V_1 is the most important factor on mechanical propertied of hot stamping products and strength increases by the increase of cooling rate; secondary cooling rate V_2 mainly takes effect in 0–20 °C/s. If V_2 is smaller, temper process is more obvious and hardness is lower; when $V_2 \leq 5$ °C/s, the self-tempering effect of materials reaches to its limit.

U-shaped hot stamping experiment is shown in Fig. 3.43. Blank thickness is 1.6 mm, die clearance is 2.06 mm. Actually, the blank used is not matched with mold because in this way a strength gradient would be created on cross-section of

Table 3.8 Coefficients of cooling rate-strength-hardness exponential index model

α	β	χ	δ	ε	a	b	b	
3.1410	−2.2585	−1.2045	−0.3769	−0.1094	0.1339	3.3227	0.0911	
ϱ	θ	ς	ξ	ψ	φ	τ	ϕ	ϑ
0.3847	1.9379	−12.4378	−0.0863	0.3844	1.9553	−13.0659	4.0600	−0.0870

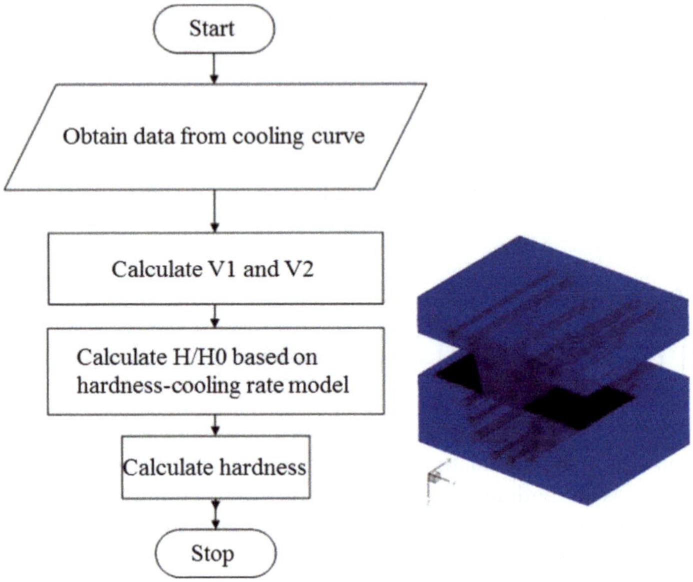

Fig. 3.41 Schematics of prediction strength procedure for hot forming and the FE model

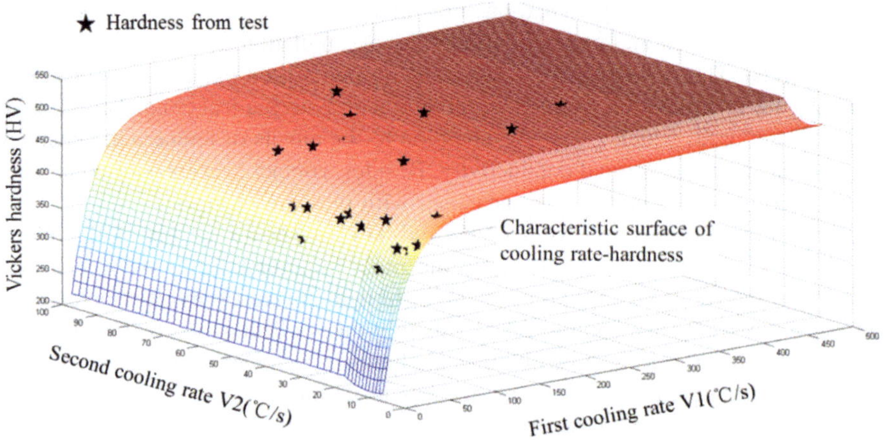

Fig. 3.42 Three-dimensional diagram of cooling rate-hardness exponential index model

U-shaped part. During the test, blank was heated to 900 °C, insulated for 6 min and cooled to 750 °C on receiver of water-cooled mold, then blank was stamped quickly and kept in mold under 20 MPa about 30 s.

Figure 3.44 shows that the simulation results are in good agreement with the experimental results. Position of upper and lower corner on blank are cooled faster, while area on sidewall and bottom are cooled slower. It might be that there is a

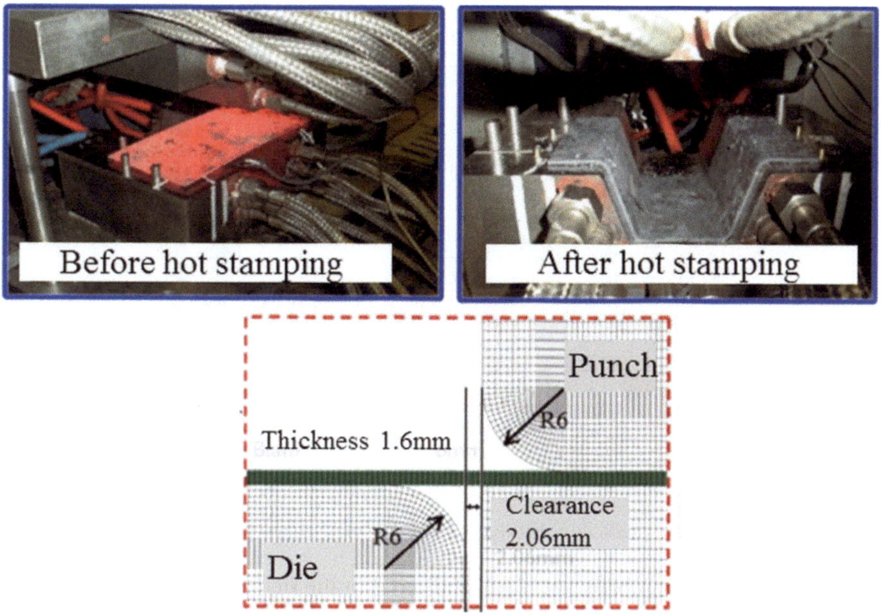

Fig. 3.43 The parameters for U-shape hot forming experiment and simulation

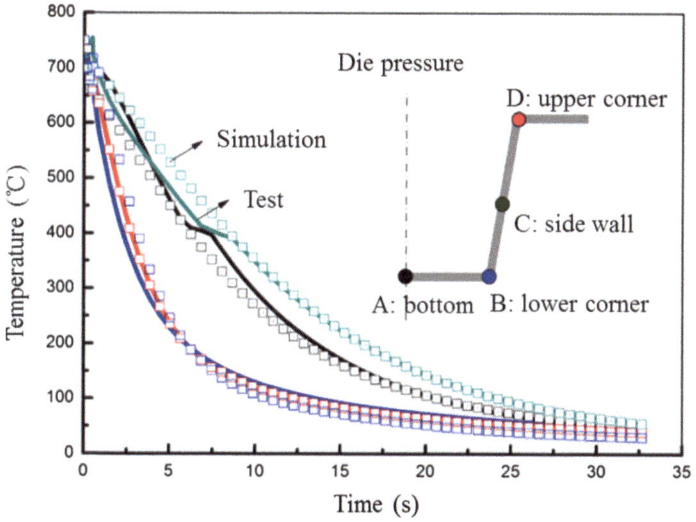

Fig. 3.44 The temperature curves of typical measured points on the blank

clearance between sidewall and mold, and the pressure is small so that the cooling rate is small. In addition, position near lower corner had small cooling rate too, this might be caused by smaller thickness under excessive stretch in mold; at the bottom

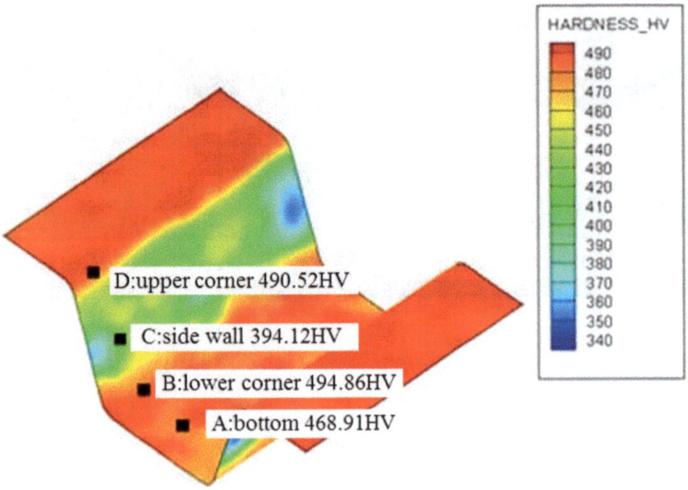

Fig. 3.45 The hardness prediction distribution contour of U-shape product

position, there is a gap between blank and mold in the first 1–2 s so that the cooling rate is small, and only when the blank is stamped completely, the cooling rate becomes bigger.

Hardness distribution of U-shaped part from simulation is shown in Fig. 3.45. It is clear that area in the side walls is not quenched completely and its hardness is only about 394.12HV, while the other three location points are fully quenched and the hardness is larger than 450HV. Hardness resulting from 4 points (A-bottom, B-lower corner, C-sidewall, D-upper corner) shows that experimental test and simulation are fitted well and the error is between 20 and 24HV which indicates that the numerical model can obtain a good prediction.

3.5 Summary

In this chapter, three major parameters about mechanical properties in hot stamping process are studied. From the experiment, hot stamping process is able to avoid the difference in thickness on boron steel if material is fully quenched. Macroscopic mechanical properties of materials (yield strength, tensile strength, hardness, elongation) under different heating temperature, holding time and cooling rate are studied through experiment. In addition, microscopic structure (martensite morphology, austenite grain size assessment) is also analyzed. All these work help to understand the behavior of material during hot stamping and the optimal process parameters are as follows: heating temperature 900–920 °C, holding time 3–5 min, cooling rate 85–175 °C/s. Based on the optimal hot stamping process, microstructure in steel is fully transformed to martensite and tensile strength of material can exceed over 1500 MPa on the other hand.

Orthogonal and integral evaluation methods are adopted in KAHN tear test. Parameters like heating temperature, stamping temperature, soaking time, are chosen to study toughness of material under quenching and tempering. Evaluation of tensile strength, elongation, tearing strength, and crack nucleation energy in unit area is used to know the influence of different parameters to material toughness at high-temperature quenching process.

In consideration of improving mechanical properties of hot stamping material, determine the heating temperature, soaking time, quenching temperature, optimum tempering temperature are 920 °C, 1 min, 650–700, 200 °C, respectively. The outcome of orthogonal analysis and variance test indicated that this technic process is able to guide actual production in hot stamping.

At last, to obtain a hot stamping component with overall high strength but lower strength locally for energy absorption, a parted flat mold is designed to study tailored properties of material. In this research, mold temperature and pressure are studied with the purpose to guide actual production. On the basis of this test, exponential model of strength, hardness, and cooling rate are established through dimensional analysis. Meanwhile, this model is integrated into KMAS/HF, a independent developed software, and a typical U-shaped part is adopted to evaluate the accuracy of simulation using 22MnB5 blank under hot stamping process. With this technic, prediction of material properties can be realized and development of hot stamping component with tailored properties can also be simulated.

References

1. Abareshi M, Emadoddin E (2011) Effect of retained austenite characteristics on fatigue behavior and tensile properties of transformation induced plasticity steel. Mater Des 32:5099–5105
2. Abbasi M, Saeed-Akbari A, Naderi M (2012) The effect of strain rate and deformation temperature on the characteristics of isothermally hot compressed boron-alloyed steel. Mater Sci Eng, A 538:356–363
3. Benyounis KY, Olabi AG (2008) Optimization of different welding processes using statistical and numerical approaches—a reference guide. Adv Eng Softw 39:483–496
4. Bin W (2004) Brief introduction to standard for no TCHED bar im PAC t testing of ASTM. Phys Testing Chem Anal (Part A Physical Testing) 7:367–371 (in Chinese)
5. Byoungchul Hwang CGLA (2010) Low-temperature toughening mechanism in thermomechanically processed high-strength low-alloy steels. Metall Mater Trans A
6. Cao R, Feng W, Peng Y et al (2010) Investigation of abnormal high impact toughness in simulated welding CGHAZ of a 8 %Ni 980 MPa high strength steel. Mater Sci Eng, A 528:631–642
7. Chang Y, Meng Z-H, Ying L et al (2011) Influence of hot press forming techniques on properties of vehicle high strength steels. J Iron Steel Res Int 18:59–63
8. Chengxi L, Cui J, Xing Z et al (2012) Microstructure prediction of high strength steel square-cup part in hot stamping. J Plast Eng 6:40–44 (in Chinese)
9. Edmonds DV, He K, Rizzo FC et al (2006) Quenching and partitioning martensite—a novel steel heat treatment. Mater Sci Eng, A 438–440:25–34

10. George R, Bardelcik A, Worswick MJ (2012) Hot forming of boron steels using heated and cooled tooling for tailored properties. J Mater Process Technol 212:2386–2399
11. Guoying Z, Liu C, Zhang H (2005) Electronic theory for the refinement mechanism of ultrafine steel. Physica Sinica 2:875–879 (in Chinese)
12. Honglin Y, Liu X, Li J et al (2013) Research status hot stamping steel coating technology. J Iron Steel Res Int 6:1–7+12 (in Chinese)
13. Hongsheng L, Xing Z, Lei C (2012) Hot formation quality of high strength steel BR1500HS for hot stamping without cooling system. Trans Nonferrous Metals Soc China 2:542–547 (in Chinese)
14. Hongyuan Y (2013) Research on hot forming process improvement of automobile high strength steel and its application. Dalian University of Technology, p 72 (in Chinese)
15. Hou H, Ying L, Wu X et al (2012) The influence of heating temperature on 22MnB5 microstructure and austenite grain. China Metalforming Equip Manuf Technol 6:89–91 (In Chinese)
16. Jingchun F, Chen L, Hu Z (2003) Comprehensive mark-giving method and its model study for bidding evaluation of water conservancy projects. J Hehai Univ (Nat Sci) 4:461–465 (in Chinese)
17. Junye C, Zhao A, Chen Y et al (2013) EBSD studies of 30MnB5 hot stamping steel temperature at different temperature. Metallurgica Sinica 2:137–145 (in Chinese)
18. Kolleck R, Veit R, Merklein M et al (2009) Investigation on induction heating for hot stamping of boron alloyed steels. CIRP Ann Manuf Technol 58:275–278
19. Lee S, Kim J, Lee S-J et al (2011) Effect of Cu addition on the mechanical behavior of austenitic twinning-induced plasticity steel. Scripta Mater 65:1073–1076
20. Lianfang H, Guoqun Z, Huiping L et al (2011) Optimization of quenching parameters for hot stamping boron steel B1500HS based on response surface methodology. J Mech Eng 47:77–82 (in Chinese)
21. Liang Y (2013) Research and application on key process experiment of high strength steel for hot forming. Dalian University of Technology, p 179 (in Chinese)
22. Liang Y, Zhang F, Dai M et al (2015) Optimization of hot forming vehicle structure with tailored strength based on side crash simulation. In: 2015 annual meeting of China Automotive Engineering Society. Shanghai, China (in Chinese)
23. 23. Liu HW (1983) On the fundamental basis of fracture mechanics. Eng Fracture Mech 17:425–438
24. Ma N (2011) Research on hot forming of high strength steel. Dalian University of Technology, p 224 (in Chinese)
25. Ma N, Hu P, Guo W (2010) Experiments and analysis of relations among heat, stress and transformation of boron steel for hot forming. Trans Mater Heat Treat 11:33–36 (in Chinese)
26. Ma N, Hu P, Yan K et al (2010) Technology and its application of high-strength boron steel hot forming. J Mech Eng 14:68–72 (In Chinese)
27. Merklein M, Lechler J (2006) Investigation of the thermo-mechanical properties of hot stamping steels. J Mater Process Technol 177:452–455
28. Mori K, Ito D (2009) Prevention of oxidation in hot stamping of quenchable steel sheet by oxidation preventive oil. CIRP Ann Manuf Technol 58:267–270
29. Nader M (2007) Hot stamping of ultra high strength steels. In:Rheinisch-Westfälischen Technischen Hochschule Aachen
30. Naderi M, Saeed-Akbari A, Bleck W (2008) The effects of non-isothermal deformation on martensitic transformation in 22MnB5 steel. Mater Sci Eng, A 487:445–455
31. Feuser P, Schweiker T, Merklein M (2011) Partially hot-formed Parts from 22MnB5-process window material characteristics and component test results. In: 10th international conference on technology of plasticity. Aachen, Germany %\ 2013–09-28 16:21:00
32. Rongfeng L, Yu L (2013) Some issues on cracking perpendicular to thickness in fractured appearance of mechanical tested steel specimen. Dev Appl Mater 3:1–8 (in Chinese)
33. Ruijiang L, Zhang Y, Wen C et al (2010) Study on the design and analysis methods of orthogonal experiment. Exp Technol Manag 9:52–55 (in Chinese)

34. Sha Q, Sun Z (2009) Grain growth behavior of coarse-grained austenite in a Nb–V–Ti microalloyed steel. Mater Sci Eng, A 523:77–84
35. Shudong Y, Shi W (2009) High temperature rapid tempering process for hardened steel. Heat Treat Met 2:92–95 (in Chinese)
36. Siying D (2013) Influence of tailored tempering factors on the mechanical properties of born steels within hot stamping. Dalian University of Technology, p 81 (in Chinese)
37. Svec T, Gruner M, Merklein M (2011) FE-simulation of the heat transfer by defined cooling conditions during the hot stamping process. Key Eng Mater 699–706 %\ 2013-2009-2028 2009:2049:2000
38. Turetta A, Bruschi S, Ghiotti A (2006) Investigation of 22MnB5 formability in hot stamping operations. J Mater Process Technol
39. Walubita LF, Simate GS, Ofori-Abebresse E et al (2012) Mathematical formulation of HMA crack initiation and crack propagation models based on continuum fracture-mechanics and work-potential theory. Int J Fatigue 40:112–119
40. Windmann M, Röttger A, Theisen W (2013) Phase formation at the interface between a boron alloyed steel substrate and an Al-rich coating. Surf Coat Technol 226:130–139
41. Xiaojun L, Jiao S, Wang C et al (2006) Effect of tempering on microstructure and properties of direct quenching steel. Hot Work Technol 8:47–50 (in Chinese)
42. Xiaoqiang H, Yu H, Ying L et al (2013) Research on Influence of Key Process Parameters of Hot Stamping on Mechanical Properties of Parts. China Metalforming Equip Manuf Technol 4:58–62 (in Chinese)
43. Ying C, Ying L, Jin F et al (2013) Effect of tempering treatment process on mechanicalproperties of hot-stamping AHSS alloy. Trans Mater Heat Treat 11:74–78 (in Chinese)
44. Ying L, Chang Y, Hu P et al (2011) Influence of low tempering temperature on fracture toughness of ultra high strength boron steel for hot forming, pp 160–165
45. Ying L, Dai M, Hu P et al (2014) Strength and hardness prediction based on cooling rate for hot forming high strength steel. J Jilin Univ (Eng Technol Edn) 6:1716–1722 (in Chinese)
46. 46. Ying L, Jia Z, Chang Y et al (2013) Optimization of strength and toughness parameters in hot stamping process for ultra high strength steel. Mater Sci Technol 6:72–77 (in Chinese)
47. Ying L, Jia Z, Hu P et al (2013) Improvement of hot stamping process for high strength steel. China Metal Forming Equip Manuf Technol 1:75–78 (in Chinese)
48. Ying L, Lu J-D, Chang Y et al (2013) Optimization evaluation test of strength and toughness parameters for hot-stamped high strength steels. J Iron Steel Res Int 20:51–56
49. Ying L, Yu Y, Dai M et al (2015) Influence of forming-quenching parameters on hardness and strength of high strength steel hot forming product. Trans Mater Heat Treat 1:205–210 (in Chinese)
50. Ying L, Yu Y, Dai M et al (2015) Influence of hot stamping RCP process on microstructure and properties of 22MnB5 sheet. Heat Treat Met 1:35–38 (in Chinese)
51. Ying L, Zhao R, Gao T et al (2015) Optimization of water-nitrogen spray quenching process parameters for 22MnB5 high-strength boron steel. Trans Mater Heat Treat 10:230–235 (in Chinese)
52. Yu W, Qian Y-J, Wu H-B et al (2011) Effect of heat treatment process on properties of 1000 MPa ultra-high strength steel. J Iron Steel Res Int 18:64–69
53. Zemin F, Jianhua M, Wei C et al (2010) Springback model for air-bending of sheet metal based on dimensional analysis. Method J Mech Eng 46:53–58 (in Chinese)
54. Zengsheng M (2011) Characterization of the mechanical properties of metallic films by nanoindentation method. Xiangtan University, p 75 (in Chinese)
55. Zhang W (2010) inspection methods of austenitic grain size in stainless. Hot Work Technol 22:66–68 (in Chinese)
56. Zhao K, Chang Y, Hu P et al (2016) Influence of rapid cooling pretreatment on microstructure and mechanical property of hot stamped AHSS part. J Mater Process Technol 228:68–75

57. Zhaohuan M (2012) Process optimization and experimental analysis of strength and toughness for hot stamping high strength steels. Dalian University of Technology, p 78 (in Chinese)
58. Zhiyu J (2013) Experimental research on thermal-mechanical-phase transformation of vehicle high strength steel hot forming process. Dalian University of Technology, p 77 (in Chinese)

Chapter 4
The Basic Theory and Constitutive Equation of High Strength Steel for Hot Forming

4.1 Multifield Coupled Relationship Among Heat, Stress and Phase Transformation

4.1.1 Theoretical Analysis

In the front section of this book, we briefly describe the basic principle of high strength steel for hot stamping technology. The special boron alloy steel is heated to make the austenitic change. Then the red-hot sheet metal is put into a mold with cooling system to form. At the same time it is quenched by mold and the steel plate organization transforms from austenite into marten site [11]. Thus the high strength structure of car is obtained.

In the quenching process, for the diffusion transformation, it can be described by a classical K-V model [7]:

$$\frac{dX}{dt} = f(G,T)f(C)f(X) \tag{4.1}$$

where X represents the volume fraction of formed phase, t represents time, $f(G,T)$ represents the effect of austenite grain size and temperature to phase transformation rate, $f(C)$ represents the effect of alloying element to phase transformation rate, $f(X)$ represents the effect of the volume fraction of formed phase to phase transformation rate.

Li et al. [10] modified the K-V model to improve the accuracy of predicting microstructure evolution while in the continuous cooling process, which can be seen in Sect. 8.4.2.

In the quenching process, for the diffusion-less martensite transformation, the relationship between temperature and phase change is shown as follows [8]:

P. Hu et al., *Hot Stamping Advanced Manufacturing Technology of Lightweight Car Body*, DOI 10.1007/978-981-10-2401-6_4

$$X_{\mathrm{M}} = 1 - \exp[-\theta(M_{\mathrm{s}} - T)] \tag{4.2}$$

where X_{M} represents the fraction of martensitic transformation: θ represents the material parameter which reflects the austenite-martensite transformation rate, and T represents temperature.

The volume fraction of remaining austenite is shown as follows:

$$X_{\mathrm{A}} = 1 - X_{\mathrm{F}} - X_{\mathrm{P}} - X_{\mathrm{B}} - X_{\mathrm{M}}. \tag{4.3}$$

where $X_{\mathrm{F}}, X_{\mathrm{P}}, X_{\mathrm{B}}$ are volume fractions of ferrite, pearlite, and bainite respectively.

The ultimate hardness of hot stamping structural component can be predicted [13] by Eq. (4.4)

$$Hv = (X_{\mathrm{F}} + X_{\mathrm{P}})Hv_{\mathrm{F+P}} + X_{\mathrm{B}}Hv_{\mathrm{B}} + X_{\mathrm{M}}Hv_{\mathrm{M}} \tag{4.4}$$

where $Hv_{\mathrm{F+P}}, Hv_{\mathrm{B}}, Hv_{\mathrm{M}}$ represent the hardness of mixture of ferrite and pearlite, bainite, and martensite respectively.

The phase change volume strain and phase change plastic strain caused in the process of phase transformation are described in Eqs. (4.5) and (4.5) respectively [3, 14]:

$$d\varepsilon_{ij}^{\mathrm{tr}} = \sum_{I}^{N} \beta_I \delta_{ij} dX_I \tag{4.5}$$

$$d\varepsilon_{ij}^{tp} = \sum_{I}^{N} 3K_I(1 - X_I)s_{ij} dX_I \tag{4.6}$$

In the upper equations, β represents the phase transformation expansion factor of phase $I; X_I$ represents the volume fraction of generated phase $I; \delta_{ij}$ represents Kronecker symbol; K_I represents the phase transformation plastic strain factor of phase I; N represents the amount of phases; s_{ij} represents deviatoric stress.

The above-mentioned equations are obtained by experiment from generally quenching process. Both large deformation and quenching happen during the hot forming process of boron steel, which integrate the traditional sheet metal forming technology and heat treatment technology together. Then boron steel will bear the effect of large deformation and large stress during quenching, and plastic heat damage and fracture phenomena are the possible results. So, the study of the action of heat, stress, phase transformation's coupled relationship under high stress in boron steel (beyond the yield stress and can achieve tensile strength) is of great significance to deeply understand hot forming mechanism and the same to subsequent numerical simulation.

By taking heat, stress, phase transformation's coupled relationship during hot stamping into consideration, we assume that the martensite transformation starting temperature M_{s}, material constants of martensite transformation rate θ and phase change plastic coefficient k, will all depend on the real equivalent stress $\bar{\sigma}$, thus the heat, force, and phase change's coupling relationship can be expressed as [5, 6, 12]:

$$X_{\mathrm{M}} = X_{\mathrm{M}}(\bar{\sigma}, T) \tag{4.7}$$

$$\dot{\varepsilon}^{\mathrm{trM}} = \dot{\varepsilon}^{\mathrm{trM}}(\bar{\sigma}, T, X_{\mathrm{M}}; \dot{\bar{\sigma}}.\dot{T}, \dot{X}_{\mathrm{M}}) \tag{4.8}$$

$$\dot{\varepsilon}^{\mathrm{tpM}} = \dot{\varepsilon}^{\mathrm{tpM}}(\bar{\sigma}, T, X_{\mathrm{M}}; \dot{\bar{\sigma}}.\dot{T}, \dot{X}_{\mathrm{M}}) \tag{4.9}$$

The above definite expression of the constitutive equations needs to be determined by means of the experiment. Then as an example, the hot forming high strength steel boron is taken to show the relationship between M_s, θ, K_{M} and equivalent stress $\bar{\sigma}$.

4.1.2 The Determination of the Parameters

In order to determine the heat, force, phase transformation coupling relationship during high strength steel plate hot forming process, the experiment scheme is designed, which contains all key characteristics in hot forming process. The experimental process is showed in Fig. 4.1: the experiment sample is heated to 950 °C with a heating speed of 10 °C/s, a heat preservation for 5 min at 950 °C, and then is cooled to room temperature with a cooling rate of more than 30 °C/s. During the cooling process, displacement load is applied when the temperature is among 750–500 °C until it rises up to room temperature. We record the experimental process force, displacement, the expansion quantity in width direction and the curve of the temperature versus time. After the experiment, the metallographic

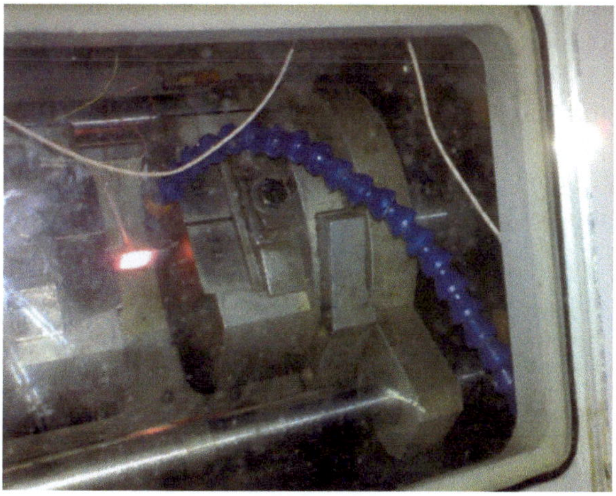

Fig. 4.1 Experimental process by forming and quenching at the same time

analysis is made after linear cutting of the central mean temperature district, to verify whether the martensite transformation is sufficient or not.

The specimen welded with thermocouple is showed in Fig. 4.2, and the thickness is 1.6 mm. In the experiment, the samples are heated through the current. In order to assure uniform heating in the region of the standard test distance of sample, it is necessary to adjust electrode contact way, fixture contact area, etc., and then to test the temperature through welding thermocouple on specimen, which is used for temperature controlling and temperature outputting. The welding machine which is used for thermocouple welding is independently developed. The interior of spot welding machine is a capacitor, and its principle is as follows: First, charge the capacitor first to achieve the capacitance of the voltage, and then click the discharge buttons to make it discharge quickly, and make use of the instantaneous oversize electric to melt the thermocouple wire on sheet metal. In this experiment, K-thermocouples is the widely used metal thermocouple at low price, and the usage amount is the summation of that of other thermocouples. K-thermocouple has many advantages, such as good linearity, big thermo-electromotive force, high sensitivity, higher stability, uniformity, strong oxidation resistance, cheap, etc., and it can be used in oxidizability inert atmosphere (Fig. 4.3).

The diagram of working room schematic of multifunction thermal-mechanical simulator is shown in Fig. 4.3. After welding thermocouples, the sample is fixed on the hydraulic clamp by the installation fixture to carry on tensile action; after thermocouples are connected with the simulator special thermocouple, the temperature online testing can be implemented; during testing process, the samples are quenched by cooling ducts which are also shown in the Fig. 4.3, and the simulator's displacement sensor and force sensors collect real-time data, and the high temperature displacement sensor records the expansion shrinkage in transverse direction of the sample at the same time. The other physical quantities can be monitored in real-time by interface as shown in Fig. 4.4.

The relation of physical parameters and starting temperature of martensite transformation is explained by the curve $\Delta d - T$ (Fig. 4.5), the relation of sheet width inflation quantity and temperate and the curve $F - T$ (Fig. 4.6), the relation of force and the temperature in a certain experiment.

The horizontal shrinkage of AB section in Fig. 4.5 is resulted from thermal-mechanical coupled effect of the thermal strain ε^{th}, elastic strain ε^{e}, and plastic strain ε^{p} (which are determined on whether tension stress exceeds the yielding limit or not). The horizontal shrinkage of BC section is caused by thermal-mechanical transformation coupled effect, such as $\varepsilon^{tr}\varepsilon^{tp}$ and $\varepsilon^{th}\varepsilon^{e}\varepsilon^{p}$, which

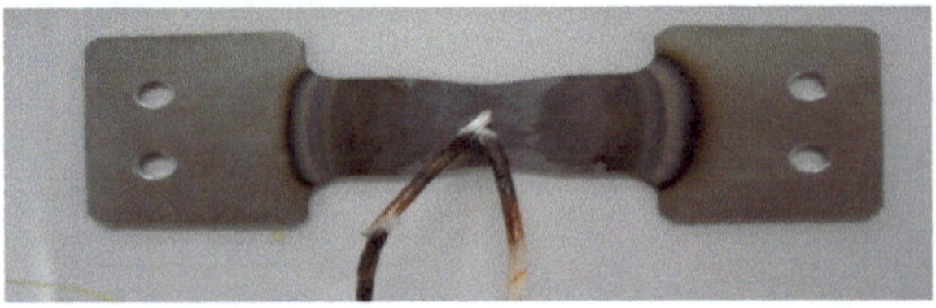

Fig. 4.2 Specimen with thermocouple for testing of coupled relations

Fig. 4.3 Working room of multifunction thermal-mechanical simulator

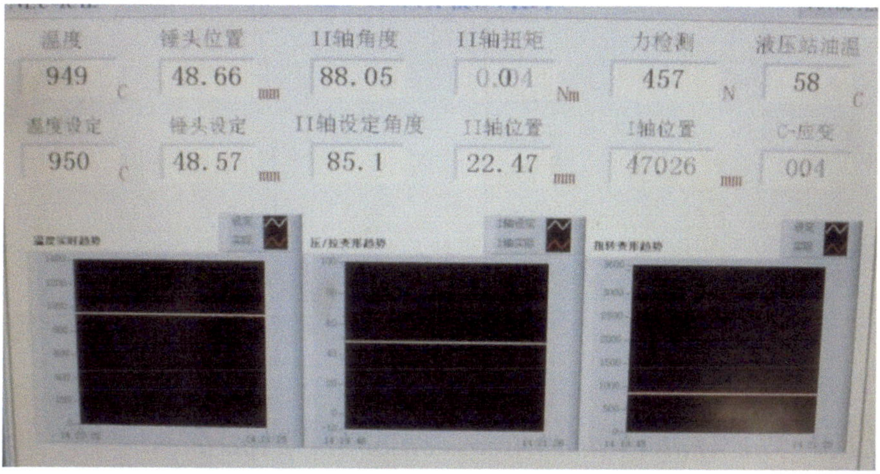

Fig. 4.4 Testing interface of multifunction thermal-mechanical simulator

is resulted from phase transformation. The horizontal shrinkage of CD section is caused by thermal-mechanical coupled effect after phase transformation such as ε^{th}, ε^{e}, and ε^{p} (which is determined on whether the stress exceeds the yielding limit or not). In this paper, the thermal-mechanical-transformation coupled parameters θ and k are determined by the phase transition process of BC.

The martensite transformation starting temperature can be determined by the Ms point of curves in Figs. 4.5 and 4.6 obviously.

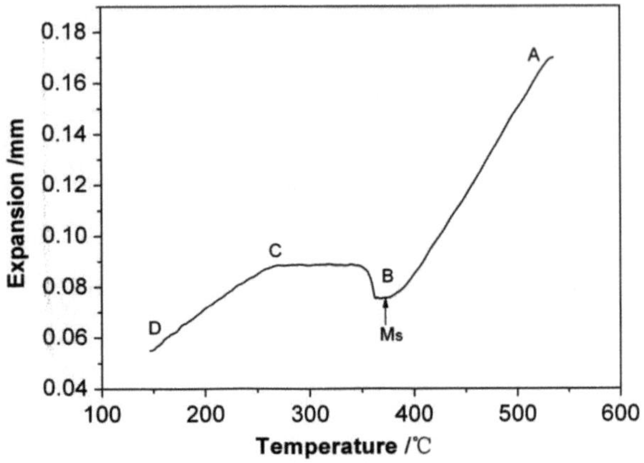

Fig. 4.5 Thermal expansion as a function of temperature

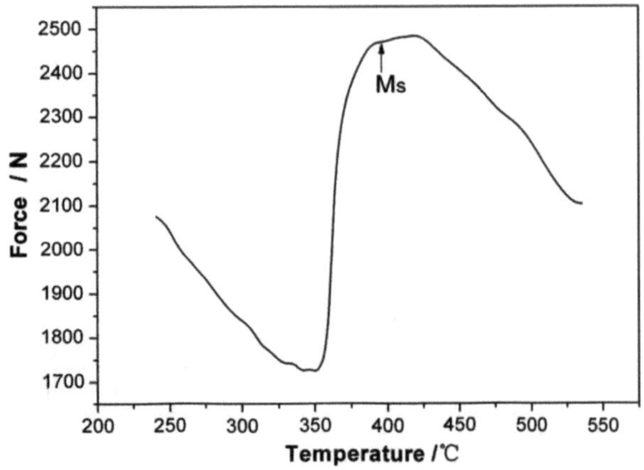

Fig. 4.6 Tensile force as a function of temperature

4.1.3 The Analysis and Discussion on the Experiment Results

The metallographic analysis of the specimen after experiment shows that it consists of more than 95 % martensite and a small amount of residual austenite.

The corresponding relation between the stress and the starting temperature of martensite transformation is shown in the Fig. 4.7. When the stress $\bar{\sigma}$ is less than a certain value, M_s is essentially constant (380 °C). When $\bar{\sigma}$ changes within a certain range, M_s increases with the increase of $\bar{\sigma}$. When $\bar{\sigma}$ is more than one particular value, M_s tends to be the certain value (565 °C). It can be seen from the curve that when the stress reaches 400 MPa, the temperature of martensite transformation increases by about 180 °C, which is important to the hot forming mechanism, production engineering and numerical simulation of boron steel [5, 6].

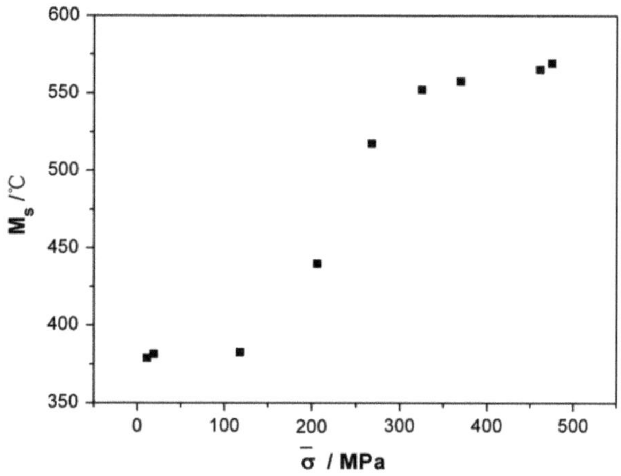

Fig. 4.7 The coupled relation between the martensite start temperature M_s and the effective stress by experiment

The martensite nucleation theory can explain the corresponding relation between stress and M_S, which is shown in Fig. 4.7. The total incremental of the Gibbs free energy of martensite nucleation ΔG can be expressed as [16]:

$$\Delta G = A\psi + V\Delta G_s - V\Delta G_v - \Delta G_d \tag{4.10}$$

where ψ is the interfacial free energy; ΔG_s is the strain energy; ΔG_v is the volume free energy; ΔG_d stands for the dislocation interaction energy; V is the volume of the core; and A is the surface area.

The tensile stress is conducive to the nucleation of martensite [2], which reflects that ΔG_v in formula (4.10) reduces the system's free energy; the plastic strain can reduce the energy barrier of nucleation, but also is conducive to the martensite nucleation [1, 17], which reflects that ΔG_d in formula (4.10) also reduces the system's free energy; however, the large plastic deformation will present the mechanical stabilization of austenite, which prevents the nucleation of martensite [4]. These theories can be used to explain the experimental results in Fig. 4.7.

Figures 4.8 and 4.9 show the relationships of martensitic transformation rate parameters θ, transformation plasticity coefficient K_M and corresponding stress respectively. As shown in the figure, when the stress $\bar{\sigma}$ is less than certain value, both θ and K_M are essentially constant. When $\bar{\sigma}$ changes within a certain range, both θ and K_M increase with the increase of $\bar{\sigma}$. When $\bar{\sigma}$ is more than a particular value, both θ and K_M tend to be a certain value; it is the same tendency as the relationship between stress and the starting temperature of martensite transformation, because the factors that affect martensite nucleation also exert an influence on the growing up and transformation amount of martensite.

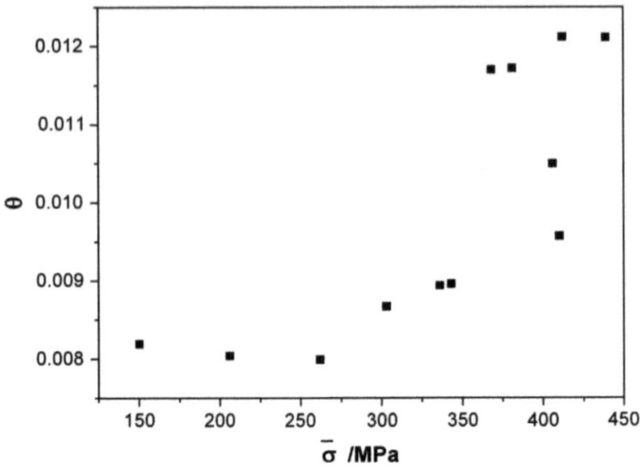

Fig. 4.8 The coupled relation between the martensite transformation rate parameter θ and the effective stress by experiment

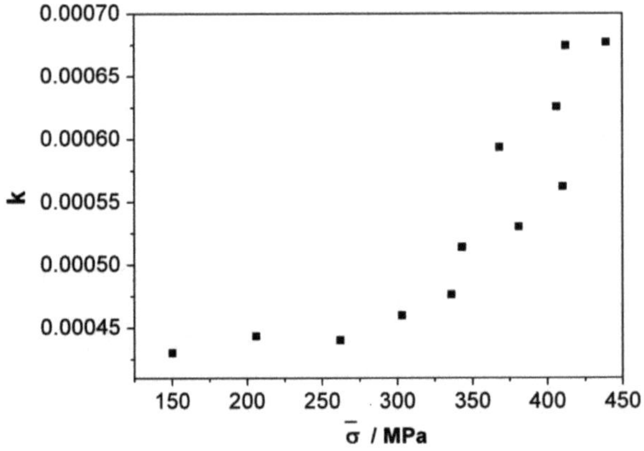

Fig. 4.9 The coupled relation between the transformation plasticity coefficient K_M and the effective stress by experiment

4.1.4 Thermal-Mechanical Transformation Coupled Constitutive Model

The relationships between boron steel starting temperature of martensite transformation M_s, martensite transformation rate parameter θ, and phase-change plasticity coefficient K_M and stress $\bar{\sigma}$ are established by combining the above experimental results, such as the formula (4.11), (4.12), and (4.13).

$$\begin{cases} M_s = M_{s0} & \bar{\sigma} \leq \sigma_{a0} \\ M_s = M_{s0} + \chi \times (\bar{\sigma} - \sigma_{a0}) & \sigma_{a0} < \bar{\sigma} < \sigma_{p0} \\ M_s = M_{s0} + \chi \times (\bar{\sigma}_p - \sigma_{p0}) & \bar{\sigma} \geq \sigma_{p0} \end{cases} \tag{4.11}$$

$$\begin{cases} \theta = \theta_0 & \bar{\sigma} \leq \sigma_a \\ \theta = \theta_0 + \lambda \times (\bar{\sigma} - \sigma_a) & \sigma_a < \bar{\sigma} < \sigma_p \\ \theta = \theta_0 + \lambda \times (\bar{\sigma}_p - \sigma_a) & \bar{\sigma} \geq \sigma_p \end{cases} \quad (4.12)$$

$$\begin{cases} K_M = K_{M0} & \bar{\sigma} \leq \sigma_a \\ K_M = K_{M0} + \gamma \times (\bar{\sigma} - \sigma_a) & \sigma_a < \bar{\sigma} < \sigma_p \\ K_M = K_{M0} + \gamma \times (\sigma_p - \sigma_a) & \bar{\sigma} \geq \sigma_p \end{cases} \quad (4.13)$$

In formula (4.11), M_{s0} is the base point temperature of martensite transformation; χ is the coupling coefficient between the phase transition temperature and stress, and σ_{a0},σ_{p0} are the stress base points for the martensitic transformation and the range of stress coupling respectively (depending on the materials and the way of cooling).

In formula (4.12), θ_0 is the base point temperature of the rate of martensitic transformation; λ is the coupling coefficient between the phase transition rate and stress, σ_a, σ_p are the stress base points for the martensitic transformation rate and the range of stress coupling respectively (depending on the materials and the way of cooling).

In formula (4.12), K_{M0} is the base point of the coefficient of phase transformation plasticity; γ is the coupling coefficient between the coefficient of phase transformation plasticity and stress, and σ_a, σ_p are the stress base points for the coefficient of phase transformation plasticity and the range of stress coupling respectively (depending on the materials and the cooling way).

The material parameters are shown in Table 4.1.

The thermal-mechanical transformation coupled constitutive model during hot forming process of boron steel can be obtained, such as Eqs. (4.14), (4.15), and (4.16).

$$X_M = 1 - \exp[-\theta(\bar{\sigma})(M_s(\bar{\sigma}) - T)] \quad (4.14)$$

$$d\varepsilon_{ij}^{trM} = \beta_M \delta_{ij} dX_M \quad (4.15)$$

$$d\varepsilon_{ij}^{trM} = 3K_M(\bar{\sigma})(1 - X_M)s_{ij}dX_M \quad (4.16)$$

where $M_s(\bar{\sigma})$, $\theta(\bar{\sigma})$, $k(\bar{\sigma})$ can be determined, respectively, by formulas (4.15), (4.17), and (4.18).

Table 4.1 Thermal-mechanical transformation coupled coefficients

M_{s0}/°C	χ	θ_0	λ	K_{M0}	γ	σ_{a0}/MPa	σ_a/MPa	σ_{p0}/MPa	σ_p/MPa
380	0.6273	7.9902E −3	2.7508E −5	4.401E −4	1.5627E −6	141	262	475	412

4.2 Hot Forming Stress and Strain Analysis

4.2.1 *Mixed Law*

Hot forming of high strength steel material is composed of multi-phase particles, and thermodynamic parameters and mechanical properties can be described by the mixed laws [9, 15]:

$$x^{\text{effective}} = \sum_{i=1}^{n} x_i \xi_i, \quad \sum_{i=1}^{n} \xi_i = 1 \tag{4.17}$$

where x_i is a single-phase material property, and ξ_i is the volume fraction of the ith phase, and $x^{\text{effective}}$ is the equivalent material property. Formula (4.17) can be applied to calculate thermal parameters, such as heat capacity, thermal conductivity, and thermal expansion coefficient of the multi-phase hybrid organization, and also used for calculating elastic modulus, yield strength, and other mechanical parameters. For example, when the coefficients of thermal expansion α_1, α_2 and the volume fractions ξ_1, ξ_2 of austenite and martensite of the two-phase composite are given, the equivalent thermal expansion coefficient $\alpha^{\text{effective}}$ of the composite can be obtained with this method:

$$\alpha^{\text{effective}} = \alpha_1 \xi_1 + \alpha_2 \xi_2, \quad \xi_1 + \xi_2 = 1 \tag{4.18}$$

The thermal parameters and mechanical properties parameters of the multi-phase composite can be calculated with this law, but some approximate linear processings are needed.

4.2.2 *Strain Analysis*

In hot forming process phase, changes will cause some changes of the material volume. For isotropic materials, transformation expansion coefficient β is the same in each direction. The transformation expansion coefficient of phase I can be denoted by β_I in heterogeneous organization. The $\varepsilon_{ij}^{\text{tr}_k}$ transformation volumetric strain can be denoted by [3] as follows:

$$\varepsilon_{ij}^{\text{tr}_I} = \beta_I \Delta X_I \delta_{ij} \tag{4.19}$$

$$\beta_{\text{I}} = \frac{1}{3} \frac{\rho_{\text{A}} - \rho_{\text{I}}}{\rho_{\text{A}}} \tag{4.20}$$

$$I = F, P, B, M \tag{4.21}$$

$$\Delta X_I = X_I^t - X_I^0 \tag{4.22}$$

Ferrite, pearlite, bainite, and martensite are denoted by F, P, B, M, respectively, in the equation. Where ΔX_I is the volume increment of phase I, X_I^t is the volume of phase I at time t; X_I^0 is the volume of phase I at the initial time before phase changes. ρ_A is the density of austenite, and ρ_I are the densities of ferrite, pearlite, bainite, and martensite respectively. The density of austenite and martensite can be expressed as a function associated with carbon content %C. The density of each phase is

$$\begin{aligned} \rho_\mathrm{A} &= 8156 - 216 \times \%\mathrm{C} \quad \mathrm{kg/m^3} \\ \rho_\mathrm{F} &= 7897 \quad \mathrm{kg/m^3} \\ \rho_\mathrm{M} &= 7897 - 248 \times \%\mathrm{C} \quad \mathrm{kg/m^3} \end{aligned} \tag{4.23}$$

The sheet total strain rate $\dot{\varepsilon}^{\mathrm{total}}$ in phase transition process and after the martensitic transformation process can be expressed as

$$\dot{\varepsilon}^{\mathrm{total}} = \dot{\varepsilon}^{\mathrm{e}} + \dot{\varepsilon}^{\mathrm{p}} + \dot{\varepsilon}^{\mathrm{th}} + \dot{\varepsilon}^{\mathrm{tr}} + \dot{\varepsilon}^{\mathrm{tp}} \tag{4.24}$$

where $\dot{\varepsilon}^{\mathrm{e}}$ is elastic strain rate caused by external force; $\dot{\varepsilon}^{\mathrm{p}}$ is plastic strain rate due to external force; $\dot{\varepsilon}^{\mathrm{th}}$ is thermal strain rate resulted from temperature; $\dot{\varepsilon}^{\mathrm{tr}}$ is transformation volumetric strain rate; $\dot{\varepsilon}^{\mathrm{tp}}$ is transformation plastic strain rate which can be obtained from Eqs. (4.6), (4.16), and (4.19).

$$\varepsilon^{\mathrm{th}} = \alpha^{\mathrm{effective}} \dot{T} \tag{4.25}$$

where $\alpha^{\mathrm{effective}}$ is the equivalent thermal expansion coefficient, and $\dot{T}$ is the change of temperature.

The total strain rate $\dot{\varepsilon}^{\mathrm{total}}$ in austenite phase and the transformation after total strain rate can be expressed as folows:

$$\dot{\varepsilon}^{\mathrm{total}} = \dot{\varepsilon}^{\mathrm{e}} + \dot{\varepsilon}^{\mathrm{p}} + \dot{\varepsilon}^{\mathrm{th}} \tag{4.26}$$

4.2.3 Stress Analysis

Compared to the thermal stress caused by temperature changes, the volume stress caused by phase change was defined: if the transformation volume deformation is constrained, the hydrostatic stress will be produced in the material. This stress will be defined as volume phase transition stress.

If the volume stress $\varepsilon_{ij}^{\mathrm{tr}_I}$ caused by phase I is constrained, the transformation volume stress, namely the hydrostatic stress is (introduce the basic assumption that plastic strain does not change the volume)

$$\sigma_{ij}^{\mathrm{tr}_I} = -\frac{E^I}{1-2\nu^I}\varepsilon_{ij}^{\mathrm{tr}_I} \tag{4.27}$$

If the volume strain $\varepsilon^{\mathrm{tr}}$ caused by organizational transformation is all constrained the transformation volume stress of the organization can be expressed as

$$\sigma_{ij}^{\mathrm{tr}} = -\frac{E^{\mathrm{effective}}}{1-2\nu^{\mathrm{effective}}}\varepsilon_{ij}^{\mathrm{tr}} \tag{4.28}$$

Substitute Eq. (4.19) into Eq. (4.28):

$$\sigma_{ij}^{\mathrm{tr}} = -\frac{E^{\mathrm{effective}}}{1-2\nu^{\mathrm{effective}}}\sum_{I=F}^{M}\beta_I\Delta X_I\delta_{ij} \tag{4.29}$$

where $E^{\mathrm{effective}}$ and $\nu^{\mathrm{effective}}$ are equivalent elastic modulus and equivalent Poisson's ratio respectively. With mixed law in Sect. 4.2.1 they can be expressed as follows:

$$E^{\mathrm{effective}} = \sum_{I=F}^{M} E_I X_I, \quad \nu^{\mathrm{effective}} = \sum_{I=F}^{M} \nu_I X_I \tag{4.30}$$

Compared with the transformation volume stress mentioned above, the transformation plastic stress caused by transformation can be defined in the same way: if the transformation plastic deformation is constrained, the deviatoric stress will be produced in the material, which is defined as transformation plastic stress.

If the transformation volume strain $\varepsilon_{ij}^{\mathrm{tp}}$ caused by organizational transformation is all constrained in the transformation plastic stress, namely deviatoric stress is (introduce the basic assumption that the plastic strain only relates to the shape distortion)

$$\sigma_{ij}^{\mathrm{tp}} = \frac{2}{3}\frac{\bar{\sigma}(\bar{\varepsilon})}{\bar{\varepsilon}}\varepsilon_{ij}^{\mathrm{tp}} \tag{4.31}$$

where $\bar{\sigma}(\bar{\varepsilon})$ and $\bar{\varepsilon}$ are the equivalent stress and equivalent strain respectively. Integrate with Eqs. (4.6), (4.16), and (4.31):

$$\sigma_{ij}^{\mathrm{tp}} = \frac{2}{3}\frac{\bar{\sigma}(\bar{\varepsilon})}{\bar{\varepsilon}}\sum_{I=F}^{M} 3K_I(\bar{\sigma})s_{ij}\left(1-\frac{1}{2}X_I\right)X_I \tag{4.32}$$

4.3 Constitutive Model of Hot Forming

For establishing the constitutive equation of high strength steel hot forming [5], the classic thermal elastic-plastic constitutive equation was modified based on the concept of transformation volumetric strain $\varepsilon^{\mathrm{tr}}$, volume transformation stress $\sigma_{ij}^{\mathrm{tr}}$, transformation plastic strain $\varepsilon^{\mathrm{tp}}$, and transformation plastic stress $\sigma_{ij}^{\mathrm{tp}}$ mentioned above. Based on the strain and stress analysis in hot forming process in Sect. 4.2, the transformation volumetric strain, stress can be scaled with temperature strain, stress in same proportion. And similarly the transformation plastic strain, stress can be transformed into physical quantity related to plastic strain, stress in hot forming process. Then the analysis can be carried out.

4.3.1 Hot Forming Constitutive Relation of Total Strain Theory

Based on the Ilyushin simple loading theorem and the definition of transformation volumetric strain, stress transformation volumetric stress, transformation plastic strain and transformation plastic stress, hot forming constitutive relation of total strain theory is

$$s_{ij}^{\mathrm{total}} = \frac{2}{3}\frac{\bar{\sigma}(\bar{\varepsilon})}{\bar{\varepsilon}}\left(e_{ij}^{\mathrm{total}} - \varepsilon_{ij}^{\mathrm{tp}}\right) \tag{4.33}$$

$$\sigma_{kk}^{\mathrm{total}} = -\frac{E^{\mathrm{affective}}}{1 - 2\nu^{\mathrm{affective}}}\left(\varepsilon_{kk}^{\mathrm{total}} - \varepsilon_{kk}^{\mathrm{th}} - \varepsilon_{kk}^{\mathrm{tr}}\right) \tag{4.34}$$

where the $\bar{\sigma}(\bar{\varepsilon}) \sim \bar{\varepsilon}$ curve is determined by high temperature tensile test in Chap. 7, Sect. 7.3.

4.3.2 Hot Forming Constitutive Relation of Incremental Theory

By introducing the transformation volumetric strain, stress and transformation plastic strain, stress into traditional thermal plastic incremental constitutive equation, we can get the following:

$$\begin{cases} de_{ij} = \dfrac{1}{2G^{\mathrm{effective}}} dS_{ij} + d\lambda S_{ij} + d\varepsilon_{ij}^{\mathrm{tp}} \\ d\varepsilon_{kk} = \left(\dfrac{1 - 2\nu^{\mathrm{effective}}}{E^{\mathrm{effective}}}\right) d\sigma_{kk} + d\varepsilon_{kk}^{\mathrm{th}} + d\varepsilon_{kk}^{\mathrm{tr}} \end{cases} \tag{4.35}$$

By substituting Eqs. (4.5), (4.6), and (4.25) into Eq. (4.35) :

$$\begin{cases} de_{ij} = \dfrac{1}{2G^{\text{effective}}} ds_{ij} + d\lambda s_{ij} + \sum\limits_{I=F}^{M} 3K_I s_{ij}(1 - X_I)dX_I \\ d\varepsilon_{kk} = \left(\dfrac{1 - 2\nu^{\text{effective}}}{E^{\text{effective}}}\right) d\sigma_{kk} + \alpha^{\text{effective}} dT\delta_{kk} + \sum\limits_{I=F}^{M} \beta_I dX_I \delta_{kk} \end{cases} \tag{4.36}$$

Considering the relation between material and temperature, the strain rate can be expressed as follows:

$$\begin{cases} de_{ij} = B^{\text{e}} ds_{ij} + d\lambda s_{ij} + \sum\limits_{I=F}^{M} 3K_I s_{ij}(1 - X_I)dX_I \\ \qquad + \left(\dfrac{\partial B^{\text{e}}}{\partial T} dT + \dfrac{\partial B^{\text{e}}}{\partial \dot{\varepsilon}} d\dot{\varepsilon}\right) s_{ij} \\ d\varepsilon_{kk} = D^{\text{e}} d\sigma_{kk} + \alpha^{\text{effective}} dT\delta_{kk} + \sum\limits_{I=F}^{M} \beta_I dV_I \delta_{kk} \\ \qquad + \left(\dfrac{\partial D^{\text{e}}}{\partial T} dT + \dfrac{\partial D^{\text{e}}}{\partial \dot{\varepsilon}} d\dot{\varepsilon}\right) \sigma_{kk} + \dfrac{\partial \alpha^{\text{effective}}}{\partial T} (T - T_0) dT\delta_{kk} \\ B^{\text{e}} = \dfrac{1}{2G^{\text{effective}}(T, \dot{\varepsilon})}, \quad D^{\text{e}} = \dfrac{1 - 2\nu^{\text{effective}}}{E^{\text{effective}}(T, \dot{\varepsilon})} \end{cases} \tag{4.37}$$

4.4 Summary

In this chapter, the basic theory in hot stamping process was explained and researched in detail. The specific form of heat, stress, and phase transformation coupled relationship was established by means of high strength boron steel hot stamping experiment.

The coupling relationship among heat, stress, and phase transformation of boron steel under hot stamping process is the basis of hot stamping theory research. It is also the key factor that will affect the veracity of numerical simulation and hot forming technology as well. In this factor, the diffusive and nondiffusive phase transformation model and hardness predictive model in boron steel hot forming process was introduced and analyzed first. At the same time, the tensile experiment under high temperature and quenching experiment of hot forming boron steel was adopted, while the hot forming process under large deformation and stress was considered. The experiment process is as follows: after the boron steel specimen is austenitized (950 °C), the heat is preserved. The drawing force is applied while continuous cooling. The force, displacement, swell increment, and temperature variation were recorded. By means of analyzing the microstructure property and transformation law of the above-mentioned parameters under different cooling rate and drawing force, the relationship between martensite phase transformation point M_s, martensite transformation rate θ, phase transformation plastic coefficient K_M and stress was established. What's more, the heat, stress, and phase transformation-coupled model of boron steel hot stamping process was derived as well. Based on the mixed law, an equivalent analysis of the multi-phases mixed microstructure thermal capacity, heat conductivity coefficient, thermal expansivity, elasticity modulus E, and poisson ratio v was completed. The composition and formation mechanism of stain after hot stamping was analyzed and the conception of transformation volume stress and transformation plastic stress was defined. Finally, the thermal elastoplastic constitutive models of total strain theory and incremental theory in hot forming process were established respectively.

References

1. Brooks JW, Loretto MH, Smallman RE (1979) In situ observations of the formation of martensite in stainless steel. Acta Metall 27:1829–1838
2. Fc F (1951) Capillary equilibria of dislocated crystals. Acta Crystallogr 497–501
3. Gür CH, Tekkaya AE (1996) Finite element simulation of quench hardening. Steel Res 67:298–306
4. Hsu X, Zuyao TY (2006) Martensitic transformation under stress. Mat Sci Eng A 438–440:64–68
5. Hu P, Ma N (2011) Development of hot forming technology for ultra high strength steel and its mechanical problem. Adv Mech 3:310–334 (in Chinese)
6. Hu P, Shi DY, Ying L et al (2013) Coupled of thermal-mechanical-transformation numerical simulation on hot stamping with static explicit algorithm. In: The 11th international conference on numerical methods in industrial forming processes: numiform 2013, pp 394–405
7. Kirkaldy JS, Venugopalan D (1983) Prediction of microstructure and harden ability in low-alloy steels. Phase Transform Ferrous Alloys 125–148
8. Koistinen DP, Marburger RE (1959) A general equation prescribing the extent of the austenite-martensite transformation in pure iron-carbon alloys and plain carbon steels. Acta Metall 7:59–60
9. Leblond JB, Mottet G, Devaux JC (1986) A theoretical and numerical approach to the plastic behaviour of steels during phase transformations—II. Study of classical plasticity for ideal-plastic phases. J Mech Phys Solids 34:411–432
10. Li MV, Niebuhr DV, Meekisho LL et al (1998) A computational model for the prediction of steel hardenability. Metall Mat Trans B 29:661–672
11. Liang Y (2013) Research and application on key process experiment of high strength steel for hot forming. Dalian University of Technology, p 179 (in Chinese)
12. Ma N (2011) Research on hot forming of high strength steel. Dalian University of Technology, p 224 (in Chinese)
13. Maynier P, Jungmann B, Dollet J (1977) Creusot–Loire system for the prediction of the mechanical properties of low alloy steel products. Hard Concepts Appl Steel 518–545

14. Nagasaka Y, Brimacombe JK, Hawbolt EB et al (1993) Mathematical model of phase transformations and elasto-plastic stress in the water spray quenching of steel bars. Metal Trans A-Phys Metal Mat Sci 24:795–808
15. Petit-Grostabussiat S, Taleb L, Jullien J-F (2004) Experimental results on classical plasticity of steels subjected to structural transformations. Int J Plast 20:1371–1386
16. Porter DA, Easterling KE (1981) Phase transformations in metals and alloys. Van Nostrand Reinhold Co
17. Sinclair R, Mohamed HA (1978) Lattice imaging study of a martensite-austenite interface. Acta Metall 26:623–628

Chapter 5
Constitutive Integration Algorithm of Crystal Thermal Deformation

The uniform stress–strain curves are used to describe the material properties in the calculation of macroscopic finite element model. However, the actual high strength steel is constructed by multiple crystals with specific orientations, and the single crystals with different orientations have diverse anisotropic properties in the finite deformation process. Thus, a more precise material model of single crystal needs to be constructed in order to obtain the integration of microscopic and macroscopic view.

The goal of this chapter is to build the constitutive calculative model of single crystal of high strength steel in the process of thermal deformation. The element is assumed to contain several single grains. The constitutive model of single crystal is combined with the macroscopic homogenization model and finite element method. Then the transient stress–strain relationship of macroscopic element scale could be obtained at any time in the hot forming process of high strength steel, which provides a more precise description of the actual hot forming process.

5.1 The Constitutive Integration Method of Single Crystal Finite Deformation at Variable Temperature Conditions

5.1.1 Elastic Deformation Gradient as Basic Variable

From the microscopic view, there are two basic types of single crystal plastic deformation: slip and twinning. In the general circumstances, the single crystal plastic deformation occurs through slip. The twinning happens when it is difficult for the crystal to slip. The slip plane is always on the atomic planes with a dense arrangement of atoms and the slip direction is towards the atomic concentrated arrangement direction of this plane. For example, there are 12 slip systems in the face center cubic (FCC) crystal, which correspond to 4 {111} planes with different

P. Hu et al., *Hot Stamping Advanced Manufacturing Technology of Lightweight Car Body*, DOI 10.1007/978-981-10-2401-6_5

directions. Each slip plane has three different concentrated arrangement directions of ⟨110⟩.

In the room temperature, the plastic deformation is considered to be caused by the dislocation motions of crystals. The elastoplastic deformation process could be divided into two continuous deformation periods. First, the plastic deformation is caused by dislocation motion, and the configuration is called the intermediate configuration. In this process, the crystal lattice and the orientation remain constant. Second, the plastic deformation occurs on the basis of the plastic deformation and the configuration is changed to the current configuration. At this period, not only the crystal lattice, but also the orientations begin to change. The thermal effect is also considered in the whole process and the single crystal kinematics [12] is shown as Fig. 5.1.

From this figure, because of the temperature variation, the initial configuration in the initial temperature θ_0 changes to the intermediate thermal configuration. Then, the relaxed configuration is obtained by uniform shear strain in the crystal slip direction, as the crystal orientation remain constant. After the crystal lattice deformation and rigid rotation, the deformed configuration could be achieved in the current temperature θ. Since the crystal orientation has already changed, the orientation variation could be described by the polar decomposition of elastic deformation gradient.

The total deformation gradient of single crystal is shown as follows [3]:

$$\mathbf{F} = \mathbf{F}^e\mathbf{F}^p\mathbf{F}^\theta \tag{5.1}$$

From this equation, the total deformation gradient is divided into three parts: elastic, plastic, and thermal deformation parts. As for the cubic crystals, the expansion characteristic is isotropic, thus the temperature deformation gradient could be described by the isotropic thermal expansion [2]:

$$\mathbf{F}^\theta = [1 + A(\theta - \theta_0)]\mathbf{I} \tag{5.2}$$

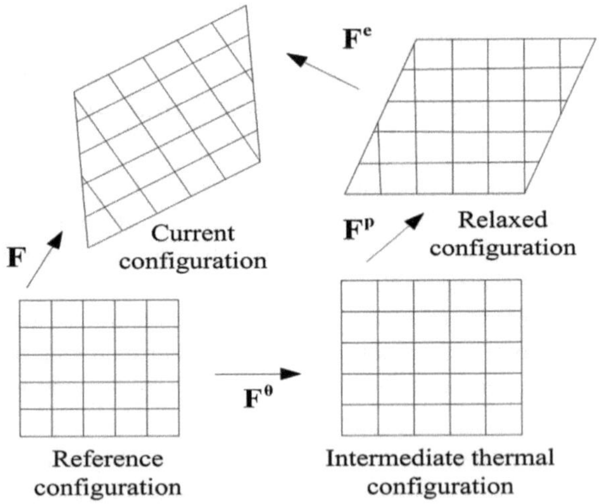

Fig. 5.1 Schematic diagram of multiplicative decomposition of the total deformation gradient

In this equation, A is the thermal expansion coefficient, θ the current temperature, θ_0 the initial temperature and $\mathbf{I}$ the two order unit vector.

From Eq. (5.1), the velocity gradient is divided into three parts:

$$\mathbf{L} = \dot{\mathbf{F}}\mathbf{F}^{-1} = \mathbf{L}^e + \mathbf{F}^e\mathbf{L}^p\mathbf{F}^{e^{-1}} + \mathbf{F}^e\mathbf{F}^p\mathbf{L}^\theta\mathbf{F}^{p^{-1}}\mathbf{F}^{e^{-1}} \tag{5.3}$$

where, $\mathbf{L}^e$ is the elastic velocity gradient. $\mathbf{L}^p$ is the plastic velocity gradient constructed by the shear strain rate $\dot{\gamma}^\alpha$ in the slip system and the corresponding Schmid tensor $\bar{\mathbf{P}}^\alpha$. Here, the unit vectors of the slip direction and the slip normal direction, $\bar{\mathbf{s}}^\alpha$ and $\bar{\mathbf{m}}^\alpha$ refer to the intermediate configuration, and the corresponding plastic velocity gradient also refers to the intermediate configuration. $\mathbf{L}^\theta$ is the thermal velocity gradient, which reflects the evolution of the relaxed configuration in the thermal expansion process.

The specific equation of these three parts of the velocity gradient is shown as below

$$\mathbf{L}^e = \dot{\mathbf{F}}^e\mathbf{F}^{e^{-1}} \tag{5.4}$$

$$\mathbf{L}^p = \dot{\mathbf{F}}^p\mathbf{F}^{p^{-1}} = \sum_\alpha \dot{\gamma}^a\bar{\mathbf{P}}^\alpha \tag{5.5}$$

$$\mathbf{L}^\theta = \dot{\mathbf{F}}^\theta\mathbf{F}^{\theta^{-1}} \tag{5.6}$$

The current velocity gradient and the temperature are given to calculate the elastic deformation gradient, stress, strain, and critical shear stress at the next increment. Here, the evolution function of the elastic deformation gradient is shown

$$\dot{\mathbf{F}}^e = \mathbf{L}\mathbf{F}^e - \mathbf{F}^e\mathbf{L}^p - \mathbf{F}^e\mathbf{F}^p\mathbf{L}^\theta\mathbf{F}^{p^{-1}} \tag{5.7}$$

The deformation history and the temperature history are written in the form of time increment, and the control equation is discretized by the backward Euler method. Thus, a implicit calculation method is obtained to solve the elastic deformation gradient at $t+\Delta t$

$$\begin{aligned}\mathbf{F}^e_{t+\Delta t} = \mathbf{F}^e_t + \Delta t\mathbf{L}_{t+\Delta t}\mathbf{F}^e_{t+\Delta t} - \Delta t\mathbf{F}^e_{t+\Delta t}\left(\sum_\alpha \dot{\gamma}^\alpha_{t+\Delta t}\bar{\mathbf{P}}^\alpha\right) \\ - \Delta t\mathbf{F}_{t+\Delta t}\mathbf{F}^{\theta^{-1}}_{t+\Delta t}\mathbf{L}^\theta_{t+\Delta t}\mathbf{F}^\theta_{t+\Delta t}\mathbf{F}^{-1}_{t+\Delta t}\mathbf{F}^e_{t+\Delta t}\end{aligned} \tag{5.8}$$

where the predefined quantities are the $\mathbf{F}^e_t$, $\mathbf{L}_{t+\Delta t}$, $\mathbf{F}_{t+\Delta t}$, $\mathbf{F}^\theta_{t+\Delta t}$ and $\mathbf{L}^\theta_{t+\Delta t}$. The temperature part of the velocity gradient could be calculated by the current temperature and the temperature rate. Here the shear strain rate of the slip system at $t+\Delta t$ is shown

$$\dot{\gamma}^{\alpha}_{t+\Delta t} = \dot{\gamma}_0 \left| \frac{\tau^{\alpha}_{t+\Delta t}(\theta)}{\hat{\tau}^{\alpha}_t(\theta)} \right|^{\frac{1}{m}} \operatorname{sign}(\tau^{\alpha}) \tag{5.9}$$

The Newton–Raphson method is used to calculate the variable, and the equation of the kth iteration is obtained

$$\mathbf{R}^K = \mathbf{F}^e_k - \mathbf{F}^e_t - \Delta t \mathbf{L} \mathbf{F}^e_k + \Delta t \mathbf{F}^e_k \left(\sum_{\alpha} \dot{\gamma}^{\alpha}_k \bar{\mathbf{P}}^{\alpha} \right) + \Delta t \mathbf{F} \mathbf{F}^{\theta^{-1}} \mathbf{L}^{\theta} \mathbf{F}^{\theta} \mathbf{F}^{-1} \mathbf{F}^e_k \tag{5.10}$$

where the subscript k is changed with the iteration, and the subscripts $t + \Delta t$ of all the variables are omitted with simplicity. The $k+1$th iteration equation of calculating $\mathbf{F}^e_{k+1}$ is shown as follows:

$$D\mathbf{R}^k(\mathbf{F}^e_k)[\Delta \mathbf{F}^e_k] = -\mathbf{R}^k \tag{5.11}$$

$$\mathbf{F}^e_{k+1} = \mathbf{F}^e_k + \Delta \mathbf{F}^e_k \tag{5.12}$$

where $D\mathbf{R}^k(\mathbf{F}^e_k)$ is a fourth-order tensor represents the derivative of residual value $\mathbf{R}^k$ by the variable $\mathbf{F}^e_k$.

As the elastic deformation gradient is obtained with the satisfied convergence condition, the strength value of the next increment is updated

$$\hat{\tau}^{\alpha}_{t+\Delta t} = \hat{\tau}^{\alpha}_t + \Delta t \dot{\hat{\tau}}^{\alpha} \tag{5.13}$$

The crystal rotation is described by the polar decomposition of the elastic deformation gradient

$$\mathbf{F}^e = \mathbf{R}^e \mathbf{U} \tag{5.14}$$

5.1.2 Plastic Deformation Gradient as Basic Variable

According to the constructed single crystal kinematics equation, the velocity gradient of the reference intermediate configuration is shown as below

$$\mathbf{L}^p = \dot{\mathbf{F}}^p \mathbf{F}^{p^{-1}} = \sum_{\alpha} \dot{\gamma}^{\alpha}(\theta) \bar{\mathbf{P}}^{\alpha} \tag{5.15}$$

$$\bar{\mathbf{P}}^{\alpha} = \bar{\mathbf{s}}^{\alpha} \otimes \bar{\mathbf{m}}^{\alpha} \tag{5.16}$$

Here, backward Euler method is used, and the Eq. (5.15) is discretized in that time increment

$$\mathbf{L}^{p}_{t+\Delta t} = \frac{1}{\Delta t}(\mathbf{I} - \mathbf{F}^{p}_{t}\mathbf{F}^{p^{-1}}_{t+\Delta t}) \quad (5.17)$$

Combined with the third item in the Eq. (5.15), this equation could be rewritten as

$$\sum_{\alpha} \dot{\gamma}^{\alpha}_{t+\Delta t}\bar{\mathbf{P}}^{\alpha} - \frac{1}{\Delta t}(\mathbf{I} - \mathbf{F}^{p}_{t}\mathbf{F}^{p^{-1}}_{t+\Delta t}) = 0 \quad (5.18)$$

Considering the incompressible volume in the plastic deformation gradient

$$\det \mathbf{F}^{p}_{t+\Delta t} = 1 \quad (5.19)$$

The control equation based on the plastic deformation gradient variable is constructed

$$\sum_{\alpha} \dot{\gamma}^{\alpha}_{t+\Delta t}\bar{\mathbf{P}}^{\alpha} - \frac{1}{\Delta t}(\mathbf{I} - \mathbf{F}^{p}_{t}\mathbf{F}^{p^{-1}}_{t+\Delta t}) + \lambda(\det \mathbf{F}^{p}_{t+\Delta t} - 1)\mathbf{I} = 0 \quad (5.20)$$

where λ is the penalty parameter and the $\lambda = 10^3$ is chosen to constrain the incompressible plastic volume. Assume that the $\mathbf{F}^p$, $\mathbf{F}$, and θ are given at t, while $\mathbf{L}$ and $\dot{\theta}$ are given at $t + \Delta t$. Here the detailed method of calculating $\mathbf{F}^p$, $\mathbf{T}$, and $\mathbf{R}^e$ at $t + \Delta t$ is introduced. The Newton–Raphson method is used in the control equation and the function of the residual value of the kth iteration

$$\mathbf{R}^{k} = \sum_{\alpha} \dot{\gamma}^{\alpha}_{k}\bar{\mathbf{P}}^{\alpha} - \frac{1}{\Delta t}(\mathbf{I} - \mathbf{F}^{p}_{t}\mathbf{F}^{p^{-1}}_{k}) + \lambda(\det \mathbf{F}^{p}_{k} - 1)\mathbf{I} \quad (5.21)$$

The Taylor expansion of the $k + 1$th iteration of the residual value $\mathbf{R}^{k+1}$ with respect to $\mathbf{F}^{p}_{k+1}$ is applied and the higher order terms are neglected

$$\mathbf{R}^{k+1} = \mathbf{R}^{k} + D\mathbf{R}^{k}(\mathbf{F}^{p}_{k})[\mathbf{F}^{p}_{k+1} - \mathbf{F}^{p}_{k}] \quad (5.22)$$

Set the left term as zero, and the $k + 1$th iteration of $\mathbf{F}^{p}_{k+1}$ could be obtained as follows:

$$D\mathbf{R}^{k}(\mathbf{F}^{p}_{k})[\Delta\mathbf{F}^{p}_{k}] = -\mathbf{R}^{k} \quad (5.23)$$

$$\mathbf{F}^{p}_{k+1} = \mathbf{F}^{p}_{k} + \Delta\mathbf{F}^{p}_{k} \quad (5.24)$$

where $D\mathbf{R}^k(\mathbf{F}_k^p)$ is a fourth-order tensor representing the derivative of the residual value by variable $\mathbf{F}_k^p$. The direct derivative by $\Delta\mathbf{F}_k^p$ is shown as follows:

$$D\mathbf{R}^k(\mathbf{F}_k^p)[\Delta\mathbf{F}_k^p] = \Delta\mathbf{G}_k^* - \frac{1}{\Delta t}\mathbf{F}_t^p\mathbf{F}_k^{p^{-1}}\Delta\mathbf{F}_k^p\mathbf{F}_k^{p^{-1}} + \lambda\mathbf{I}\det(\mathbf{F}_k^p)\mathrm{tr}(\mathbf{F}_k^{p^{-1}}\Delta\mathbf{F}_k^p) \quad (5.25)$$

$$\Delta\mathbf{G}_k^* = \xi_k[\Delta\mathbf{T}_k^*] \quad (5.26)$$

$$\xi_k = \sum_\alpha \frac{d\dot{\gamma}_k^\alpha}{d\tau_k^\alpha}(\bar{\mathbf{P}}^\alpha \otimes \bar{\mathbf{P}}^\alpha) \quad (5.27)$$

$$\begin{aligned}\Delta\mathbf{T}_k^* &= (-\mathbf{F}_k^{p^{-\mathrm{T}}}\Delta\mathbf{F}_k^{p^{\mathrm{T}}}\mathbf{C}_k^{p\theta} - \mathbf{C}_k^{p\theta}\Delta\mathbf{F}_k^p\mathbf{F}_k^{p^{-1}})\mathbf{T}_k^e \\ &+ \mathbf{C}_k^{p\theta}\mathbf{C}\left[\frac{1}{2}(-\mathbf{F}_k^{p^{-\mathrm{T}}}\Delta\mathbf{F}_k^{p^{\mathrm{T}}}\mathbf{C}_k^{p\theta} - \mathbf{C}_k^{p\theta}\Delta\mathbf{F}_k^p\mathbf{F}_k^{p^{-1}})\right]\end{aligned} \quad (5.28)$$

$$\begin{aligned}\mathbf{C}_k^{p\theta} &= \mathbf{F}_k^{p^{-\mathrm{T}}}\mathbf{C}^\theta\mathbf{F}_k^{p^{-1}} \\ \mathbf{C}^\theta &= \mathbf{F}^{\theta^{-\mathrm{T}}}\mathbf{F}^{\mathrm{T}}\mathbf{F}\mathbf{F}^{\theta^{-1}}\end{aligned} \quad (5.29)$$

Considering the equations from (5.23) to (5.29), the incremental value of the plastic deformation gradient $\Delta\mathbf{F}_k^p$ is calculated, thus the next iteration $\mathbf{F}_{k+1}^p$ is obtained. This process is repeated until the bound norm of the residual value is under the regulated error value. Because the plastic deformation is much larger than the elastic deformation, a reasonable predicted value should be given in the direct calculation of the plastic deformation increment [13]. According to Eq. (5.1), it could be transformed into

$$\mathbf{F}_k^p = \mathbf{F}_t^{e^{-1}}\mathbf{F}_{t+\Delta t}\mathbf{F}_{t+\Delta t}^{\theta^{-1}} \quad (5.30)$$

After the satisfied plastic deformation gradient $\mathbf{F}_{t+\Delta t}^p$ is deduced, variables such as $\hat{\tau}_{t+\Delta t}^\alpha$, $\mathbf{F}_{t+\Delta t}^e$, and $\mathbf{R}^e$ is updated.

5.1.3 Stress as the Basic Variable in the Algorithm

The calculation method of the crystal plasticity based on the second P-K stress is introduced in this chapter. The second P-K stress could be related to the Cauchy stress

$$\mathbf{T} = \mathbf{F}^{e^{-1}}\{(\det\mathbf{F}^e)\boldsymbol{\sigma}\}\mathbf{F}^{e^{-\mathrm{T}}} \quad (5.31)$$

where is the Cauchy stress tensor. The Green strain work conjugated to the second P-K stress is defined as follows:

$$\mathbf{E} = \frac{1}{2}(\mathbf{F}^{e^{\mathrm{T}}}\mathbf{F}^{e} - \mathbf{I}) \tag{5.32}$$

The elastic constitutive relationship is shown as below

$$\mathbf{T} = \varphi(\theta) : \mathbf{E} \tag{5.33}$$

where $\varphi(\theta)$ is the fourth-order elastic tensor related to the temperature.

The power-law model is used to describe the shear strain rate $\dot{\gamma}^{\alpha}$ in each slip system [9]

$$\dot{\gamma}^{\alpha} = \dot{\gamma}_0 \left| \frac{\tau^{\alpha}(\theta)}{\hat{\tau}^{\alpha}(\theta)} \right|^{\frac{1}{m}} \mathrm{sign}(\tau^{\alpha}) \tag{5.34}$$

where $\hat{\tau}^{\alpha}(\theta)$ is the critical resolved shear stress and m is the rate-dependent index.

In the hardening model, Voce-type model is revised

$$\dot{\hat{\tau}}^{\alpha} = H_0 \left\{ \frac{\hat{\tau}_{\mathrm{s}}(\dot{\gamma}, \theta) - \hat{\tau}^{\alpha}(\theta)}{\hat{\tau}_{\mathrm{s}}(\dot{\gamma}, \theta) - \hat{\tau}_0} \right\} \dot{\gamma} \tag{5.35}$$

where H_0 is the reference hardening rate coefficient and $\hat{\tau}_0$ is the initial critical resolved shear stress. Meanwhile, $\dot{\gamma}$ is the total shear strain rate in all the slip systems and $\hat{\tau}_{\mathrm{s}}$ is the hardening value when the material reaches the stress saturation state

$$\dot{\gamma} = \sum_{\alpha} \dot{\gamma}^{\alpha} \tag{5.36}$$

$$\hat{\tau}_{\mathrm{s}}(\dot{\gamma}, \theta) = \hat{\tau}_{\mathrm{s}}^{0} \left(\frac{\dot{\gamma}}{\dot{\gamma}_{\mathrm{s}}} \right)^{m'} \tag{5.37}$$

where $\hat{\tau}_{\mathrm{s}}^{0}$, $\dot{\gamma}_{\mathrm{s}}$ and m' are the material parameters.

The evolution of plastic velocity gradient is solved by the following flow rule:

$$\dot{\mathbf{F}}^{p} = \mathbf{L}^{p}\mathbf{F}^{p^{-1}} \tag{5.38}$$

Using the expression of plastic velocity gradient (5.5), the evolution equation of $\mathbf{F}^{p}$ (5.38) could be revised as the incremental fully implicit [10]:

$$\mathbf{F}^{p}_{t+\Delta t} = \exp(\Delta t \mathbf{L}^{p}_{t+\Delta t})\mathbf{F}^{p}_{t} \tag{5.39}$$

The Taylor expansion is used to simplify this equation

$$\mathbf{F}^p_{t+\Delta t} = (\mathbf{I} + \Delta t \mathbf{L}^p_{t+\Delta t})\mathbf{F}^p_t \tag{5.40}$$

Thus

$$\mathbf{F}^{p^{-1}}_{t+\Delta t} = \mathbf{F}^{p^{-1}}_t \left(\mathbf{I} - \Delta t \sum_\alpha \dot{\gamma}^\alpha \bar{\mathbf{P}}^\alpha\right) \tag{5.41}$$

$$\mathbf{F}^{p^{-\mathrm{T}}}_{t+\Delta t} = \left(\mathbf{I} - \Delta t \sum_\alpha \dot{\gamma}^\alpha \bar{\mathbf{P}}^{\alpha^{\mathrm{T}}}\right)\mathbf{F}^{p^{-\mathrm{T}}}_t \tag{5.42}$$

According to the stress updating method of Kalidindi [5–7], the time increment of second P-K stress is shown as below

$$\mathbf{T}^e_{t+\Delta t} = \varphi : \left[\frac{1}{2}(\mathbf{F}^{p^{-\mathrm{T}}}_{t+\Delta t}\mathbf{F}^{\theta^{-\mathrm{T}}}_{t+\Delta t}\mathbf{F}^{\mathrm{T}}_{t+\Delta t}\mathbf{F}_{t+\Delta t}\mathbf{F}^{\theta^{-1}}_{t+\Delta t}\mathbf{F}^{p^{-1}}_{t+\Delta t} - \mathbf{I})\right] \tag{5.43}$$

Set $\mathbf{x} = \mathbf{F}^{p^{-\mathrm{T}}}_{t+\Delta t}\mathbf{F}^{\theta^{-\mathrm{T}}}_{t+\Delta t}\mathbf{F}^{\mathrm{T}}_{t+\Delta t}\mathbf{F}_{t+\Delta t}\mathbf{F}^{\theta^{-1}}_{t+\Delta t}\mathbf{F}^{p^{-1}}_{t+\Delta t}$.
Substitute (5.41) and (5.42) into this setting

$$\begin{aligned}\mathbf{x} = {} & \mathbf{F}^{p^{-\mathrm{T}}}_t\mathbf{F}^{\theta^{-\mathrm{T}}}_{t+\Delta t}\mathbf{F}^{\mathrm{T}}_{t+\Delta t}\mathbf{F}_{t+\Delta t}\mathbf{F}^{\theta^{-1}}_{t+\Delta t}\mathbf{F}^{p^{-1}}_t - \sum_\alpha \Delta\gamma^\alpha \bar{\mathbf{P}}^{\alpha^{\mathrm{T}}}\mathbf{F}^{p^{-\mathrm{T}}}_t\mathbf{F}^{\theta^{-\mathrm{T}}}_{t+\Delta t}\mathbf{F}^{\mathrm{T}}_{t+\Delta t}\mathbf{F}_{t+\Delta t}\mathbf{F}^{\theta^{-1}}_{t+\Delta t}\mathbf{F}^{p^{-1}}_t \\ & - \mathbf{F}^{p^{-\mathrm{T}}}_t\mathbf{F}^{\theta^{-\mathrm{T}}}_{t+\Delta t}\mathbf{F}^{\mathrm{T}}_{t+\Delta t}\mathbf{F}_{t+\Delta t}\mathbf{F}^{\theta^{-1}}_{t+\Delta t}\mathbf{F}^{p^{-1}}_t \sum_\alpha \Delta\gamma^\alpha \bar{\mathbf{P}}^\alpha + o(\Delta\gamma^{\alpha^2})\end{aligned} \tag{5.44}$$

where

$$\Delta\gamma^\alpha = \Delta t \dot{\gamma}^\alpha(\mathbf{T}_{t+\Delta t}, \hat{\tau}^\alpha_{t+\Delta t}) \tag{5.45}$$

Simplify the Eq. (5.44)

$$\mathbf{x} = \mathbf{A} - \sum_\alpha \Delta\gamma^\alpha \mathbf{B}^\alpha \tag{5.46}$$

where

$$\mathbf{A} = \mathbf{F}^{p^{-\mathrm{T}}}_t\mathbf{F}^{\theta^{-\mathrm{T}}}_{t+\Delta t}\mathbf{F}^{\mathrm{T}}_{t+\Delta t}\mathbf{F}_{t+\Delta t}\mathbf{F}^{\theta^{-1}}_{t+\Delta t}\mathbf{F}^{p^{-1}}_t \tag{5.47}$$

$$\mathbf{B}^\alpha = \mathbf{A}\bar{\mathbf{P}}^\alpha + \bar{\mathbf{P}}^{\alpha^{\mathrm{T}}}\mathbf{A} \tag{5.48}$$

The second P-K could be expressed as

$$\mathbf{T}_{t+\Delta t} = \mathbf{T}^{\mathrm{tr}} - \sum_{\alpha} \Delta\gamma^{\alpha}(\mathbf{T}_{t+\Delta t}, \hat{\tau}^{\alpha}_{t+\Delta t})\mathbf{C}^{\alpha} \tag{5.49}$$

where

$$\mathbf{T}^{\mathrm{tr}} = \phi : \left[\frac{1}{2}(\mathbf{A} - \mathbf{I})\right] \tag{5.50}$$

$$\mathbf{C}^{\alpha} = \varphi : [\frac{1}{2}\mathbf{B}^{\alpha}] \tag{5.51}$$

In addition, the implicit integration method is used to update the hardening model, and its evolution equation is deduced

$$\hat{\tau}^{\alpha}_{k+1} = \hat{\tau}^{\alpha}_{k} + \Delta t H_0 \left\{ \frac{\hat{\tau}_{\mathrm{s}}\left[\dot{\gamma}(\mathbf{T}_{n+1}, \hat{\tau}^{\alpha}_{k+1}), \theta\right] - \hat{\tau}^{\alpha}(\theta)}{\hat{\tau}_{\mathrm{s}}\left[\dot{\gamma}(\mathbf{T}_{n+1}, \hat{\tau}^{\alpha}_{k+1}), \theta\right] - \hat{\tau}_0} \right\} \dot{\gamma}(\mathbf{T}_{n+1}, \hat{\tau}^{\alpha}_{k+1}) \tag{5.52}$$

Therefore, a two-level iteration is used to solve the updating of the second P-K and the critical resolved shear stress. Firstly, the second P-K is calculated as the critical resolved shear stress is given a suitable value

$$D\mathbf{R}^{n}(\mathbf{T}_n)[\Delta\mathbf{T}_n] = \mathbf{I}_4 + \sum_{\alpha} \mathbf{C}^{\alpha} \otimes \frac{\partial}{\partial \mathbf{T}_n} \Delta\gamma^{\alpha}(\mathbf{T}_n, \hat{\tau}^{\alpha}_{k}) \tag{5.53}$$

$$\frac{\partial \Delta\gamma^{\alpha}(\mathbf{T}_n, \hat{\tau}^{\alpha}_{k})}{\partial \mathbf{T}_n} = \Delta t \frac{\partial \dot{\gamma}^{\alpha}(\mathbf{T}_n, \hat{\tau}^{\alpha}_{k})}{\partial \mathbf{T}_n} \tag{5.54}$$

The relationship between the second P-K and resolved shear stress is inserted

$$\frac{\partial \dot{\gamma}^{\alpha}(\mathbf{T}_n, \hat{\tau}^{\alpha}_{k})}{\partial \mathbf{T}_n} = \frac{d\dot{\gamma}^{\alpha}(\mathbf{T}_n, \hat{\tau}^{\alpha}_{k})}{d\tau^{\alpha}} D\tau^{\alpha} \tag{5.55}$$

where

$$\frac{d\dot{\gamma}^{\alpha}(\mathbf{T}_n, \hat{\tau}^{\alpha}_{k})}{d\tau^{\alpha}_{n}} = \frac{\dot{\gamma}_0}{m\hat{\tau}^{\alpha}_{k}} \left| \frac{\tau^{\alpha}_{n}}{\hat{\tau}^{\alpha}_{k}} \right|^{\frac{1}{m}-1} \tag{5.56}$$

The expression of the resolved shear stress in the slip system

$$\tau^{\alpha} = \mathbf{F}^{p^{-\mathrm{T}}} \mathbf{F}^{\theta^{-\mathrm{T}}} \mathbf{F}^{\mathrm{T}} \mathbf{T}_n \mathbf{F}^{p^{-\mathrm{T}}} \mathbf{F}^{\theta^{-\mathrm{T}}} \mathbf{F}^{\mathrm{T}} \mathbf{F}^{\theta^{\mathrm{T}}} \mathbf{F}^{p^{\mathrm{T}}} : \bar{\mathbf{p}}^{\alpha} \tag{5.57}$$

Then

$$D\tau^{\alpha} = \mathbf{F}^{p^{-\mathrm{T}}}\mathbf{F}^{\theta^{-\mathrm{T}}}\mathbf{F}^{\mathrm{T}}\mathbf{F}^{p^{-\mathrm{T}}}\mathbf{F}^{\theta^{-\mathrm{T}}}\mathbf{F}^{\mathrm{T}}\mathbf{F}^{\theta^{\mathrm{T}}}\mathbf{F}^{p^{\mathrm{T}}} : \bar{\mathbf{p}}^{\alpha} \tag{5.58}$$

The configuration updating in the calculation

$$\mathbf{s}^{\alpha}_{t+\Delta t} = \mathbf{F}^{e}(t+\Delta t)\mathbf{s}^{\alpha}_{0} \tag{5.59}$$

$$\mathbf{m}^{\alpha}_{t+\Delta t} = \mathbf{F}^{e}(t+\Delta t)\mathbf{m}^{\alpha}_{0} \tag{5.60}$$

This is deduction of the crystal plasticity model based on second P-K stress.

5.2 Comparison Between Two Deformation Gradient Algorithms

5.2.1 Elastic and Plastic Deformation Gradient

The comparison between these two gradient algorithms is analyzed in the algorithm construction, control equation building and the initial value of the iteration

1. Starting from the thermal part of the velocity $\mathbf{L}^{\theta}$ and the elastic deformation gradient $\mathbf{F}^{e}$, the control equation of variation $\mathbf{F}^{e}$ is obtained by solving the Green strain $\mathbf{E}^{e}$ and other variables. The control equation of variation of elastic deformation gradient $\mathbf{F}^{e}$ could also be obtained by beginning at the definition equation of $\mathbf{F}^{e}$. Combining these two formulas, a closed solving system is constructed to form the final calculation model. Where, $\mathbf{L}^{\theta}$ represents the history variation of temperature and the thermal effect is considered in the whole process. Similarly, the plastic deformation gradient is set as the control variable of the control equation in the intermediate reference configuration to form the ultimate solving system.
2. In the $\mathbf{F}^{p}$ model, the direct penalty function is used to insert the incompressible plastic volume into the control Eq. (5.20). However, in the control Eq. (5.7) of $\mathbf{F}^{e}$ model, the incompressible plastic property is not considered. Thus, the component of $\mathbf{F}^{e}$ should be constrained to ensure the incompressible plastic volume in solving the convergence of $\mathbf{F}^{e}$.
3. To choose the initial value of iteration, the elastic deformation is generally small and the convergence value of $\mathbf{F}^{e}$ in the last increment could be used as the initial iteration value. However, in the $\mathbf{F}^{p}$ model, the Eq. (5.30) should be considered to predict the initial value in order to guarantee a better convergence rate. In short, these two kinds of implicit integration methods are different in the building process. They are similar in the solving method, but have their own characters in the detailed calculation process.

5.2.2 The Implicit and Explicit Algorithms

Although large increment step could be used in the implicit algorithm, the Jacobian matrix should be updated in the iteration, not to mention its complicated deduction process. The forward Euler integration method is used to build the explicit calculation model combined with the crystal plasticity theory and the thermal kinetics.

The current velocity gradient $\mathbf{L}$ and thermal velocity gradient $\mathbf{L}^{\theta}$ are given. It starts from the elastic deformation gradient at time t. The basic process is introduced as follows:

$$\mathbf{F}^{e}(t) = \mathbf{F}(t)\mathbf{F}^{\theta^{-1}}(t)\mathbf{F}^{p^{-1}}(t) \tag{5.61}$$

$$\mathbf{E}(t) = \frac{1}{2}[\mathbf{F}^{e^{\mathrm{T}}}(t)\mathbf{F}^{e}(t) - \mathbf{I}] \tag{5.62}$$

$$\mathbf{T}(t) = \mathbf{C}(\theta) : \mathbf{E}(t) \tag{5.63}$$

From these three equations, the second P-K stress $\mathbf{T}(t)$ is solved at time t, and the resolved shear stress $\tau^{\alpha}(t)$ and the shear strain rate $\dot{\gamma}^{\alpha}(t)$ are calculated

$$\tau^{\alpha}(t) = \mathbf{F}^{e^{\mathrm{T}}}(t)\mathbf{F}^{e}(t)\mathbf{T}(t) : \bar{\mathbf{P}}^{\alpha} \tag{5.64}$$

$$\dot{\gamma}^{\alpha}(t) = \dot{\gamma}_0 \left|\frac{\tau^{\alpha}(\theta)}{\hat{\tau}^{\alpha}(\theta)}\right|^{\frac{1}{m}} \operatorname{sign}(\tau^{\alpha}) \tag{5.65}$$

The plastic deformation gradient at time τ is calculated, the components of which are constrained using the incompressible plastic volume condition by Eq. (5.5):

$$\mathbf{F}^{p^{-1}}(\tau) = \mathbf{F}^{p^{-1}}(t)\left[\mathbf{I} - \Delta t\left(\sum_{\alpha} \dot{\gamma}^{a}(t)\bar{\mathbf{P}}^{\alpha}\right)\right] \tag{5.66}$$

Using Eq. (5.6), the thermal deformation gradient is updated, and $\mathbf{T}(\tau)$ is obtained in the next step

$$\mathbf{F}^{e}(\tau) = \mathbf{F}(\tau)\mathbf{F}^{\theta^{-1}}(\tau)\mathbf{F}^{p^{-1}}(\tau) \tag{5.67}$$

$$\mathbf{E}(\tau) = \frac{1}{2}[\mathbf{F}^{e^{\mathrm{T}}}(\tau)\mathbf{F}^{e}(\tau) - \mathbf{I}] \tag{5.68}$$

$$\mathbf{T}(\tau) = \mathbf{C}(\theta) : \mathbf{E}(\tau) \tag{5.69}$$

The critical shear stress is updated as follows:

$$\dot{\hat{\tau}}^{\alpha}(t) = H_0\left\{\frac{\hat{\tau}_s(\dot{\gamma}, \theta) - \hat{\tau}^{\alpha}(\theta)}{\hat{\tau}_s(\dot{\gamma}, \theta) - \hat{\tau}_0}\right\}\dot{\gamma}(t) \tag{5.70}$$

$$\hat{\tau}^{\alpha}(\tau) = \hat{\tau}^{\alpha}(t) + \Delta t\dot{\hat{\tau}}^{\alpha}(t) \tag{5.71}$$

The crystal rotation is calculated by Eq. (5.14), and the calculation continues as this process in the next increment

5.3 The Constitutive Integration Method of Polycrystalline

The ordinary metal materials are polycrystalline constructed by single crystals with their own orientations. The FCC metal structure of single crystal is described by a local coordinate system based on three crystal axes ([001], [010], [001]). The relationship between the local and the global coordinate systems is shown as Fig. 5.2

After determination of the global and the local crystal coordinate system, the crystal orientation is described by the position of the local system reflected in the global system. The rotation matrix **R** is used to describe the three Euler angle of single crystal in the global system. If $\mathbf{D}_g$ represents a random tensor in the global system and $\mathbf{D}_c$ represents a tensor in the local crystal system, the relationship between them is shown as below

$$\mathbf{D}_c = \mathbf{R}\mathbf{D}_g\mathbf{R}^T \tag{5.72}$$

$$\mathbf{D}_g = \mathbf{R}^T\mathbf{D}_c\mathbf{R} \tag{5.73}$$

The Euler angles (θ, φ, γ) are described in the way of Bunge [1], which is obtained by rotating the three axes in a certain order shown as Fig. 5.3.

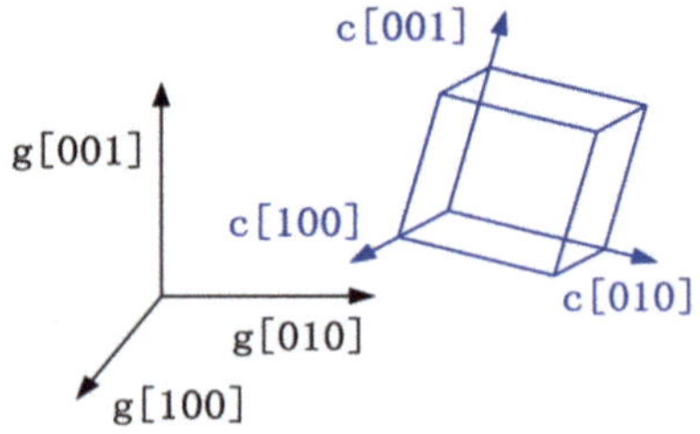

Fig. 5.2 The local crystal coordinate system and global system

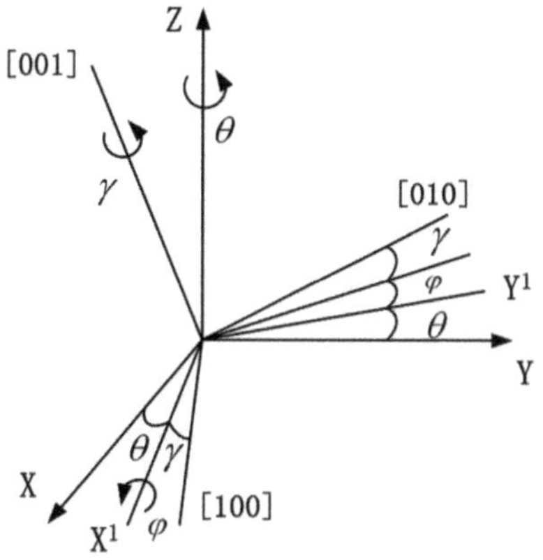

Fig. 5.3 The orientation of the Euler rotation

The rotation matrix **R** obtained by Euler rotation is shown as below

$$\mathbf{R} = \begin{pmatrix} \cos\theta\cos\gamma - \sin\theta\sin\gamma\cos\varphi & \sin\theta\cos\gamma + \cos\theta\sin\gamma\cos\varphi & \sin\gamma\sin\varphi \\ -\cos\theta\sin\gamma - \sin\theta\cos\gamma\cos\varphi & -\sin\theta\sin\gamma + \cos\theta\cos\gamma\cos\varphi & \cos\gamma\sin\varphi \\ \sin\theta\sin\varphi & -\cos\theta\sin\varphi & \cos\varphi \end{pmatrix} \tag{5.77}$$

The value range of three Euler angles θ, φ, γ are presented in Table 5.1 [7]. In the cubic crystal system, because of its high symmetry, the value range of Euler angle is set between 0° and 90° to ensure its uniqueness.

5.3.1 The Construction of Taylor Model

The macroscopic stress is obtained by the volume-average stress of each crystal. The expression is shown as follows:

Table 5.1 Value range of Euler angle for different crystal symmetries

Crystal structure	Crystal symmetry		Sample symmetry		
			Orthorhombic	Monoclinic	Triclinic
	φ	γ	θ	θ	θ
Cubic	90°	90°			
Tetragonal	90°	90°			
Orthorhombic	90°	180°			
Hexagonal	90°	60°	90°	180°	360°
Trigonal	90°	120°			
Monoclinic	90°	360°			
Triclinic	180°	360°			

$$\boldsymbol{\sigma}_{\text{avg}} = \sum_{K=1}^{N} w_k \boldsymbol{\sigma}_k \tag{5.78}$$

where $\boldsymbol{\sigma}_{\text{avg}}$ is the volume average stress, N is the total crystal number, and w_k is the volume fraction of each grain.

The calculation procedure of Taylor polycrystalline model is conducted as follows

1. Circulate the orientation of each grain to rotate the velocity gradient from the global to the local coordinate system

$$\mathbf{L}_{\text{c}} = \mathbf{R}\mathbf{L}_{\text{g}}\mathbf{R}^{\text{T}} \tag{5.79}$$

2. The stress is obtained from the constructed single crystal model and the stress is rotated from the local to the global coordinate system

$$\boldsymbol{\sigma}_{\text{g}} = \mathbf{R}^{\text{T}}\boldsymbol{\sigma}_{\text{c}}\mathbf{R} \tag{5.80}$$

3. The rotation matrix is updated in the next increment

$$\mathbf{R} = \mathbf{R}^{e}\mathbf{R} \tag{5.81}$$

4. The macroscopic stress of the polycrystal is solved by Eq. (5.78). Continue to calculate until all the grains are solved.

5.3.2 *The Multiscale Finite Element Model*

The constitutive model of the single crystal in the finite deformation is combined with the tangent slope of the stress–strain curves in the elastic-plastic matrix of the macroscopic finite element model. The microscopic information of the single crystal is contained in the single element, which provides a more precise description of the stress–strain relationship of each single element in the finite element model.

The hot stamping process of the high strength steel is constructed by complex multi-physic field coupling, such as thermal, mechanical, and phase fields. The total strain rate $\dot{\varepsilon}_{ij}$ of the hot stamping process could be divided into five parts [4]

$$\dot{\varepsilon}_{ij} = \dot{\varepsilon}_{ij}^{\mathrm{e}} + \dot{\varepsilon}_{ij}^{\mathrm{p}} + \dot{\varepsilon}_{ij}^{\mathrm{th}} + \dot{\varepsilon}_{ij}^{\mathrm{tr}} + \dot{\varepsilon}_{ij}^{\mathrm{tp}} \tag{5.82}$$

where $\dot{\varepsilon}_{ij}^{\mathrm{e}}$ is the elastic strain rate, $\dot{\varepsilon}_{ij}^{\mathrm{p}}$ the plastic strain rate, $\dot{\varepsilon}_{ij}^{\mathrm{th}}$ the thermal strain rate, $\dot{\varepsilon}_{ij}^{\mathrm{tr}}$ the phase transition strain rate, $\dot{\varepsilon}_{ij}^{\mathrm{tp}}$ the phase transition plastic strain rate.

$$\dot{\varepsilon}_{ij}^{\mathrm{th}} = \alpha_{ij}\dot{T} + (T - T_0)\dot{\alpha}_{ij} \tag{5.83}$$

$$\dot{\varepsilon}_{ij}^{\mathrm{tr}} = \sum_{I}^{N} \beta_I \delta_{ij} \dot{X}_I \tag{5.84}$$

$$\dot{\varepsilon}_{ij}^{\mathrm{tp}} = \sum_{I}^{N} 3K_I(1 - X_I)s_{ij}\dot{X}_I \tag{5.85}$$

where I could be replaced by F, P, B, M representing ferrite, pearlite, bainite, and martensite, respectively.

Based on the dynamic explicit algorithm of the thermal large deformation, the thermal, mechanical and phase transition are coupled in the incremental constitutive formula. The Jaumann stress (derivative of the Cauchy stress) is used with objectivity

$$\sigma_{ij}^{\nabla J} = \dot{\sigma}_{ij} - \sigma_{ik}\omega_{kj} - \sigma_{kj}\omega_{ki} \tag{5.86}$$

where, $\dot{\sigma}_{ij}$ is the material derivative of the Cauchy stress. ω_{ij} is the rotation rate tensor.

The tensor form of the constitutive equation is constructed as follows:

$$\sigma_{ij}^{\nabla J} = D_{ijkl}^{\mathrm{e}}(\dot{\varepsilon}_{kl} - \dot{\varepsilon}_{kl}^{\mathrm{p}} - \dot{\varepsilon}_{kl}^{\mathrm{th}} - \dot{\varepsilon}_{kl}^{\mathrm{tr}} - \dot{\varepsilon}_{kl}^{\mathrm{tp}}) + \dot{D}_{ijkl}^{\mathrm{e}}(\varepsilon_{kl} - \varepsilon_{kl}^{\mathrm{p}} - \varepsilon_{kl}^{\mathrm{th}} - \varepsilon_{kl}^{\mathrm{tr}} - \varepsilon_{kl}^{\mathrm{tp}}) \tag{5.87}$$

where D_{ijkl}^{e} is the elastic-plastic constitutive matrix. $\dot{\varepsilon}_{kl}$, $\dot{\varepsilon}_{kl}^{\mathrm{th}}$, $\dot{\varepsilon}_{kl}^{\mathrm{tr}}$, $\dot{\varepsilon}_{kl}^{\mathrm{tp}}$ are the Green strain rate tensor, thermal strain rate tensor, phase transition strain rate tensor and phase transition plastic strain rate tensor, respectively in the adjacent reference configuration.

$$\sigma_{ij}^{\nabla J} = \left(D_{ijkl}^{\mathrm{e}} - \frac{D_{ijkl}^{\mathrm{e}} S_{kl} S_{ij} D_{ijkl}^{\mathrm{e}}}{L}\right)(\dot{\varepsilon}_{kl} - \dot{\varepsilon}_{kl}^{\mathrm{th}} - \dot{\varepsilon}_{kl}^{\mathrm{tr}} - \dot{\varepsilon}_{kl}^{\mathrm{tp}}) + \dot{M}_{ij} \tag{5.88}$$

where

$$\begin{aligned} \dot{M}_{ij} = {} & \left(s_{ij} - \frac{\frac{2}{3} D_{ijkl}^{e} S_{kl} \sigma_y^2}{L}\right)\frac{\dot{\mu}}{\mu} + \frac{\frac{2}{3}\sigma_y D_{ijkl}^{e} S_{kl}}{L}\left(\frac{\partial \sigma_{y0}}{\partial T} + \frac{\partial H}{\partial T}\bar{\varepsilon}^p\right)\dot{T} \\ & + \left(1 - \frac{D_{ijkl}^{e} S_{kl} S_{ij}}{L}\right)\frac{\sigma_{kk}}{3}\frac{\dot{K}}{K}\delta_{ij} \end{aligned} \tag{5.89}$$

$$L = S_{ij} D^{e}_{ijkl} s_{kl} + \frac{4}{9} H \sigma_y^2 \tag{5.90}$$

$$H = \frac{E_{\mathrm{T}} E}{E - E_{\mathrm{T}}} \tag{5.91}$$

In the equation, E is the elastic tensor. E_{T} is the tangent slope of the stress–strain curves, which could reflect the microscopic information. The slope is provided in the single crystal algorithm, and therefore a micro–macro relationship is constructed.

The stress tensor with the microscopic information is inserted in the dynamic explicit finite element model

$$\mathbf{M}\ddot{\mathbf{u}} + \mathbf{C}\underline{\mathbf{u}} = \mathbf{P} - \mathbf{F} \tag{5.92}$$

where $\mathbf{M}$ is the mass matrix. It is defined as:

$$\mathbf{M} = \sum \int_{\mathbf{V}_e} \rho \mathbf{N}^{\mathrm{T}} \mathbf{N} \mathrm{d}\mathbf{V} \tag{5.93}$$

$\mathbf{C}$ is the damping matrix. It is defined as

$$\mathbf{C} = \alpha \mathbf{M} \tag{5.94}$$

$\mathbf{P}$ is the nodal internal force vector

$$\mathbf{P} = \sum \int_{V_e} \mathbf{N}^{\mathrm{T}} \mathbf{p} \mathrm{d}V + \sum \int_{S_e} \mathbf{N}^{\mathrm{T}} \mathbf{q} \mathrm{d}S \tag{5.95}$$

$\mathbf{F}$ is the nodal force vector

$$\mathbf{F} = \sum \int_{V_e} \mathbf{B}^{\mathrm{T}} \boldsymbol{\sigma} \mathrm{d}V \tag{5.96}$$

where $\boldsymbol{\sigma}$ reflects the stress in the elastic-plastic deformation, thermal expansion, phase transition and the microscopic structure of the material (crystal orientation).

5.4 The Numerical Calculation and Experimental Verification of Thermal Tensile of the High Strength Steel

5.4.1 The Thermal–Mechanical Coupling Tensile Experiment

As the most commonly used hot stamping steel, 22MnB5 is produced by adding a certain amount of B in the C-Mn steel. The B of solution solid state segregates at austenite grain boundaries, which delays the nucleation process of the ferrite, pearlite and bainite to further increase the material strength. Before hot stamping, most components of the steel are ferrite and pearlite and its tensile strength is about 600 MPa. After hot stamping, the internal component is transformed into martensite and its tensile is increased to 1500 MPa.

Under isothermal condition, the tensile experiment is set as Fig. 5.4a. The specimen is heated to 900 °C and the temperature remains for 5 min. Then the temperature is cooled to the test temperature by more than 40 °C/s and also the temperature remains constant. The experimental data is collected through the

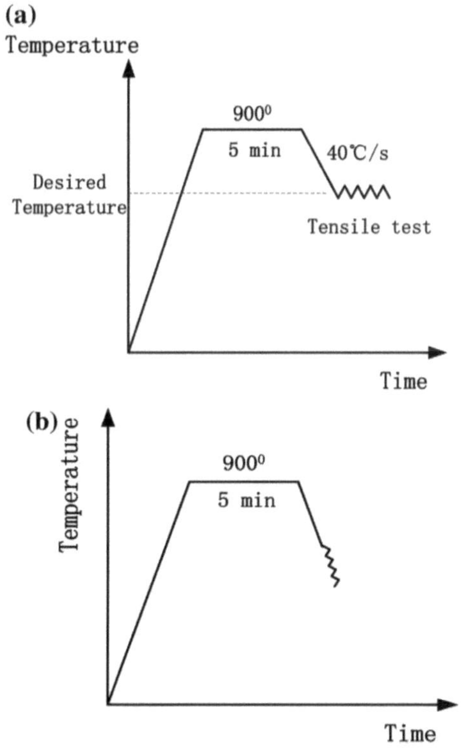

Fig. 5.4 **a** Experimental process of isothermal tensile deformation, **b** experimental process of non-isothermal tensile deformation

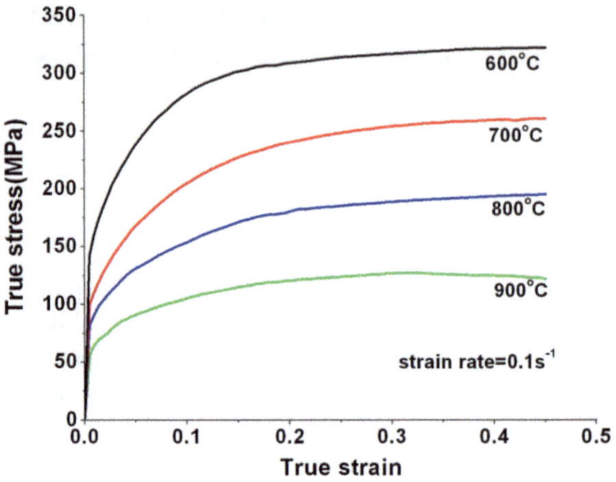

Fig. 5.5 Experimental data of 22MnB5 in tensile deformation at 600–900 °C and strain rate 0.1 s^{-1}

thermal tensile test afterwards. The temperature variation in the tensile test is shown in Fig. 5.4b. The difference between (a) and (b) is that the cooled temperature is not kept constant in the temperature variation test. This experiment studies the effect of different cooled tensile temperatures on the flow behavior of 22MnB5 in the temperature cooling process.

Figure 5.5 shows the stress–strain curves at different temperatures when the strain rate remains 0.1 s^{-1}. It could be seen that the temperature has a strength influence on the mechanical behavior of the material. As the temperature increases,

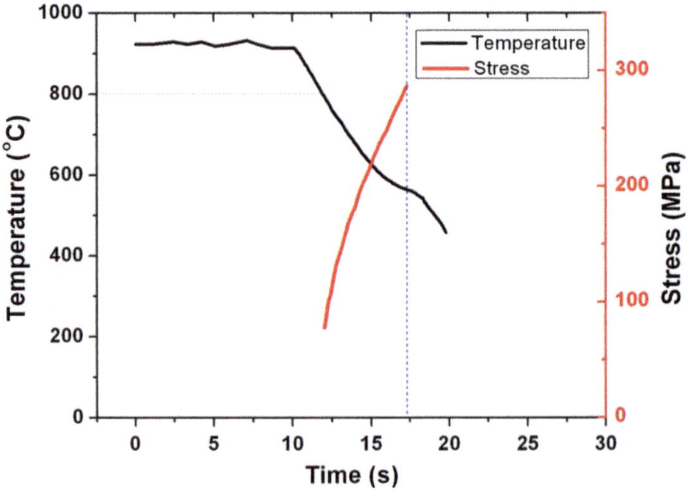

Fig. 5.6 The stress and temperature evolution during non-isothermal tensile test (the starting temperature is 800 °C)

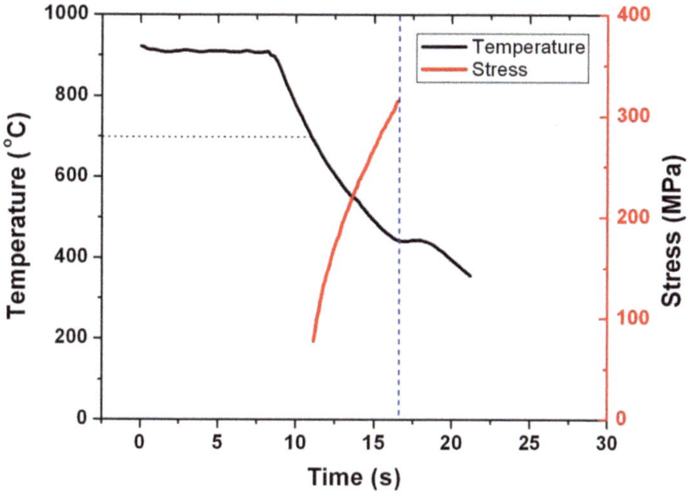

Fig. 5.7 The stress and temperature evolution during non-isothermal tensile test (the starting temperature is 700 °C)

the yield stress and the strengthening ability are both decreased. In the temperature of 600 or 700 °C, the stress–strain curves have an obvious strengthening state. As the temperature goes higher, the flow stress under high temperature undergoes a certain increase and quickly approaches the horizontal asymptote. The stress level gradually tends to be stable. This phenomenon is mostly caused by the dynamic recrystallization and recovery in the hot deformation materials [8].

Figure 5.6 shows the stress and temperature tensile curves versus time in the non-isothermal conditions. The figure reflects that the beginning temperature of the thermal tensile is 800 °C. After the cooling stage of 5.3 s, the slope of the declining temperature curve starts to change. This slope change means that new component different from the austenite phase is generated. The material structure change could cause the change of material properties such as mechanical and thermos-mechanical behaviors. At this time, the phase transition temperature is about 570 °C. As the phase transition has occurred, the microstructure of the material also changes. Thus, the phase transition point has to be controlled to avoid that in the forming stage, the early phase transition causes the hardness of the material to change early, further leading to the crack of components. In the Fig. 5.7, the tensile begins at 700 °C after cooling from the 900 °C. The temperature goes down to 445 °C after 5.5 s and the phase transition begins. The stress–strain curves of 22MnB5 in the thermal tensile at 800 and 700 °C are given in Fig. 5.8. As the temperature goes down, the elastic modulus and the strength ability are enhanced. The flow stress therefore tends to show the linear strengthening behavior. The temperature elevation at the beginning of tensile test causes the obvious decrease of the stress level and strengthening ability.

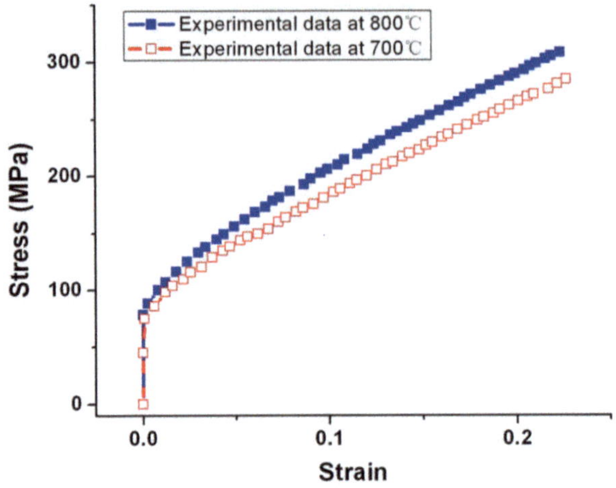

Fig. 5.8 Experimental data of stress–strain curves in cooling process

5.4.2 Comparison Analysis of the Numerical Simulation

According to the experimental material, the equipment and the experimental process introduced before, the isothermal and non-isothermal condition are considered in the thermal tensile experiment of the high strength steel 22MnB5. In this section, the polycrystalline calculation model is used to conduct the numerical simulation and experimental verification on thermo-mechanical behavior of the high strength steel 22MnB5.

According to the method of determining the model parameters, 200 grains with random orientations are circulated by their own positions and the macroscopic velocity gradient is rotated into the local coordinate system to calculate the stress. Then the local stress is transformed to the global coordinate system The Taylor assumption is used to average the stress by volume. Finally, the parameters of the crystal plasticity model on the macroscopic material are shown in Table 5.2.

Table 5.2 Parameters for model [11]

θ (°C)	600	700	800	900
m	0.05	0.06	0.06	0.06
$\dot{\gamma}_0$ (s^{-1})	1.0	1.0	1.0	1.0
τ^0_{cr} (MPa)	58.2	43.2	36.7	25.7
H_0 (Mpa)	184.0	110.8	60.0	40.1
$\hat{\tau}^\alpha$ (Mpa)	121.8	106.5	83.7	58.7
$\dot{\gamma}_s$ (s^{-1})	5.0×10^{10}	5.0×10^{10}	5.0×10^{10}	5.0×10^{10}
m'	0.005	0.006	0.008	0.011

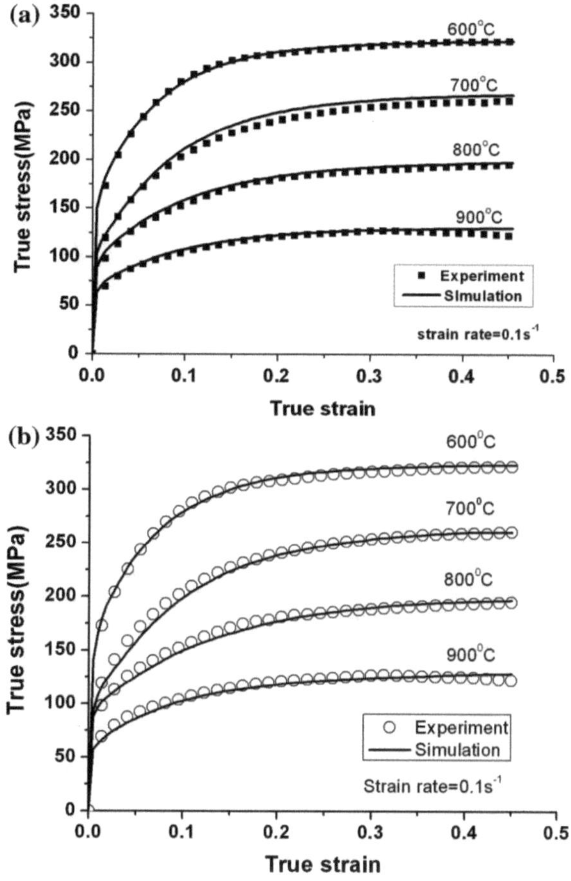

Fig. 5.9 Comparison between experimental data and simulation results at different temperatures and at strain rate 0.1 s^{-1}, **a** $\mathbf{F}^e$ model, **b** $\mathbf{F}^p$ model

The implicit algorithm based on deformation gradient is used to simulate the flow behavior of 22MnB5 at different temperatures and strain rates, the result of which is compared to the experimental data in Fig. 5.9. This Figure shows the stress–strain response of $\mathbf{F}^e$ model when the temperature is between 600 and 900 °C and the strain rate is 0.1 s^{-1}. From this figure, the temperature has a great effect on the flow behavior of 22MnB5. At constant strain rate, the flow stress of the material decreases while the temperature goes up. Figure 5.9b shows the results of $\mathbf{F}^p$ model. In the numerical simulation, the difference between two implicit polycrystalline algorithms is mostly the construction of the algorithm with two different integration variables to calculate the stress update in single crystal model. The calculation efficiency and the precision of these two models are basically the same in calculating the thermal tensile behavior of 22MnB5.

In the temperature variation condition, the flow behavior of the thermal tensile of high strength steel 22MnB5 is also calculated by the polycrystalline model. Figure 5.10 shows the comparison between the numerical simulation of $\mathbf{F}^p$ model

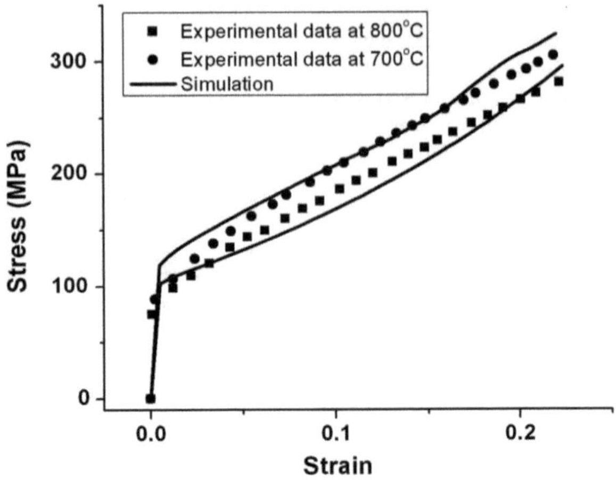

Fig. 5.10 Stress–strain correlations between experimental data and implicit computational model in cooling process

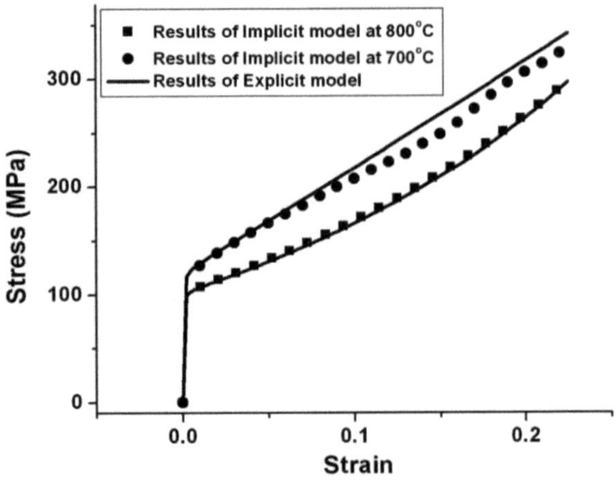

Fig. 5.11 Comparison of computational results in cooling process of implicit model and explicit model

and the experimental data. In the temperature cooling process, the flow stress of the material shows the tendency of continuous strengthening. In beginning of the tensile test, the elevation of temperature (from 700 to 800 °C) causes the decrease of the stress level and the strengthening ability. The numerical simulation of the implicit model agrees well with experimental data. The implicit and explicit models are also compared in the Fig. 5.11. It shows that the results of these two models are basically the same. The constructed algorithm could well describe the flow behavior of high strength steel 22MnB5 in the temperature variation condition.

5.5 Summary

This chapter introduces the constitutive integration algorithm in finite strain framework. The application is practically in the warm forming process. First, the constitutive model of single crystal is introduced. And then the algorithm is constructed based on second P-K stress, plastic deformation gradient, and elastic deformation gradient. The Newton–Raphson method is introduced to solve the time-discretized crystal plasticity formulation and an implicit time integration method based on finite strain is built. Under the Taylor model homogenization, the microscopic response of single crystal is associated with the macroscopic response. Thus, the synchronous mechanical response of high strength steel blank in hot forming can be described. In the simulation compared to experiment, it proves that in the crystal plasticity constitutive algorithm the macroscopic stress–strain response can predict the temperature-varying curves in different strain rates and elevated temperatures. This constitutive algorithm provides a unique tool for research on material behavior and prediction of new material mechanical response in temperature-varying condition.

References

1. Bunge HJ, Morris PR (1982) Texture analysis in materials science: mathematical methods. Butterworth, London
2. Dan Z, Zhu Y, Hu P et al (2013) Constitutive model for single crystal thermal finite deformation. Chin J Solid Mech 34:266–271 (in Chinese)
3. Ganapathysubramanian S, Zabaras N (2005) Modeling the thermoelastic-viscoplastic response of polycrystals using a continuum representation over the orientation space. Int J Plast 21:119–144
4. Hu P, Ma N, Liu L et al. (2012) Theories, methods and numerical technology of sheet metal cold and hot forming: analysis, simulation and engineering applications. Springer Jan 15 2013
5. Kalidindi SR, Bronkhorst CA, Anand L (1992) Crystallographic texture evolution in bulk deformation processing of FCC metals. J Mech Phys Solids 40:537–569
6. Liu Y, Zhu Y, Ying L et al (2015a) Experimental and numerical simulation of 6061 Al tensile behavior at high temperature. IDDRG
7. Liu Y, Zhu Y, Ying L et al (2015b) Numerical simulation and experimental verification of aluminum alloy tensile behavior at elevated temperature. In: ICSHU 2015
8. Merklein M, Lechler J (2006) Investigation of the thermo-mechanical properties of hot stamping steels. J Mater Process Technol 177:452–455
9. Roters F, Eisenlohr P, Hantcherli L et al (2010) Overview of constitutive laws, F, homogenization and multiscale methods in crystal plasticity finite-element modeling: theory, experiments, applications. Acta Mater 58:1152–1211
10. Weber G, Anand L (1990) Finite deformation constitutive equations and a time integration procedure for isotropic, hyperelastic-viscoplastic solids. Comput Methods Appl Mech Eng 79:173–202
11. Zhao D, Zhu Y, Hu P et al (2013a) Modeling of tensional behavior of 22MnB5 boron steel at isothermal conditions by extended crystal plasticity. In: AIP conference proceedings, pp 304–310

12. Zhao D, Zhu Y, Ying L et al (2013b) Modeling and experimental verification of thermo-mechanical coupled behavior of face-centered-cubic polycrystals. Mat Des 52:289–294
13. Zhao D, Zhu YG, Hu P et al (2013c) A new integration algorithm for finite deformation of thermo-elasto-viscoplastic single crystals. Acta Mech Sin 29:709–717

Chapter 6
Heat Transfer in Hot Stamping Process of High-Strength Steel

6.1 Heat Transfer Theory and Behavior Analysis

6.1.1 Basic Principle [1, 2]

1. Basic law and differential equation

(1) Basic law

Fourier law is known as the basic law of heat conduction [3], and it can be described as,

$$\mathbf{q} = -\lambda \text{grad}T = -\lambda \frac{\partial T}{\partial n}\mathbf{n} \quad (6.1.1)$$

where, $\mathbf{q}$ is heat flux vector, λ is thermal conductivity, grad$T$ is temperature gradient of some point and $\mathbf{n}$ is normal unit vector on the isothermal line through that point.

Fourier law can be expressed as follows: in the process of heat conduction, the heat through a given section per unit time is proportional to the temperature gradient perpendicular to that section and the corresponding sectional area, but the direction of heat transfer is opposite to the temperature rising direction.

(2) Differential equation of heat transfer

According to the Fourier law and energy conservation law, a partial differential equation is established to describe the heat transfer process which is applicable to any heat transfer problem except nanoscale heat transfer process.

The differential equation of heat transfer in Cartesian coordinate system is given as

$$\rho c \frac{\partial T}{\partial \tau} = \frac{\partial}{\partial x}\left(\lambda \frac{\partial T}{\partial x}\right) + \frac{\partial}{\partial y}\left(\lambda \frac{\partial T}{\partial y}\right) + \frac{\partial}{\partial z}\left(\lambda \frac{\partial T}{\partial z}\right) + \dot{Q} \quad (6.1.2)$$

P. Hu et al., *Hot Stamping Advanced Manufacturing Technology of Lightweight Car Body*, DOI 10.1007/978-981-10-2401-6_6

where ρ is density, c is specific heat, λ is thermal conductivity and $\dot{Q}$ is internal heat source. All these parameters can take the values associated with temperature.

When λ is a constant, Eq. (6.1.2) is simplified as

$$\frac{\partial T}{\partial \tau} = a\left(\frac{\partial^2 T}{\partial x^2} + \frac{\partial T^2}{\partial y^2} + \frac{\partial T^2}{\partial z^2}\right) + \frac{\dot{Q}}{\rho c} \tag{6.1.3}$$

where $a = \lambda/\rho c$ is thermal diffusion coefficient.

(3) Definite conditions

Heat conduction differential equation presents the general mathematical expression of heat transfer process. For obtaining the temperature distribution of a specific problem, some additional conditions, i.e., definite conditions which are composed by initial conditions and boundary conditions should be provided.

There are three kinds of boundary conditions of heat conduction problems. One of the easiest ways to specify the thermal conditions on a surface is to specify the temperature. In addition, when there is sufficient information about energy interactions at a surface, it may be possible to determine the rate of heat transfer and thus the heat flux on that surface and this information can be used as one of the boundary conditions. Moreover, convection is probably the most common boundary encountered in practice since most heat transfer surfaces are exposed to an environment at a specified temperature. Three boundary conditions mentioned above are also called Dirichlet, Neumann, and Robin conditions respectively.

2. Heat transfer mechanisms

Heat can be transferred in three different ways: conduction, convection, and radiation [4]. All modes of heat transfer require the existence of a temperature difference, and all modes of heat transfer are from the high-temperature medium to a lower temperature one. The detailed introduction is given as follows.

(1) Heat conduction

Heat conduction is the transfer of energy from the more energetic particles of a substance to the adjacent less energetic ones as a result of interactions between the particles. Conduction can take place in solids, liquids, or gases. In gases and liquids, conduction is due to the collisions and diffusion of the molecules during their random motion. In solids, it is due to the combination of vibrations of the molecules in a lattice and the energy transported by free electrons. Heat conduction follows the Fourier law shown in Eq. (6.1.1).

(2) Heat convection

Heat convection is the mode of energy transfer between a solid surface and the adjacent liquid or gas that is in motion, and it involves the combined effects of conduction and fluid motion. The faster the fluid motion, the greater the convectional heat transfer.

Despite the complexity of convection, the rate of convectional heat transfer is observed to be proportional to the temperature difference, and is conveniently expressed by Newton's law of cooling as follows:
fluid is heated,

$$q = h(T_w - T_f) \tag{6.1.4}$$

fluid is cooled,

$$q = h(T_f - T_w) \tag{6.1.5}$$

where T_w and T_f are the wall temperature and fluid temperature, respectively, h is the convectional heat transfer coefficient in W m^{-2} K^{-1}.

(3) Heat radiation

Heat radiation is the energy emitted by matter in the form of electromagnetic waves as a result of the changes in the electronic configurations of the atoms or molecules.

The heat radiation of matter is given by the Stefan–Boltzmann law as

$$q = \frac{Q}{A} = \varepsilon\sigma\left(T_w^4 - T_\infty^4\right) \tag{6.1.6}$$

where ε is the emissivity of the matter. The property emissivity, whose value is in the range $0 \leq \varepsilon \leq 1$, is dependent on the material property, surface condition and surface temperature. σ is the Stefan–Boltzmann constant, and the value is 5.67e10^{-8} W m^{-2} K^{-4}. T_w is the surface temperature, and T_∞ is the surrounded temperature.

6.1.2 Heat Transfer Behavior Analysis

Hot stamping process is composed by three stages: heating stage, transferring stage and stamping and quenching stage [5, 6]. Regardless of the heating stage in Fig. 6.1, the hot blank is cooled in the air mainly in the form of heat radiation and heat convection during the transferring stage. In the stamping and quenching stage, the blank heat passes to tool then the heat is taken away by the circulating water [7], and the main forms of heat transfer are conduction, convection with clearance, and forced convection with cooling water.

In the actual hot stamping process, the conduction, convection, and radiation do not work alone, but influence and interweave each other. Heat transfer coefficient can be thought as a comprehensive evaluation in the whole process of heat transfer. The method for determining the integrative heat transfer coefficient is specifically described in next section.

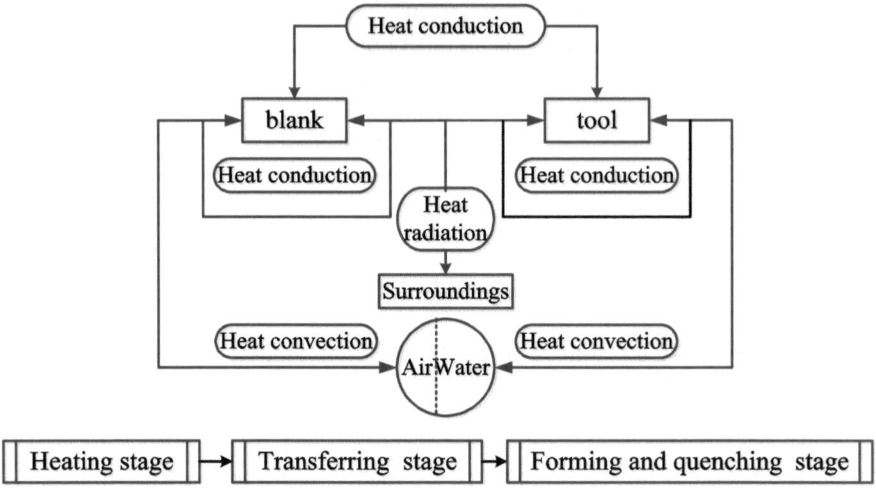

Fig. 6.1 Heat transfer forms in hot stamping process

6.2 Determination of Heat Transfer Coefficient in Hot Stamping Process

6.2.1 Method

Heat transfer coefficient not only depends on the physical properties of contact bodies as well as the contact surface shape, size, and layout, but also has a close relationship with factors such as temperature, pressure, and flow rate.

So far, four methods are widely used to calculate and estimate the interfacial heat transfer coefficient. They are empirical formula method, heat balance method, Beck inverse technique, and FEM optimization method [8]. The interfacial heat transfer coefficient related to pressure one by one is directly obtained by the empirical formula, but the experience formula is generally generalized based on certain experimental conditions which are always given bigger prediction value. If the mold temperature is constant and Biot Number is smaller, the heat transfer problem can be solved by Newton cooling formula, i.e., heat balance method [9, 10], which is applicable to heat transfer process with a constant tool temperature, but not to the process with a sharp change of tool temperature. According to the known initial conditions and the temperature curves of specified points measured from experiment, the heat transfer coefficient is got by solving the partial differential equation with repeatedly updating the trial and error heat flux boundary, and this method is called inverse heat conduction problem [11–13]. The solution accuracy of inverse heat conduction problem is mainly determined by the temperature curves measured in experiment. For minimizing the delay and lag of heat conduction in itself, the calibration of temperature collection and location precision is put forward higher requirements. Beck inverse technique is a widely used nonlinear estimation method [14–16]. Due to the development of computer power and the enhancement of FEM simulation accuracy of temperature field, an optimization process which integrates the I-Sight and Abaqus is established to inverse estimate the heat transfer coefficient. The principle of this method is to make the simulation temperature field in accordance with the experimental value by continuously adjusting the heat transfer coefficient parameter in FEM model. FEM optimization method is welcomed by the industry but normally it can only get an effective interface heat transfer coefficient value in the process and it is unable to describe the characteristic changing with time and temperature in the hardening process.

There are three methods to get the convectional heat transfer coefficient between tool and cooling water. They are empirical formula method, experimental mensuration, and CFD simulation method. According to the experience correlations of turbulent forced convectional heat transfer in the tube slot, the convectional heat transfer coefficient is calculated at a certain velocity and ratio of channel length and diameter. However, the empirical formula is only suitable for a simple model, and only a heat transfer coefficient will be obtained in a whole heat transfer system. Based on a single cooling pipe experiment, the convectional heat transfer coefficients associated with different mass flow rate, diameter, and surface heat flux are got. Besides, when appropriate boundary conditions and mesh model are provided, it is easy to obtain the convectional heat transfer coefficient along the channel wall by CFD simulation, but the results accuracy mainly depends on the CFD model.

This section will focus on the experimental measuring methods of the interfacial heat transfer coefficient between blank and tool and forced convectional heat transfer coefficient between tool and cooling water [16–18].

6.2.2 Determination of Interfacial Heat Transfer Coefficient Between Blank and Tool

The general idea of the estimation method is to calculate the interfacial heat transfer coefficient with an inverse model by matching the experiment data and model data.

1. Beck inverse technique

The Beck inverse technique is the most classic and widely applied in handling inverse heat conduction problem. The objective function of Beck inverse method is

$$\min{:}\, f(q) = \sum_{i=1}^{n} (T_{\mathrm{est}} - T_{\mathrm{mea}}) \tag{6.2.1}$$

The solving process is summed up as follows: assume the initial heat flux is q at t_0 moment, then calculate the temperature of measurement points at 1, 2 … r moment. After that, assess the heat flux sensitivity ratio depending on temperature and modify the initial heat flux until the heat flux rate meets the requirements of convergence. The heat flux sensitivity ratio is expressed as

$$\phi = \frac{\partial T}{\partial q} = \frac{T_{\text{est}}(q * (1 + e)) - T_{\text{est}}(q)}{q * e} \tag{6.2.2}$$

The revised value of heat flux is given as

$$\Delta q = \frac{\sum_{i=1}^{r} (T_{\text{mea}} - T_{\text{est}})\phi_i}{\sum_{i=1}^{r} \phi_i^2} \tag{6.2.3}$$

The heat flux at each moment is iterative solving the one-dimensional heat transfer partial differential equation by PDE tool of MATLAB. The flowchart of Beck inverse method is shown in Fig. 6.2.

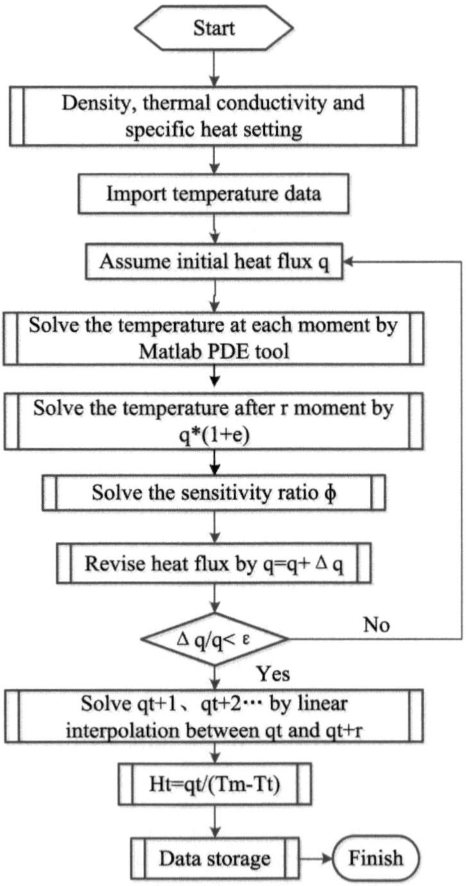

Fig. 6.2 The flowchart of Beck inverse method

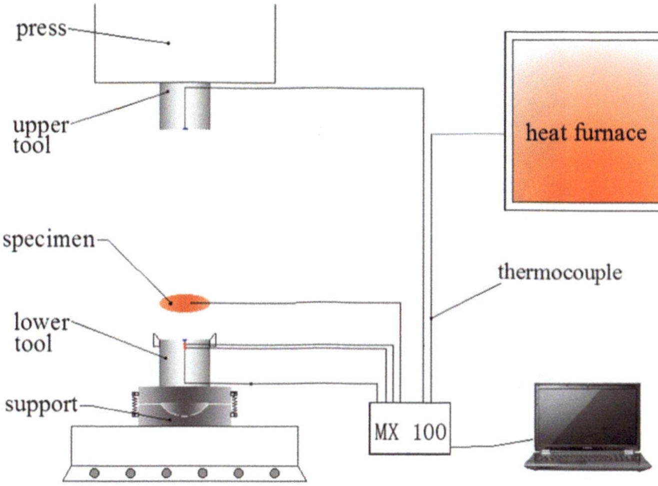

Fig. 6.3 Schematic diagram of experimental setup

2. Hot stamping experiment [1, 2]

The blank is 22MnB5 steel which has the diameter of 75 mm and thickness of 2.0 mm. A thermocouple is inserted into a depth of 20 mm hole along the radial direction. Schematic diagram of experimental setup is illustrated in Fig. 6.3. The setup is composed of hydraulic press, upper and lower dies, pressure balance support plate, furnace and temperature acquisition instrument. The maximum nominal pressure of hydraulic press is 400 KN which is adjusted by pressure control system. The tool is 45 steel with a same size as blank. The pressure balance support plate is designed to ensure the pressure of blank is uniformly distributed during the hot stamping process. The thermocouple is fusion welded on the tool surface, and then the tool surface should be polished smooth on the premise that does not destroy the nugget. For assuming the measured temperature as the approximate tool surface temperature, the welding technology is required to not only ensure the connection property of welding nugget but also obtain the nugget small enough.

The blank is heated up to 900 °C in a furnace and preserved about 4 min to achieve homogenous austenitic microstructure. Then the blank is quickly transferred to the work place and simultaneously the upper die is dropped to close and the pressure is held for 30 s. After that martensite is formed in blank which can greatly improve the strength and hardness. The temperatures of blank and tool are monitored by acquisition instrument during the whole process. According to the loading rage in industry and hydraulic press range, 0.05, 2, 5, 20, and 40 MPa are selected to operate the experiment (Fig. 6.4).

For the convenience of comparison and analysis of temperature, the initial stamping temperature of blank is controlled at 800 °C. The temperature curves of blank and tool at different holding pressures are displayed in Fig. 6.5. It is noted

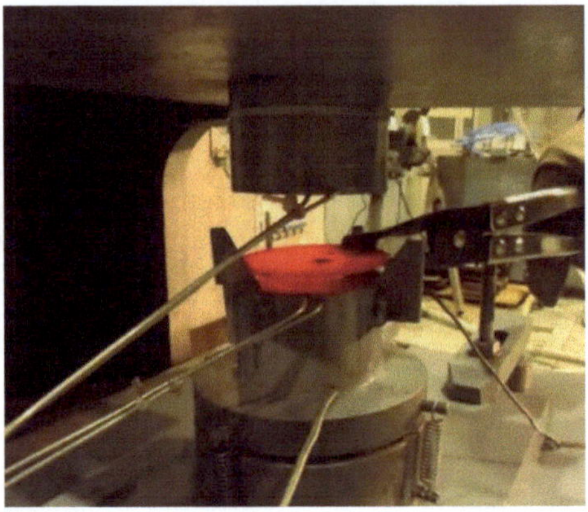

Fig. 6.4 Experiment process

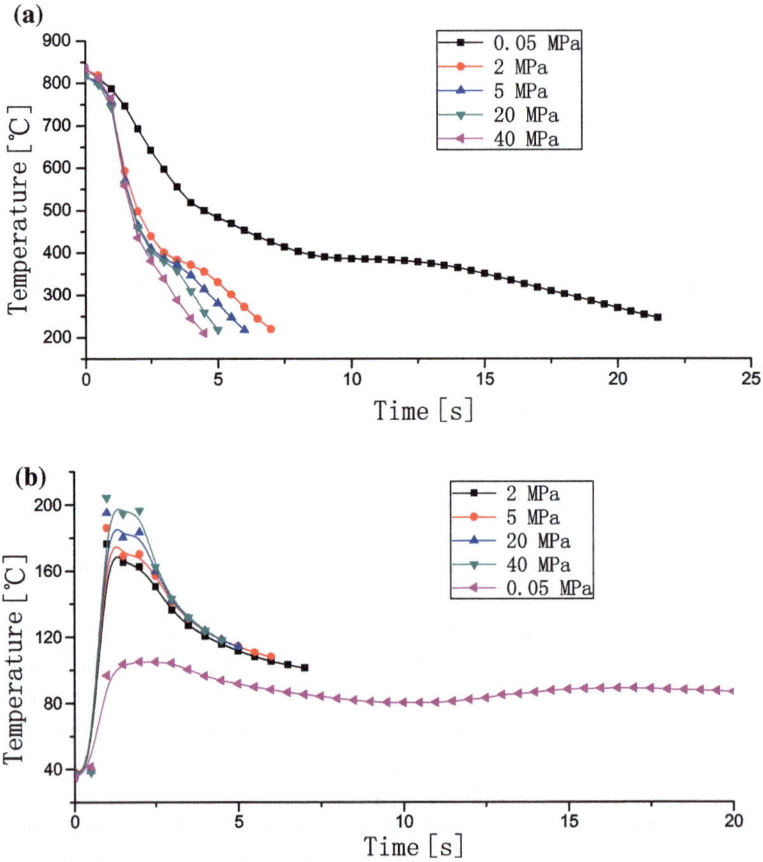

Fig. 6.5 Temperature curves at different holding pressures **a** blank; **b** tool surface

that within the temperature range from 800 to 400 °C even at the pressure of 0.5 MPa, the cooling rate of blank also can reach 50 °C/s which has exceeded the critical cooling rate of martensite transformation. Besides, within the same region mentioned above, it is not prominent that the cooling rate is kept in the range of 200 to 300 °C/s when the holding pressure is more than 2 MPa.

3. Determination of the interfacial heat transfer coefficient

According to the former introduction of Beck inverse technique and experiment setup, the temperature curves at the points below the tool surface center with distances of 2 and 4 mm are shown in Fig. 6.6. The interfacial heat transfer coefficient curves at different holding pressures which are calculated by the routine written with PDE tool in MATLAB are displayed in Fig. 6.7.

From high to low temperature, the interfacial heat transfer coefficient calculated by Beck inverse technique rises slowly first and sharply when the temperature reach 400 °C. This is because the martensite phase transformation of boron alloy steel will happen at 400 °C which will produce the latent heat to promote the increase of coefficient.

6.2.3 Determination of Convectional Heat Transfer Coefficient Between Tool and Cooling Water

1. Experiment principle [19]

According to the Newton's law of cooling, the convectional heat transfer coefficient is defined by the temperature difference between channel wall and cooling water,

$$q = h(T_{\mathrm{W}} - T_{\mathrm{B}}) \tag{6.2.4}$$

where q is heat flux, h is convectional heat transfer coefficient, T_{W} is the channel wall temperature, and T_{B} is the average temperature of flow section. Due to the globality of energy transformation, the average temperature of water is set as standard.

It is difficult to measure the T_{B} directly by experiment, but it can be determined by the following formula:

$$T_{\mathrm{B}} = T_{\mathrm{C}} + \frac{7}{96}\frac{ur_0^2}{\alpha}\frac{\partial T}{\partial x} \tag{6.2.5}$$

where T_{C} is the temperature of channel center which is measured by thermocouples installed in the channel center, u is velocity of cooling water, r_0 is radius of cooling channel, α is heat dissipating coefficient, $\frac{\partial T}{\partial x}$ is the temperature gradient along the axial direction of channel. Assume that the heat flux of the channel wall is constant

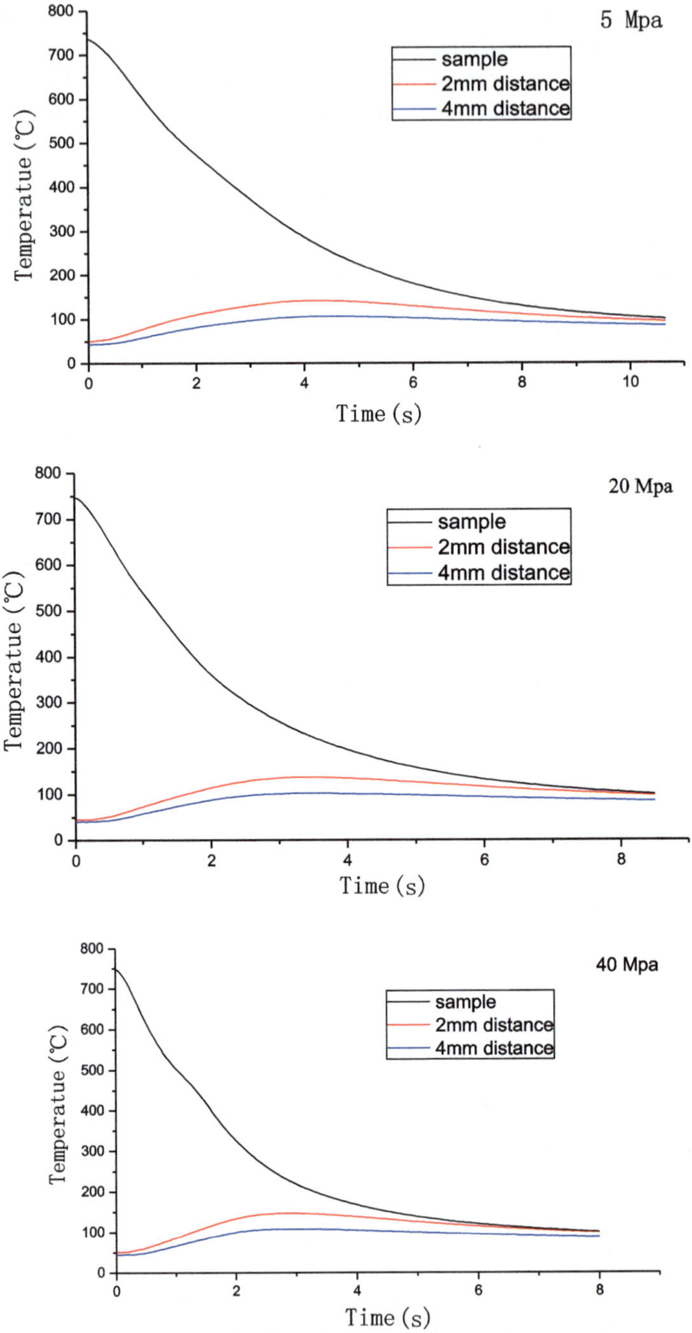

Fig. 6.6 Temperature curves at different holding pressures (5 MPa, 20 MPa, 40 MPa)

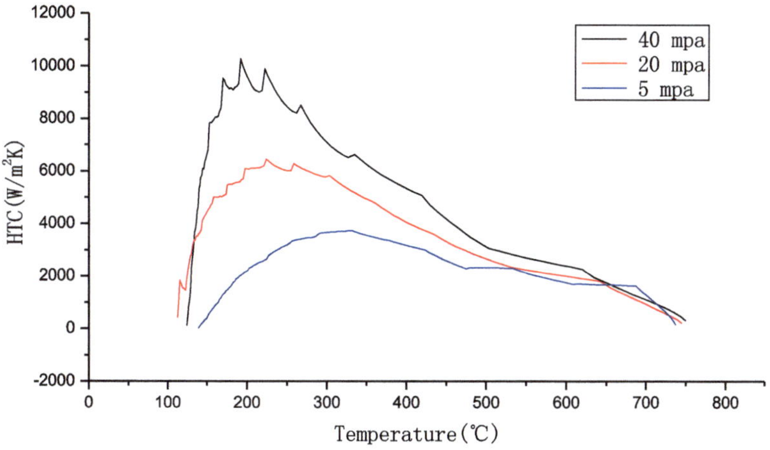

Fig. 6.7 Interfacial heat transfer coefficient curves at different holding pressures calculated by Beck inverse technique

and the average temperature of cooling water is linear increasing along the axial direction, the $\frac{\partial T}{\partial x}$ is also seen as constant. $\frac{\partial T}{\partial x}$ is calculated by the inlet and outlet temperature.

As shown in Fig. 6.8, the heat flux at any radius of r is expressed by

$$q = \frac{\lambda}{r} \frac{T_1 - T_2}{\ln(r_2/r_1)} \tag{6.2.6}$$

where λ is thermal conductivity of cylindrical tool, T_1 and T_2 are the temperature at the radius of r_1 and r_2, respectively, which are measured by thermocouples. The temperature T_w of channel inner wall is calculated by Eq. (6.2.7).

Fig. 6.8 Schematic of cylindrical heat conduction problem

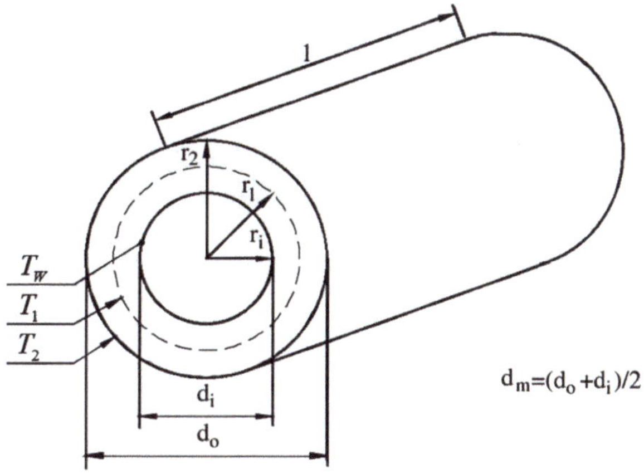

$$\frac{T_w - T_2}{T_1 - T_2} = \frac{\ln(r_i/r_2)}{\ln(r_1/r_2)} \tag{6.2.7}$$

Put in the Eqs. (6.2.5)–(6.2.7) to Eq. (6.2.4), the convectional heat transfer coefficient h will be obtained.

2. Experiment process

In the basis of experiment principle, the experiment device is shown in Fig. 6.9. The tool material is H11 tool steel with a length of 500 mm and a diameter of 60 mm. The cooling channel is drilled with the diameter of 8, 10, and 12 mm according to the experimental sequence. The roughness of drilled channel wall is the same as actual tool channel wall. For pipe installation, a flowmeter is installed at the channel inlet to control mass flow with ball valve. The cylindrical tool is fixed in a temperature controlled and uniform heated tube electric resistance furnace. The heat flux entering the cylindrical tool is controlled by adjusting furnace temperature.

Three groups of, in total nine, K-type thermocouples are arranged at the inlet, outlet, and middle positions of the cylindrical tool. Each group at each position has three thermocouples which are used to measure the temperature of the cylindrical surface, internal and channel center. The thermocouples of cylindrical tool surface and internal are welded connection and the thermocouple for water temperature is inserted into the channel center. All of the thermocouples are monitored by temperature acquisition instrument MX100. The temperature measured at channel middle is used to analyze the tool internal temperature field and calculate the average water temperature. The temperature of inlet and outlet are used to calculate and monitor the temperature change of cooling water.

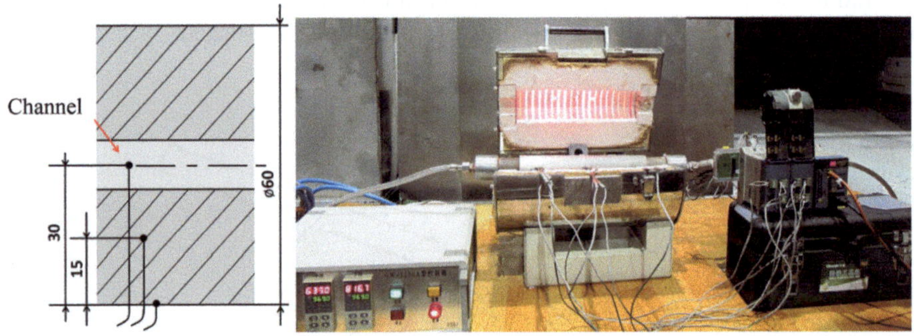

Fig. 6.9 Thermocouple layout and experiment device

Considering the steady state of experiment, the mass flow is in a range of 0.8–12.5 kg/min, the heat flux is controlled in (1.5–5.0) × 10^5 W/m^2 and three diameters of 8, 10, 12 mm are realized by three different tools.

3. Determination of convectional heat transfer coefficient

According to the experiment principle and device, the effects of mass flow rate, heat flux and channel radius on the convectional heat transfer coefficient of circular pipe are discussed. The cooling medium is pure water and the tool material is H11 steel.

(1) The effects of mass flow rate and channel radius

The effects of mass flow rate on convectional heat transfer coefficient at different channel diameters are presented in Fig. 6.10. The heat flux is (3.5 ± 0.1) × 10^5 W/m^2 and the inlet temperature is 20 °C. Obviously, mass flow rate is an important factor that affects the convectional heat transfer coefficient. It is found that the greater the mass flow rate, the greater the convectional heat transfer coefficient. This is because the mass flow rate directly affects the water velocity in internal pipe which will lead to stronger turbulence, so that the turbulent flow between adjacent layers are enhanced mixed under the condition of forced convection.

It is found from Fig. 6.10 that the convectional heat transfer coefficient increases with the decrease of channel diameter. Obviously, at the same flow rate, the smaller the channel diameter, the faster flow velocity and the higher turbulence intensity. Besides, the smaller diameter the increase trend of convection heat transfer coefficient is more obvious with the decrease of channel diameter. When the diameter is

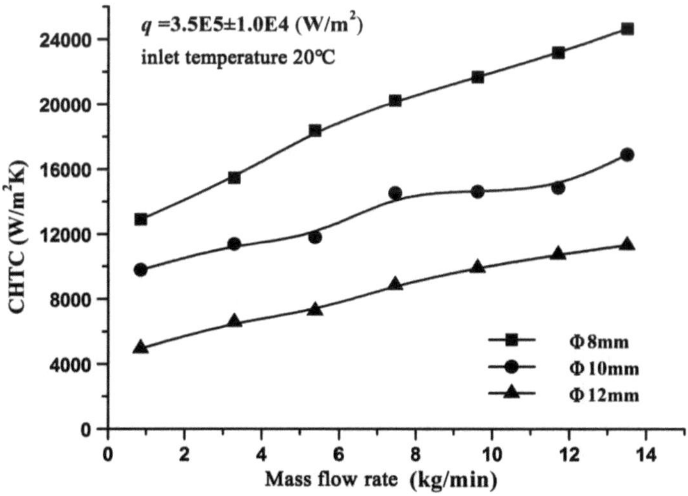

Fig. 6.10 Effects of mass flow rate on convectional heat transfer coefficient at different channel diameters

12 mm, the convectional heat transfer coefficient increases from 4900 to 11300 W/m^2 K with a 6400 W/m^2 K range ability. However, for a cooling channel diameter of 8 mm, the coefficient rises from 12900 to 24600 W/m^2 K with a 24600 W/m^2 K range ability which is apparently higher than the result from the diameter of 12 mm. This is because with the increase of mass flow rate, the Reynolds number increases more obviously and the degree of turbulence is stronger with smaller channel diameter.

(2) The effect of heat flux

The effect of heat flux on the convectional heat transfer coefficient is shown in Fig. 6.11. The channel diameter is 12 mm and the inlet temperature is 20 °C. The heat flux is the other important factor that affects the convectional heat transfer coefficient. It is noted that changing the heat flux will significantly change the convection heat transfer coefficient. The trend is the greater the heat flux, the greater the convection heat transfer coefficient. This is due to the heat flow increase is bound to cause water temperature increase which will change the thermal physical properties of water. Moreover, the change of fluid physical properties can directly affect the heat transfer process, especially the variation of viscosity with temperature. Due to the higher temperature corresponding to the relatively low viscosity coefficient, the Reynolds number increases and the degree of turbulence is stronger.

(3) Results comparison of CFD simulation and empirical formula

The fully development of forced convection heat transfer has the following classical empirical formula:

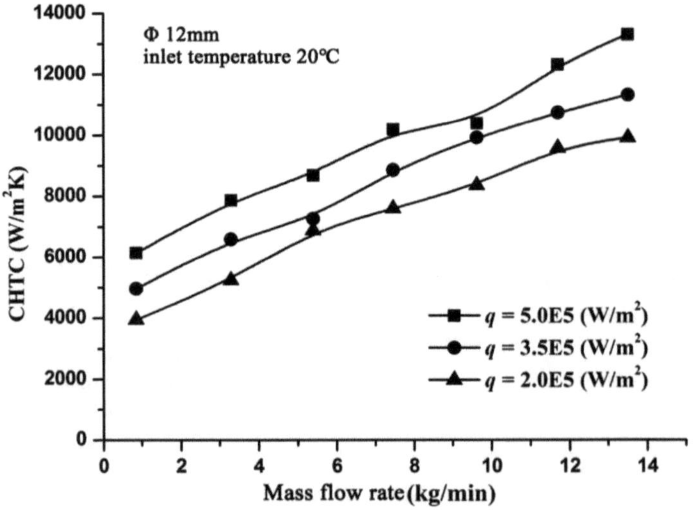

Fig. 6.11 The effect of mass flow rate on the convectional heat transfer coefficient at different heat flux

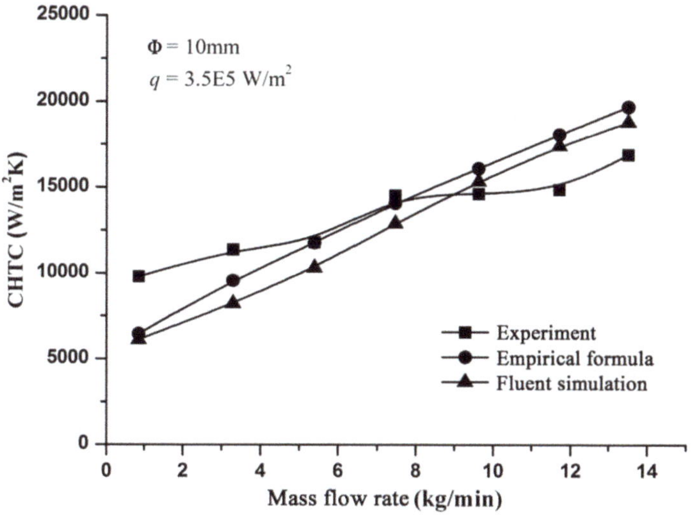

Fig. 6.12 The results comparison of experiment, empirical formula, and fluent simulation

$$Nu_d = \frac{(f/8)\mathrm{Re}_d \,\mathrm{Pr}}{1.07 + 12.8\left(\mathrm{Pr}^{2/3} - 1\right)(f/8)^{1/2}} \tag{6.2.8}$$

where f is Darcy–Weisbach friction coefficient looked up a special table according to the relative roughness and Reynolds number. Prandtl number Pr which reflects the effects of fluid physical properties on the convective heat transfer process and measures the thickness ratio of two kinds of boundary layers is also looked up a special table.

Fig. 6.13 Simulation contour of tool temperature by fluent

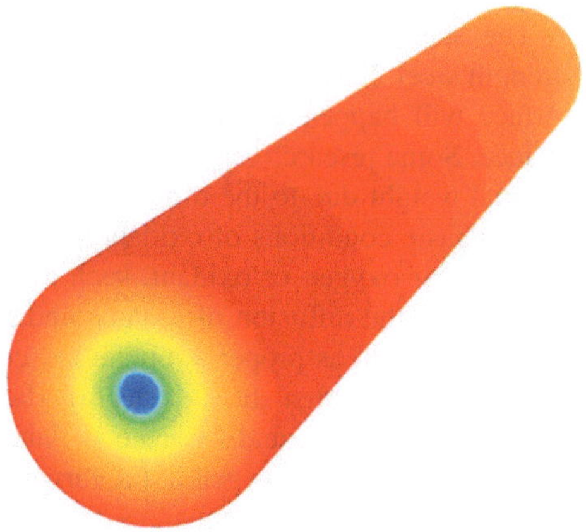

The results comparison of experiment, empirical formula and fluent simulation is exhibited in Fig. 6.12. The parameters used in fluent is the same as that used in experiment and the simulation contour is shown in Fig. 6.13.

In Fig. 6.12, the channel diameter is 10 mm and the heat flux is 3.5×10^5 W/m^2. It is obviously noted that the convectional heat transfer coefficient increases with the increase of mass flow rate. The convectional heat transfer coefficients of empirical formula and fluent simulation have the same linear upward trend, and the simulation results are just slightly smaller than that of empirical formula. Due to the particularity of tool material and the difference between experimental conditions and ideal situation, compared with empirical formula and simulation, the convectional heat transfer coefficients of experiment present slow growing with the increase of mass flow rate.

6.3 The Other Factors Influencing the Heat Transfer Coefficient of Hot Stamping Process

6.3.1 The Effect of High-Temperature Oxidized Scale

The hot stamping process is: make the boron alloy steel heat up to austenitic temperature in a furnace, and then transfer the hot blank into the stamping tool with cooling system to form and quench. After uniform quenching, the microstructure of blank is transferred from austenite to martensite which can obtain the ultrahigh-strength steel. The blank will be heated up to austenitic temperature and preserved several minutes, so the oxidation decarburization behavior occurs inevitably on the hot blank surface during the hot stamping process. Subsequently, the thickness and strength are reduced. Besides, the oxidation decarburization behavior will influence the heat transfer performance of interfacial surface and further affects the stability of component quality [17].

1. Oxidation mechanisms of high-strength steel

The steel is heated directly in a high-temperature furnace. When the high-strength steel 22MnB5 is put in furnace, the oxidation and decarburization of steel surface will happen and the scale and decarburization layer are attached on the surface. Some research reports that the oxidation burning loss weight is 0.5–3 % of original weight due to the different varieties of steel and heating technology [20].

The basic conditions of oxidation for hot stamping steel are as follows: (1) the existence of oxygen or oxidant, such as carbon dioxide, vapor, oxygen and sulfur dioxide; (2) the diffusion of oxygen and iron; (3) meet a certain chemical reaction conditions, such as proper temperature, chemical concentration, and reaction time.

Generally, the first stage of steel oxidation process is the forming of ionic oxide on the steel surface, and then the $O^{2-}$and Fe^{2+} promote the further growth of the oxide film. After the oxygen in the furnace is attached on the oxide film of steel, the

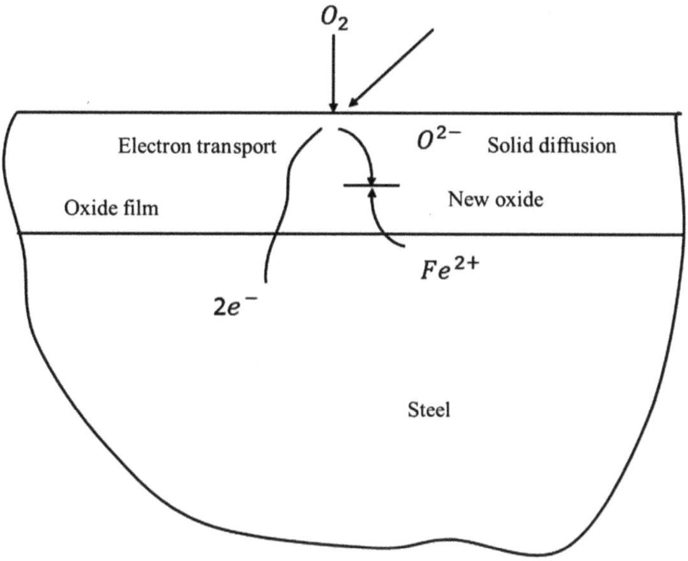

Fig. 6.14 The ion diffusion mechanism of oxide film growth

O^{2-} is formed by getting electrons, and then the O^{2-} is spread on the steel surface. However, Fe^{2+} and electrons are arrived at the steel surface by diffusion, and new oxide will be generated once the O^{2-} and Fe^{2+} meet on the surface membrane. The oxidation mechanism is shown in Fig. 6.14. Due to the different chemical composition of steel, the diffusion law of O^{2-} and Fe^{2+} has a certain difference with different oxidation properties and oxidation products. Therefore, for achieving the purpose of oxidation burning loss reduction, the oxidation characteristics and heating technology of different steels should be studied systematically.

According to the domestic and foreign studies on oxidation characteristics of different steels, it is noted that the oxidation kinetics under different experimental temperature satisfy the following parabolic rule:

$$(\Delta G/S)^2 = K_p t + C \tag{6.3.1}$$

where K_p increases quickly with temperature increase, and the increase law meets the Arrhenius equation,

$$K_p = Ae^{-Q/(RT)} \tag{6.3.2}$$

where K_p is oxidation rate constant, A is frequency factor, Q is reaction activation energy, R is the molar gas constant and T is thermodynamic temperature.

To sum up, the higher the temperature, the faster the reaction speed, the thicker the generated oxide film.

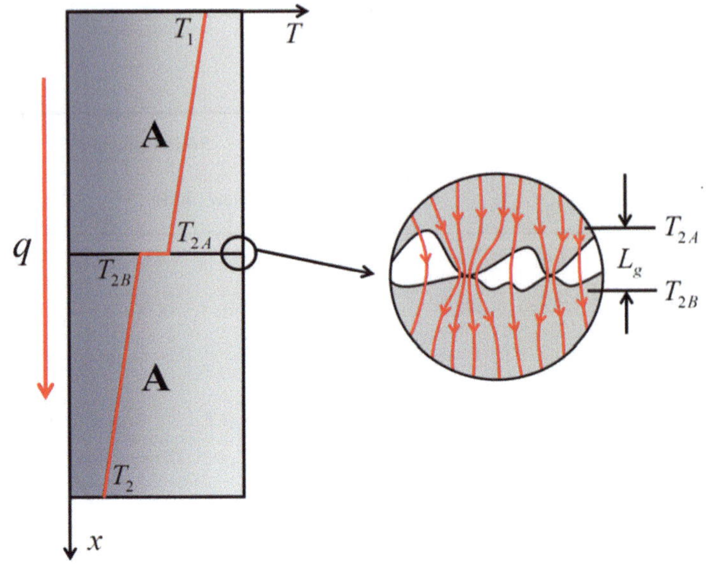

Fig. 6.15 The heat transfer contact surface of solid to solid

2. The relationship of scale thickness and heat transfer coefficient

The oxidation behavior of high steel has an important influence on the heat transfer performance in hot stamping process [21]. The effect of scale on heat transfer is important in actual production. The thermal conductivity (3 W/m^2 K) of scale is much less than that of blank and tool (42–52 W/m^2 K). The interfacial heat transfer coefficient is introduced to describe the heat transfer capability, and the equivalent interfacial heat transfer coefficient is inversely proportional with the scale thickness. The factors such as thermal physical properties of material, mechanical properties, surface properties, load, and temperature have great influence on the interfacial heat transfer coefficient. The research shows that two factors of them are primary. The first is the heat transfer between surface asperities and the second is the gas heat transfer results from contact clearance [21]. The contact surface of solid to solid is depicted in Fig. 6.15. The oxidation decarburization layer is the primary factor that affects the heat transfer in quenching stage whose essence is the reduction of interfacial heat transfer coefficient due to the gas clearance. Although the contact surface between substrate and oxide scale is not the ideal plane, it is closely linked together without air and interval. Therefore, the effect of oxide scale on the blank temperature can be neglected. But there is a discontinuous gap between the tool and oxide scale which will affect the interfacial heat transfer coefficient a lot [1].

However, with the increase of stamping load, the real contact area will increase and the contact thermal resistance will reduce. It is proved that larger interfacial heat transfer coefficient can be obtained when the contact condition is improved.

Table 6.1 The chemical composition of boron steel

Element	C	Mn	Cr	Si	B	P	S	Al
Chemical composition	0.24	1.29	0.165	0.24	0.0037	0.016	0.006	0.04

Therefore, in order to realize the good quenching performance in hot stamping production, the appropriate load should be selected to reduce the influence of oxidation and achieve good heat exchange.

Although Krzyzanowski et al. have reported that a small change of chemical composition has a great influence on oxidation [22], few people study the oxidation properties of 22MnB5. The relationship of oxide scale thickness and work time has been studied in this research. The boron steel of B1500HS with a thickness of 2.0 mm supplied by Bao Steel is used as experiment material and the chemical composition is listed in Table 6.1.

Furnace is heated to 900 °C. The samples are put into the furnace in 5 min. After the sample is heated to 900 °C, in order to get the corresponding heating time with corresponding thickness of oxide layer, there are eight different time periods (0, 3, 6, 10, 15, 30, 45, 60 min) for sample to be warmed.

The oxide layer of boron steel 22MnB5 often comes off after taking it out from the furnace. In order to avoid this kind of problem, the following methods are adopted [23]. (1) After taking out sample from the furnace, press it by clean flat die steel pieces H-11 immediately. (2) After the sample cooling to room temperature, gently lift the steel piece and put the steel piece into the metallographic mosaic machine. Pour into resin and completely coat samples to protect the oxide layer. Make them with pressing and heating. (3) Remove the resin after solidification and cut the cross section. Next, the prepared samples are observed with digital inverted

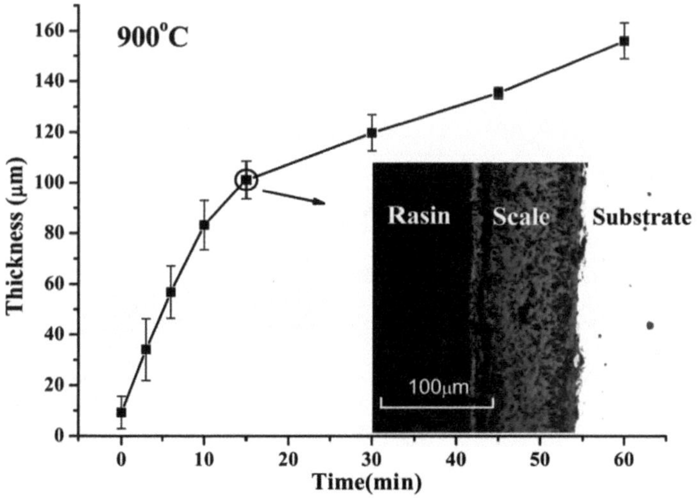

Fig. 6.16 The change of oxide layer thickness along with the heating time

metallurgical microscope. Every time corresponds to the three specimen metallographic tests and for each specimen, take five points along the direction of equidistance surface and then average them. Through observation, along the specimen surface oxide layer thickness is not completely consistent and the deviation is 8 %. But because interface heat transfer coefficient is the macroscopic quantities, partial oxidation layer thickness deviation is ignored. The relations between thickness of oxide layer and oxidation time are shown in Fig. 6.16.

In the initial stage of oxidation, the defects of intergranular have resulted in the high oxidation rate. As time goes on, the oxide layer formed initially covers the surface of samples and stops the sample matrix being exposed to air. It can be seen from Fig. 6.16 after 15 min, oxidation rate decreases obviously. Considering the contact interface heat transfer coefficient of sheet metal oxide skin, the experiment still adopts the experiment described in the Sect. 6.2.2. Oxide layer thickness and pressure are the two main factors which impact the interface heat transfer coefficient related to the temperature. This experiment adopts five stress level (8, 16, 25, 34 and 42 MPa) and eight heating duration (3 min, 6 min, 9 min, 13min, 18min, 30 min, 45 min and 60 min). After the specimen is heated to 900 °C and insulated, rapidly shift the specimen to the middle of die positioning pin. In order to ensure that each experiment has the same initial conditions, the beginning temperature of experiment is 800 °C. Later, let upper die down and make the setting pressure maintain 20 s to ensure that the specimen temperature drops to below 100 °C. In the experiments, the cooling rate of the specimens is greater than 27 °C/s, which is martensitic transformation temperature. Upper and lower molds are fully cooled

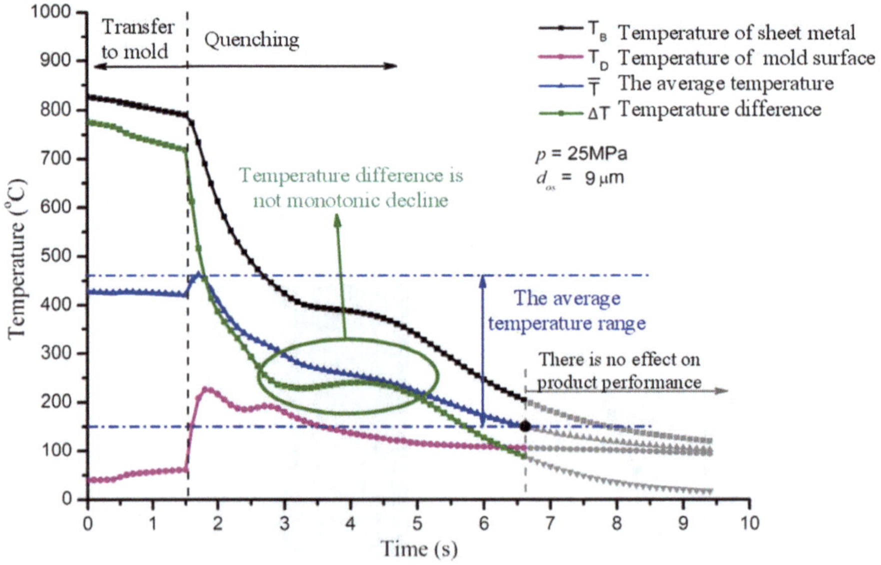

Fig. 6.17 The specimens, lower die surface temperature curve, and the corresponding average temperature and temperature difference

between the two experiments. Do each condition of experiment three times and get the final value by averaging them. In order to minimize the uncertainty of the experiment, do experiments by adopting the method of design of experiment (DOE). The interface heat transfer coefficient related to the temperature in the research of this section is defined as the functions related to average temperature between sheet metal and die surface instead of the temperature difference between them. As shown in Fig. 6.17, temperature difference between sheet metal and die is not monotonic decline, especially in the key stage of latent heat release. So it is possible to make a difference in temperature value corresponding to two or more interface heat transfer coefficient, which is not the one-to-one correspondence. For the average temperature, in addition to the specimen in contact with the mold just in a short area, in other areas average temperature declines monotonically. So in order to avoid "one-to-many" dilemma, average temperature is chosen as the physical quantities corresponding with interface heat transfer coefficient. According to Fig. 6.17, the upper limit of the average temperature is set to 450 °C, and the lower limit is set to 150 °C, because martensitic transformation is completely finished and it no longer has an impact on performance of product when the average temperature is less than 150 °C.

This estimation procedure in the study is based on optimization method and FEM. Through the guidance of commercial optimization software I-Sight, establish simulation model in the FEM software ABAQUS, and make the simulation temperature field in accordance with the experimental value by adjusting the interface heat transfer coefficient values. Interface heat transfer coefficient value related to temperature is design variable, objective function is established through the experiment and simulation temperature. It is represented as follows:

$$\min\delta_{\mathrm{B}} = \min\sum_{i=1}^{N}\left(T_{i,\mathrm{B}}^{\mathrm{FEM}} - T_{i,\mathrm{B}}^{\mathrm{EXP}}\right)^2 \tag{6.3.3}$$

$$\min\delta_{D} = \min\sum_{j=1}^{N}\left(T_{j,\mathrm{D}}^{\mathrm{FEM}} - T_{j,\mathrm{D}}^{\mathrm{EXP}}\right)^2 \tag{6.3.4}$$

δ_{B} is the error function of sheet metal temperature. N is the number of sampling points on the time scale, $\min\delta_{\mathrm{B}}$ is the first objective function, $T_{i,\mathrm{B}}^{\mathrm{EXP}}$ and $T_{i,\mathrm{B}}^{\mathrm{FEM}}$ are the sheet metal temperature of experiment and simulation, respectively. In order to balance the calculation accuracy and computational efficiency, the time scale of temperature sampling interval is set as 0.1 s. The values in formula (6.3.4) are the corresponding values of the die surface and the $\min\delta_{\mathrm{D}}$ is the second objective function. Considering the importance and different measurement precision of sheet metal and die surface, as a multi-objective optimization problem, the weight of $\min\delta_{\mathrm{B}}$ and $\min\delta_{\mathrm{D}}$ is 5:1. It is stricter with $\min\delta_{\mathrm{B}}$ in the process of optimization.

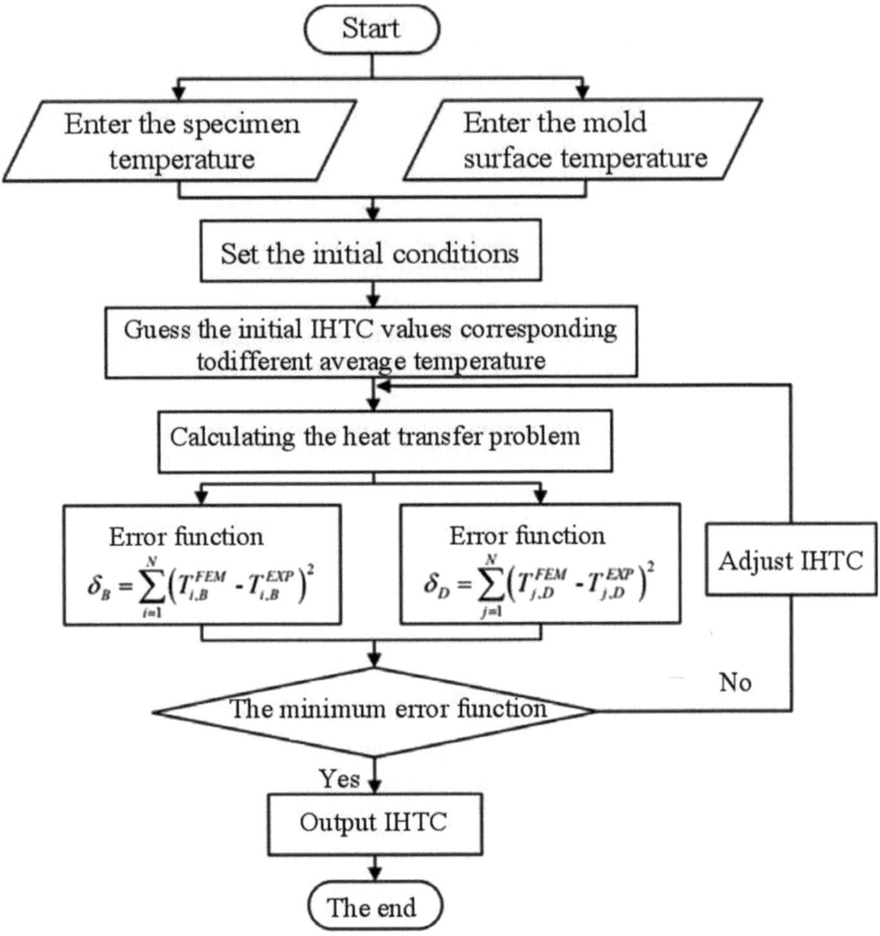

Fig. 6.18 The process flowchart of optimization

"Pointer" method is used as the optimization method in this analysis. It is a kind of automatic optimization engine which controlling four standard optimization methods (genetic algorithms, Nelder & Mead downhill algorithm, NLPQL and linear algorithm). When the optimization is start, according to the internal control parameters (step length and the number of iterations), system select the most appropriate optimization method in order to achieve maximum efficiency. Figure 6.18 is the process flowchart of optimization.

Pressure and thickness of oxide layer are taken in different values. The temperature of specimen and mold is recorded to calculate the interface heat transfer coefficient. The contrast between experimental and numerical results of temperature curve is shown in Fig. 6.19. As you can see, sheet metal temperature curve fit is very well and it also illustrates the interface heat transfer coefficient of calculated

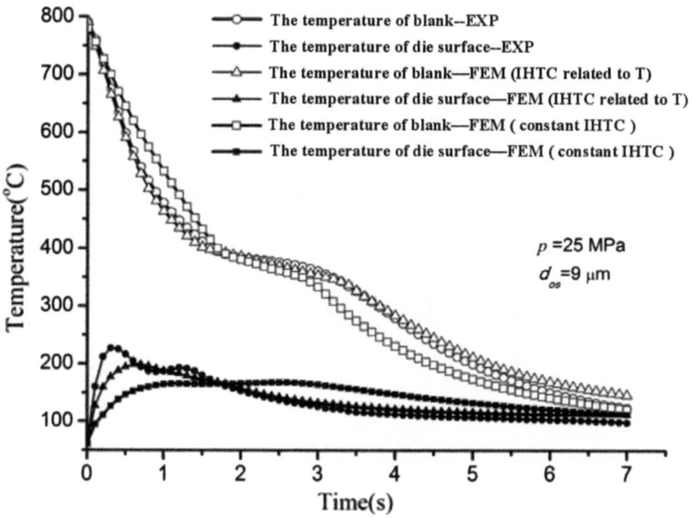

Fig. 6.19 The contrast between the specimen and die temperature of experiment and simulation

value and reliable. Mold temperature during the first 2 s has a deviation, because of the inevitable experiment error and the relative small proportion of the above-mentioned objective function.

Figure 6.20 shows the interface heat transfer coefficient changed with the average temperature under different pressure in the scope of this research. In each pressure condition, the interface heat transfer coefficient of the peak rises from 3986 W/m^2 K(8 MPa) to 6288 W/m^2 K (42 MPa). This deformation of the contact surface caused by pressure micro convex point is the mainly result and it causes to an increase of the actual solid–solid contact area. Also because of the pressure increase, it enhances the clearance of air mobility and enhances heat conduction ability. The average interface heat transfer coefficient under each pressure condition is shown in Fig. 6.20b. It can be seen that as the pressure increases, the interface heat transfer coefficient increases significantly.

It can be seen very clearly that interface heat transfer coefficient changes with the change of average temperature $\bar{T}$. From the start interface heat transfer coefficient decreases with the decrease of $\bar{T}$ and reduces to a minimum at $\bar{T} = 250\,^\circ\mathrm{C}$. When $\bar{T} = 250\,^\circ\mathrm{C}$, the corresponding sheet metal temperature is 380 °C, the martensitic transformation begins and the latent heat starts to release. According to the research of Naderi, along with the transformation of martensite material expansion is going to happen [24]. Inflation makes specimens pressed tighter on the up and down mold, oxide layer is more and more dense, contact between the specimen and die is also better. At the same time, the release of latent heat increases the temperature difference between the specimen and die (as shown in Fig. 6.17) and the movement of air molecules in the clearance is more intense. All these reasons make stronger heat transfer driving force.

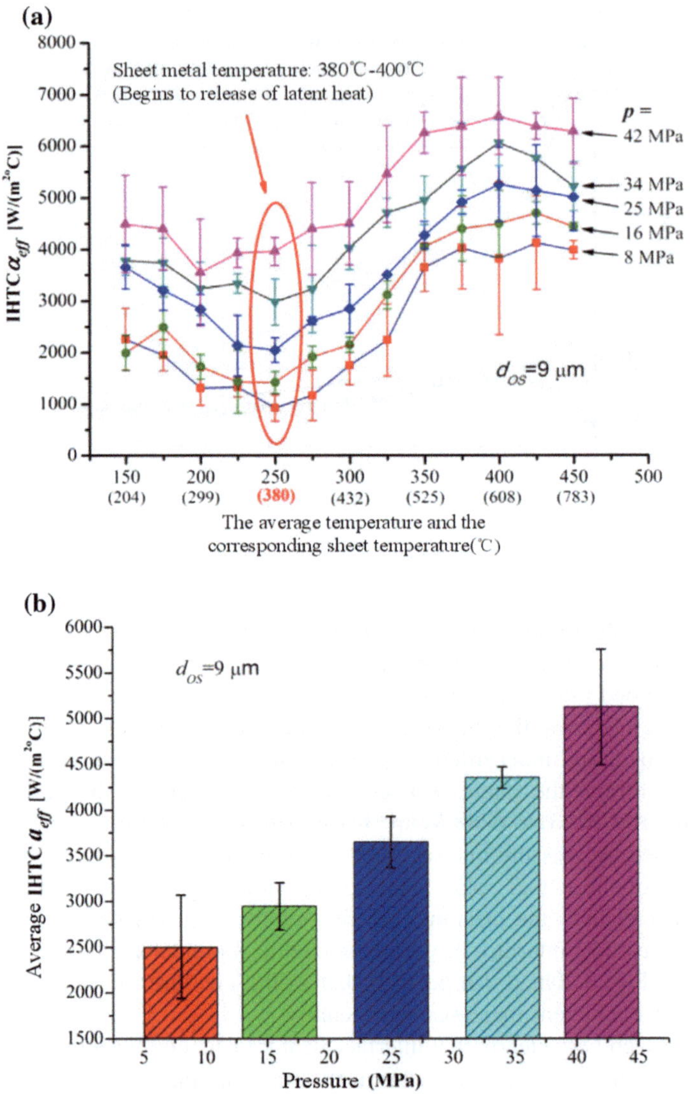

Fig. 6.20 The influence of pressure on the IHTC **a** under the condition of different pressure the change equivalent of a α_{eff} with $\overline{T}$ (**b**) the change of the average equivalent IHTC α_{eff} with different pressure

Measurement and calculation at the moment of $\bar{T} > 375\,°C$ is more difficult, because at that time under and upper dies just close and the experiment system does not reach stability. There is a volatility about interface heat transfer coefficient when $\bar{T} < 200\,°C$, because in this temperature range appropriate and accurate phase transformation research is lacked as a material condition of numerical model.

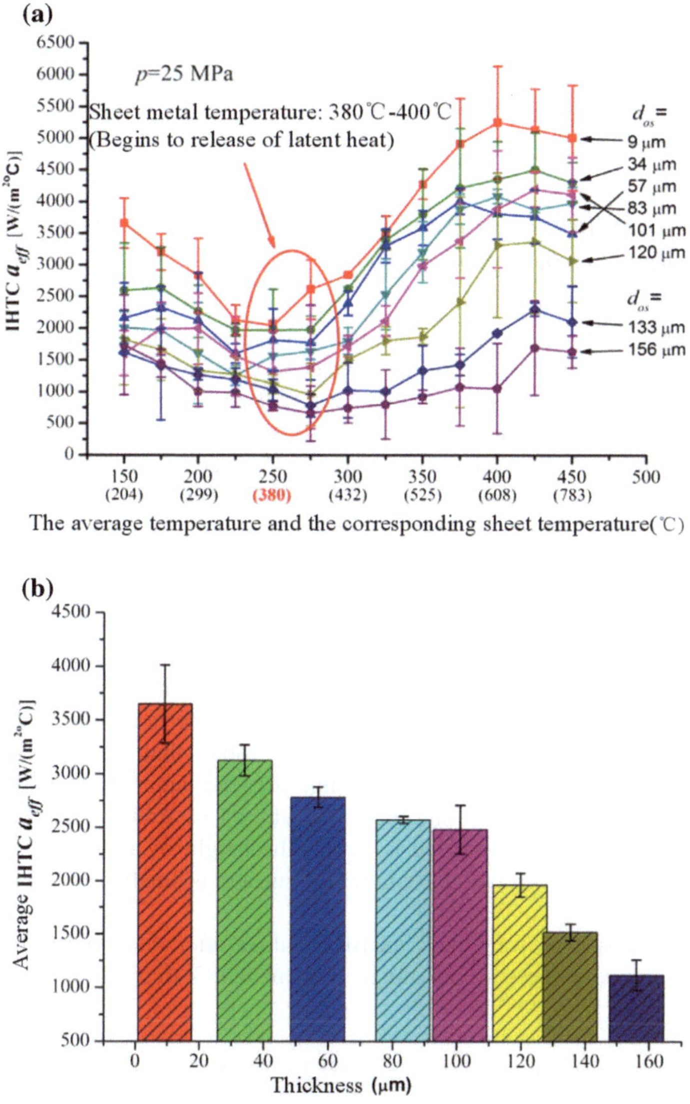

Fig. 6.21 The influence of oxide layer thickness on the IHTC (**a**) under the condition of different oxide layer thickness the change of equivalent a α_{eff} with $\overline{T}$ (**b**) the change of the average equivalent IHTC α_{eff} with different thickness

Different heating time led to different thickness of oxide layer. The equivalent interface heat transfer coefficient $\alpha_{\text{eff}}(\bar{T})$ about different thickness of oxide layer (from 9 to 156 μm) is shown in Fig. 6.21. It can be seen from the Fig. 6.21a, the interface heat transfer coefficient first falls then rises with the decline of the average temperature. For the thin oxide layer, the fluctuation is more obvious. Under the condition of 25 MPa, the maximum can reach 5253 W/m^2 K. It can be seen in Fig. 6.21b that the average of interface heat transfer coefficient and thickness of oxide layer d_{OS} show the opposite trend.

6.3.2 The Influence of Heat Transfer Coefficient About Steel Sheet Surface Roughness

The roughness values between sample and mold surface can be used to characterize the concave and convex extent of the contact surface. In actual production, of both samples and mold, the surface roughness will increase due to the machining inaccuracy or manufacturing error. Using experiment device introduced in Sect. 6.2.2, in the premise of maintaining the 45# steel mold roughness and using polishing machine with different types of sand paper (80#, 180#, 240#, 600#) polish the surface of the sample to obtain a uniform roughness gradient (as shown in Fig. 6.22) and explore interface heat transfer coefficient influenced by roughness in the hot forming process [25, 26].

Figure 6.23a–c are samples polished by 180#, 600#, and 800# sandpaper After 15 MPa pressure hot stamping, surface pattern will be filmed by the LEXT OLS4000. White-highlighted parts are stamping contact areas. The dark part is not contact area. Get the corresponding microscopic surface morphology after magnifying 4000 times to highlight area. The study found that the greater sandpaper model used and the smaller roughness value are, the larger the number of bright spots in observation area per unit and the total area is. And the number of micro convex body effectively contacted is larger, the height of a single micro convex body is smaller, the concave and convex of surface is smaller and the roughness value is smaller.

The sample surface roughness measure values after the sanding is shown in Fig. 6.24a, the roughness values range is from 0.483 to 3.542 μm. After 1 and

Fig. 6.22 The samples with different roughness

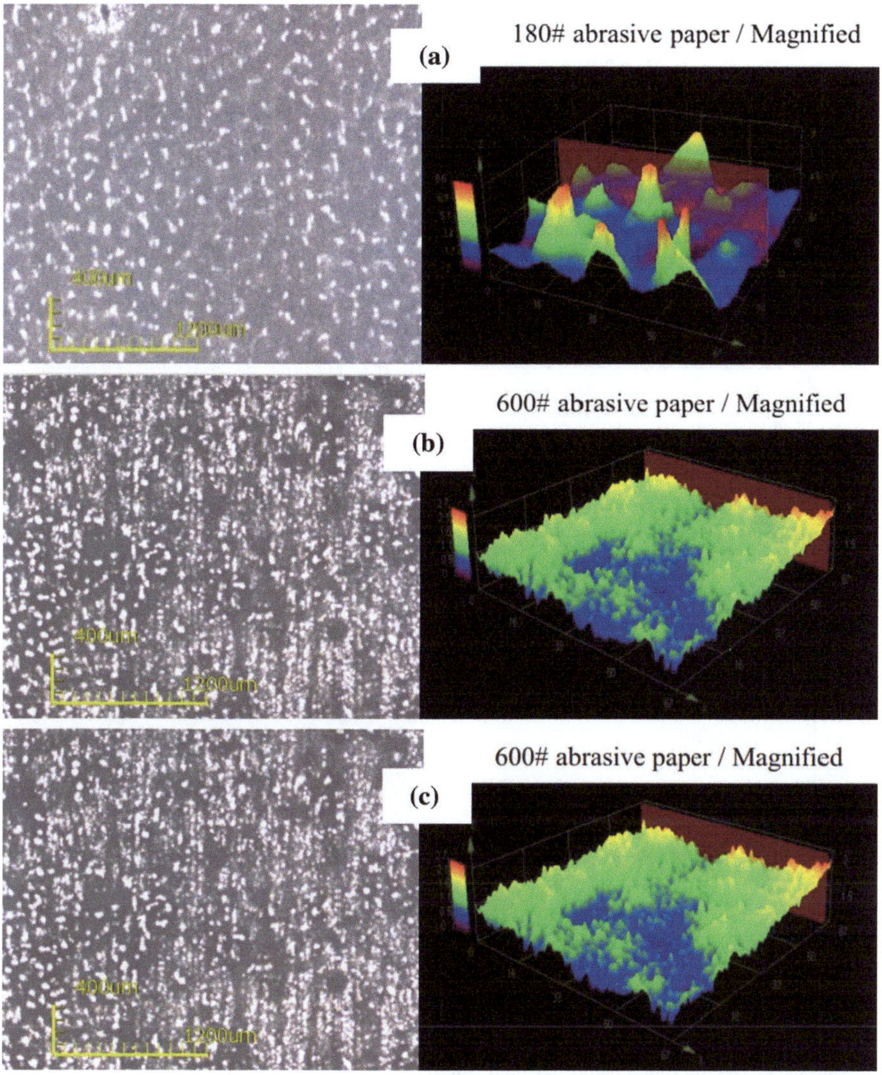

Fig. 6.23 The contact surface and slightly convex body state of boron steel burnished by different types of sand paper after stamping (**a**) the contact surface and slightly convex body state of boron steel burnished by 180# sand paper after stamping (**b**) the contact surface and slightly convex body state of boron steel burnished by 600# sand paper after stamping (**c**) The contact surface and slightly convex body state of boron steel burnished by 800# sand paper after stamping

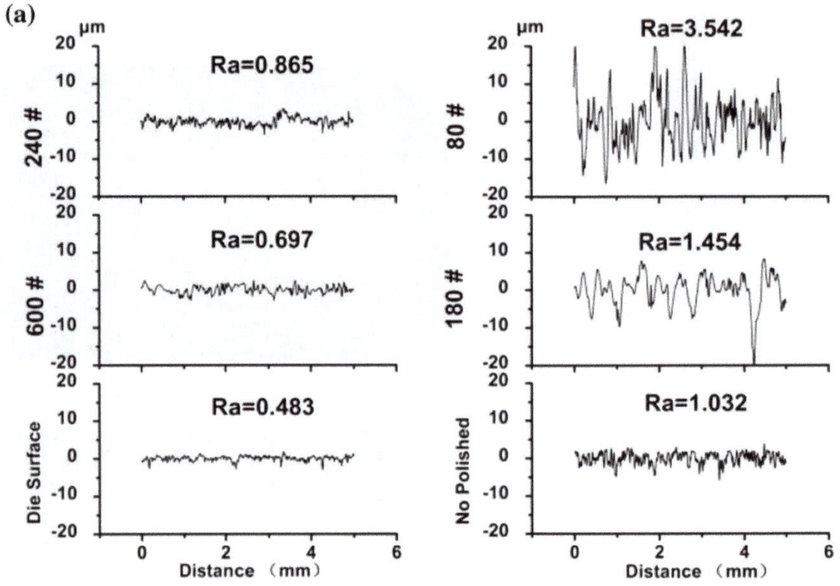

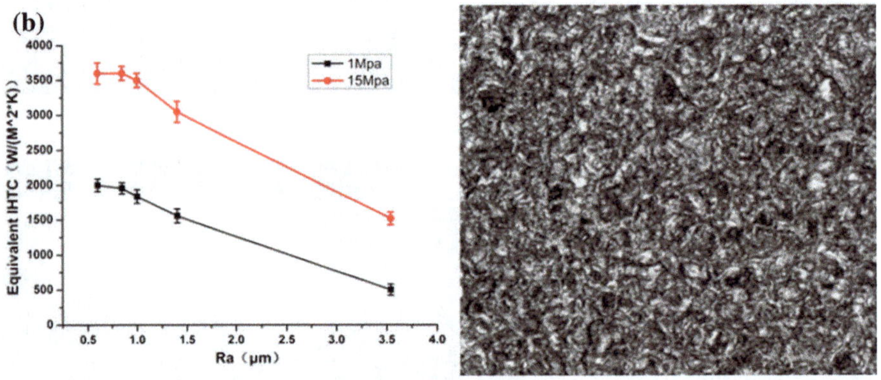

Fig. 6.24 The IHTC influenced by roughness (**a**) surface roughness burnished by different types of sand paper (**b**) the relationship between roughness and equivalent IHTC

15 MPa pressure stamping and quenching with 45# steel stamping die, the relationship between roughness value and the equivalent interface heat transfer coefficient calculated is shown in Fig. 6.24b. From the figure we can find that when the roughness is less than 1 μm, interface heat transfer coefficient will slow down the increase speed with the decrease of the roughness value, and even tend to be valued and no longer increases. This is mainly due to that the concave and convex contour is damaged by the oxide coating on the surface of samples after high-temperature heating (shown in microscopic images Fig. 6.24). When the roughness is greater than 1 μm, the interface heat transfer coefficient values will decrease along with the

increase of roughness, and with other conditions unchanged; under the condition of the stamping pressure, the greater the interface heat transfer coefficient value increases the greater rate of change is increases.

6.4 Summary

First the heat transfer theory involved in hot stamping process is introduced, including heat conduction differential equation and the definite condition, three forms of heat transfer, and heat conduction problem of three kinds of different boundary conditions. And heat transfer phenomena involved in the different stages in the process of hot stamping are also analyzed. Second the hot stamping process of sheet metal and die, and cooling heat transfer coefficient of the experimental and numerical inverse method are introduced. The temperature of hot stamping is collected by frustum of a cone design experiments, and the inverse calculation of Beck method is combined with in order to obtain the interface heat transfer coefficient about the pressure and temperature. The temperature of different locations has been gathered through the single pipe cooling experiment. The single tube flow is simplified to one-dimensional flow model for the conversion of convection heat transfer coefficient in theory, and the results of experience formula and CFD numerical simulation values are compared in order to demonstrate the effectiveness of the experimental results. Finally, the sheet metal in high-temperature scale and the influence of surface roughness on the interface heat transfer coefficient are considered. Further, the rotary table experiment results are applied, the interface heat transfer coefficient is calculated by FEM method; the relationship between oxide coating thickness and the interface heat transfer coefficient is gained indirectly. Through different levels of the steel plate surface with grinding sheet with different roughness, the interface heat transfer coefficient related to sheet metal roughness can be obtained from the calculation by testing the rotary table experiment again. In conclusion, the phenomenon of heat transfer involved in the hot stamping process is illuminated in this chapter, and the interface heat transfer coefficient and convection heat transfer coefficient, two key parameters of the experimental method, are introduced in detail, at the same time the influence on interface heat transfer coefficient about the scale and the roughness is revealed.

References

1. Li Y (2013) The optimization design of cooling system in hot stamping dies. In: Automobile engineering. Dalian University of Technology, Dalian, p 76 (in Chinese)
2. Liao Z (2013) Estimation and experimental study of heat transfer coefficient in hot forming of ULtra high strength steels. In: Automobile engineering. Dalian University of Technology, p 76 (in Chinese)
3. Hj P (2010) Heat transfer. McGraw-Hill, New York

4. Wang L, Yang X, Lu J (2006) Development of high strength steel sheets for lightweight automobile. Iron Steel 9:1–8 (in Chinese)
5. Liang Y (2013) Research and application on key process experiment of high strength steel for hot forming. Dalian University of Technology, p 179 (in Chinese)
6. Ma N, Hu P, Yan K et al (2010) Research on boron steel for hot forming and its application. Chin J Mech Eng 14:68–72 (in Chinese)
7. He B, Ying L, Hu P et al (2012) Design of water cooling tool for hot stamping process of high strength steel. China Metal Forming Equip Manuf Technol 6:62–65 (in Chinese)
8. Ying C, Tang X, Wang B et al (2015) Comparison of methods for calculating interfacial heat transfer coefficient in hot forming process. J Central South Univ (Sci Technol) 4:1208–1214 (in Chinese)
9. Bosetti P, Bruschi S, Stoehr T et al (2010) Interlaboratory comparison for heat transfer coefficient identification in hot stamping of high strength steels. Int J Mater Form 3:817–820
10. Merklein M, Lechler J, Stoehr T (2009) Investigations on the thermal behavior of ultra high strength boron manganese steels within hot stamping. Int J Mater Form 2:259–262
11. Fieberg C, Kneer R (2008) Determination of thermal contact resistance from transient temperature measurements. Int J Heat Mass Transf 51:1017–1023
12. Beck JV, Blackwell B, Clair Jr CR (1985) Inverse heat conduction. Wiley, New York
13. Zhipeng G (2007) Development of an inverse heat transfer model and its application in the prediction of the interfical. Metall Sin 43:607–611 (in Chinese)
14. Xinghui T (2014) Exploration on IHTC and influence factors of 22MnB5 steel in hot stamping process. Dalian University of Technology, p 78 (in Chinese)
15. Ying C, Lu J, Jin F et al (2015) Influence of thermal properties on KIHTC in hot forming. J Harbin Inst Technol 3:97–102 (in Chinese)
16. Zhengwei L, Ying L, Hu P et al (2013) Estimation of interface heat transfer coefficient of ultra high strength steels in hot forming. Trans Mat Heat Treat S1:167–172 (in Chinese)
17. Hu P, Ying L, Li Y et al (2013) Effect of oxide scale on temperature-dependent interfacial heat transfer in hot stamping process. J Mater Process Technol 213:1475–1483
18. Zhengwei L, Ying L, Hu P et al (2013) Basic study on heat transfer of high strength sheet during hot stamping. Hot Working Technol 5:1–4+7 (in Chinese)
19. Ye L, Ying L, Hu P et al (2013) Experimental investigation of convectional heat transfer coefficient in circular cross-section tubes. Hot Working Technol 19:15–18 (in Chinese)
20. Zheng Y (2012) Research on oxidation and decarburization of automotive high strength steel plate. Jilin University, p 63 (in Chinese)
21. Ying C, Li Y, Ying L et al (2013) Effect of oxidation behavior on heat transfer dynamatics performance of hot stamping steel. J Jilin Univ (Eng Technol Ed) 6:1551–1557 (in Chinese)
22. Krzyzanowski M, Beynon JH, Farrugia DC (2010) Oxide scale behavior in high temperature metal processing. Wiley-VCH
23. Wei DB, Huang JX, Zhang AW et al (2009) Study on the oxidation of stainless steels 304 and 304L in humid air and the friction during hot rolling. Wear 267:1741–1745
24. Naderi M, Saeed-Akbari A, Bleck W (2008) The effects of non-isothermal deformation on martensitic transformation in 22MnB5 steel. Mater Sci Eng, A 487:445–455
25. Chang Y, Li S, Tang X et al (2015) Effect of boron steel 22MnB5 interfacial morphology on interfacial heat transfer coefficient. Trans Mat Heat Treat 3:226–231 (in Chinese)
26. Chang Y, Tang X, Zhao K et al (2016) Investigation of the factors influencing the interfacial heat transfer coefficient in hot stamping. J Mater Process Technol 228:25–33

Chapter 7
The Formability of High-Strength Steel for Hot Stamping

7.1 The Concepts of Plasticity and Deformation Resistance

Metal is the most important material in modern industry, widely used in agriculture, industry, national defense industry, and other sections. Since the selection of materials for most mechanical equipments is based on its mechanical performance, the mechanical properties of metals should first be learned. This section will focus on the plasticity and deformation resistance of metal materials and their influencing factors.

Plasticity, deformation resistance, and other terms often appear in traditional stamping technology, and they are defined as following:

1. Plasticity is plastic deformation that takes place without fracture in response to external forces.
2. Deformation resistance is the unit deformation force which induces plastic deformation under given load condition, deformation temperature and rate.

The plasticity is related to not only the type of deformation object, but also the deformation mode (stress strain state) and deformation conditions (deformation temperature and deformation rate). For example, the lead generally has very good plasticity; however, it often shows brittle rupture like a brittle material without plastic deformation under three-dimensional compressive stress. And the brittle marble can develop great plastic deformation while the material bears three-dimensional compressive stress. These two examples have fully proved that: the plasticity of material is not an unchangeable property, it is related to the material type, the deformation mode, and the deformation condition.

The plasticity of materials is measured by the tensile test. Plasticity is generally indicated by elongation and reduction of area. Plasticity is an important index of metal material. Generally, the larger the elongation δ and reduction of area ψ are, the better the plasticity will be. Both the δ and ψ of plastic material are larger than

P. Hu et al., *Hot Stamping Advanced Manufacturing Technology of Lightweight Car Body*, DOI 10.1007/978-981-10-2401-6_7

that of brittle material. For example, the elongation δ of pure iron is up to 50 %, while the δ of common pig iron is lower than 1 %. Therefore, the plasticity of pure iron is much better than that of the common cast iron.

The ability of plastic deformation for metal material is very important to the manufacturability and usability of components. For example, metal material with high plasticity will be easily deformed without defects in a simple stamping process. In stamping production, how to take advantage of the metal plasticity to obtain the desired component with the minimum deformation resistance is very important. There are many factors influencing the plasticity and deformation resistance of metal. In this chapter, the influences of physical factors and stress—strain on plasticity and deformation resistance are introduced.

7.2 Factors Influencing Plasticity and Deformation Resistance of Hot Stamping Steel

7.2.1 Chemical Composition

Chemical composition will exert very complicated influence on deformation resistance. For a variety of pure metal, deformation resistance is different due to the different interactions between atoms. Even the same metal, the higher purity results in lower deformation resistance. Deformation resistance is also related to metallographic structure. For example, the high-strength steel sheet has different microstructure under high temperature such as austenite, pearlite and ferrite, which result in different deformation resistance. The deformation resistance of high-strength steel is also affected by the nature and the internal distribution of impurities. Solid solution composed of impurity atoms and matrix element leads to basic element lattice distortion, which increases the deformation resistance. The hardening effect of impurities becomes more intense when impurity elements in the periodic table are farther away from the matrix, thus deformation resistance increases dramatically. However, if the impurity elements diffusely distribute within the grain or between the grains in the form of individual inclusions, their influences on deformation resistance of metal is relatively small. If the impurity elements form brittle inclusions with reticular formation, the deformation resistance of metal decreases accordingly.

There are a variety of high-strength steel for hot stamping and the most typical HSS is boron steel. At present, chemical composition for the boron steel 22MnB5 is shown in Table 7.1 [1, 2]. The boron steel is widely used in hot stamping industry in the world in recent years. Such kind of steel is based on C-Mn steel with a certain mass fraction of boron. Nucleation of ferrite, pearlite, and bainite is delayed by the segregation of boron as solid solute on the grain boundary of austenitic, which enhances the steel strength.

Table 7.1 Chemical composition for the boron steel 22MnB5

22MnB5	C	Mn	Cr	Si	B	P	Al	Ti	Mo	S	Cu	Ni
Min	0.220	1.200	0.110	0.200	0.002	–	0.020	0.020	–	–	–	–
Max	0.250	1.400	0.200	0.300	0.005	0.020	0.050	0.050	0.100	0.005	0.100	0.100

Boron has a strong affinity with oxygen and nitrogen. If deoxidation and denitrification cannot be well achieved during steel making process, the formation of boron oxide and boron nitride will seriously affect the hardenability of steel. Therefore, we must consider the "effective boron." The so-called "effective boron" is the boron dissolved in the austenite (or chemical free) other than that is synthesized oxides or nitrides. The amount of nitrogen, oxygen in steel is different depending on the steel composition, smelting and steel making operations. 22MnB5 is fully deoxidized by adding silicon, aluminum, and other powerful deoxidizing agent to properly protect it from the oxidation of boron [3].

7.2.2 Metallic Structure

The metallic structure has a significant effect on metal plasticity, such as the lattice, grain orientation and grain boundary characteristics of component. Face centered cubic lattice (such as Al, Ni, Pb, Au, Ag, etc.) has the best plasticity compared to body centered cubic lattice (such as Fe, Cr, W, Mo, etc.), and hexagonal close packed lattice (such as Zr, Hf, Ti, etc.). Additionally, plasticity of BCC lattice is better than that of HCP lattice.

Most single crystal metals have a high plasticity at room temperature. However, polycrystalline has lower plasticity. This is due to nonuniform polycrystalline grain size, different grain orientation, weak grain boundary strength, etc., under normal circumstances. Fine grain size means large grain boundary area and high grain boundary strength, and deformation is more concentrated in the intragranular, which results in good plasticity. Ultrafine grain shows great plasticity because of its nearly spherical structure and the slide of grain boundary at low deformation rate. Coarse grain presents low plasticity due to nonuniform grain size and weak strength on grain boundary that easily results in stress concentration and crack.

7.2.3 Deformation Temperature and Work Hardening

The cold plastic deformation process of the metal material is carried out at room temperature, and the material strength and hardness increase with the increasing of deformation, and vice verse. This phenomenon is known as work hardening.

Different materials and different deformation conditions (deformation temperature, deformation rate) result in distinct degree of hardening. Material work hardening not only increases deformation force but also has a great impact on the forming process, favorable or unfavorable. Working hardening limits the further deformation of the blank sheet and reduces the maximum deformation if work hardening enlarges the deforming force. Sometimes annealing process is required to eliminate hardening effect and improve the forming limit; the material hardening on the hole edge may easily arouse cracking in the flanging process. However, sometimes the hardening is beneficial for deformation, for example, hardening along the stretch forming can make the deformation more uniform, which increases the degree of maximum deformation. Thus, in order to deal with the practical problems of sheet stamping, determine the various parameters and analyze the force condition of blank deformation zone, the laws of material hardening and its effect on the stamping process need to be investigated.

Generally speaking, metal plasticity improves with the increase of temperature since the energy of atomic thermal motion increases and those plastic deformation institutions with obvious diffusion characteristics (intergranular sliding bodies, amorphous bodies, dissolved sediment bodies) have played a role. At the same time, recrystallization softening occurs with the hardening elimination process during deformation, which increases the possibility of repairing the damage and microdefects due to plastic deformation; in addition, the appearance of new slip systems means an improvement of the plastic deformation capability.

On the other hand, the binding force between the metal atoms reduces with the increasing of temperature, which results in a decrease of critical shear stress of metal slip. Almost all metals' and alloys' deformation resistance decrease as temperature increases. However, for high-strength steel, austenite to martensite phase transformation happens when the high temperature declines to a specific value, and deformation resistance significantly increases. The detailed information will be described in the later chapters.

7.2.4 Deformation Rate

Deformation rate has a complex impact on the metal plasticity. When the deformation rate is low, plasticity reduces with the increasing of deformation rate; however, when the deformation rate is high, the plasticity will be improved with the increasing of deformation rate. The exact quantitative relationship between plasticity and deformation rate is not emerging. The heat generation rate per unit time increases with the increasing of deformation rate, which benefits the material softening and decreases the deformation resistance. On the other hand, the deformation time reduces due to the increase of the deformation rate, causing insufficient

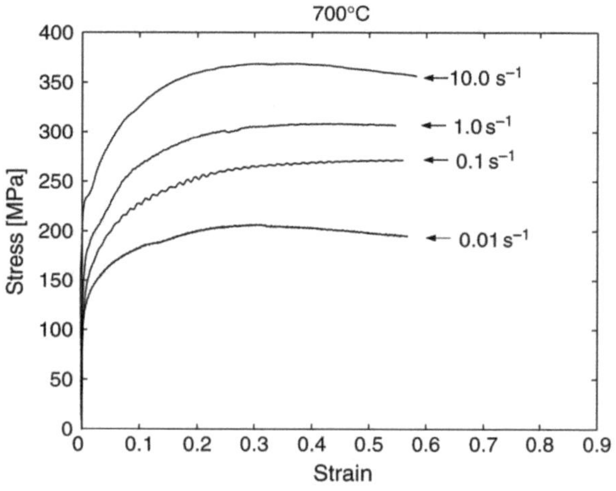

Fig. 7.1 Stress–strain curves of 22MnB5 steel at evaluated temperatures

dislocation during deformation which contributes to the increasing of deformation resistance. In general, the resistance of metals and alloys increases as the deformation rate increases. However, the degree of increase is closely related with the deformation temperature. The increasing of deformation rate makes resistance increase, and the resistance is not very sensitive to deformation rate during cold deformation. The increasing of deformation rate causes resistance to improve significantly in hot stamping. The experimental research on 22MnB5 high-strength steel was presented by some scholars; the influence of different deformation rates on the mechanical properties of hot stamping at the temperature of 700 °C was studied [4–6], as shown in Fig. 7.1. When the tensile strain rate range increased from 0.01 to 10/s, the material yield strength increased by two times. In the hot stamping process, the strain rate is usually less than 10/s, so the material rate dependent or rate-independent model can be chosen according to the different strain rates of different positions in the process of forming.

It is worth noting that the deformation rate of explosively forming is over 1000 times faster than the general sheet metal forming. In such a deformation rate, even the difficult to machining metallic iron and heat-resistant alloys can be well formed, which means the explosive forming of metals and alloys can greatly improve the plasticity with much energy saved.

Why can high-speed forming save energy and avoid breakdown in the metal formation? Robert worked on the assumption:

Assuming that the strain hardening is related with time, there is a special time limit—for a metal or alloy at a specific temperature, a "residence time" of the deformation hardening. The plastic deformation takes place in the possibly shortest time, which reduces the deformation energy consumption down to a very low level, so that the deformation process completes before crack propagation.

This hypothesis can be used to explain the reason for the explosive forming and the good effect of high-speed forging.

7.2.5 Cooling Rate

In the early 1940s, Grnage and Kiefer [7] studied the transformation process of steel under continuous cooling conditions, and proposed a continuous cooling transformation curve (CCT), which was used widely at that time; in the early 1970s they began to propose numerical simulation of the phase transformation, and CCT curve was first selected as the theoretic basis on simulation, but it only showed the phase transformation products under different cooling rates, not the process of phase transformation (transition from the beginning to the end between the transition process) which was essential for the numerical simulation [8]. CCT curve of 22MnB5 is shown in Fig. 7.2, it can be seen from the figure that the cooling rate should be over 30 °C/s to gain full martensite through hot stamping process [9, 10].

7.2.6 Deformation Degree

The impact of deformation for plasticity associates with the hardening and cracking tendencies resulting from the development of plastic deformation in processing.

For cold deformation, because of the absence of the repair process, the plasticity decreases with deformation increasing. From the perspective of the plasticity processing, there is no definite conclusion about the most appropriate degree of deformation between the two annealing in the cold deformation process and further study needs to be done. However, deformation is closely related to the property of the metal. Metals and alloys with high hardening should be given a small

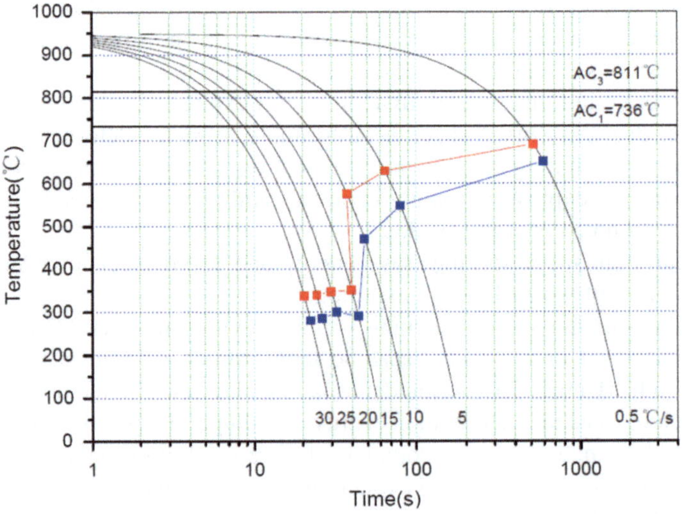

Fig. 7.2 CCT curve of boron alloy steel for hot forming

deformation before entering the next intermediate annealing to restore its plasticity; while for the low hardening metals and alloys, a great deformation can be given between two annealing processes.

On the other hand, no matter how the temperature changes, as long as recovery and recrystallization processes have not begun, the deformation increase inevitably results in working hardening and an increase of the deformation resistance. Usually when the deformation is less than 30 %, the deformation resistance increases significantly. When the deformation is larger, the deformation resistance increases slowly. Because a further increase in deformation leads to the energy of lattice distortion increase, promotes the occurrence and development of recovery and recrystallization, and also enhances the thermal effect of deformation.

7.2.7 Size Factor

The basic law of size factor for the plasticity of hot stamping steel is that the plasticity declines as the volume of work piece increases. Actually, the metal contains a large number of defects in unit volume, so the greater the volume is, the more nonuniform the deformation shows. Defects in the organization are likely to cause stress concentration which results in crack source and a decrease in plasticity.

7.3 Material Properties of High-Strength Steel at Elevated Temperature

7.3.1 Uniaxial Tensile Experiment of High-Strength Steel at Elevated Temperature

After high-strength steel material is heated to austenitizing temperature, the material microstructure transforms into austenite organization from the ferrite and pearlite hybrid organization and, an obvious shift in mechanical properties of material takes place. The blank must be transferred as quickly as possible from the furnace to the press. The temperature of austenitic materials changes at 950–600 °C. The material performance changes dramatically with the change of the temperature. In order to investigate the optimal hot stamping parameter and perform effective simulation of hot stamping process, a tensile test for hot forming material in the temperature range mentioned above is necessary to reveal the deformation rule of sheet under high temperature [11].

The elevated temperature tensile test is carried on the Multifunction thermo-mechanical simulator as shown in Fig. 7.3. The machine is a complex system with electricity, gas and liquid, mainly composed by mechanical systems, hydraulic systems, heating systems, cooling systems, vacuum systems, pneumatic systems,

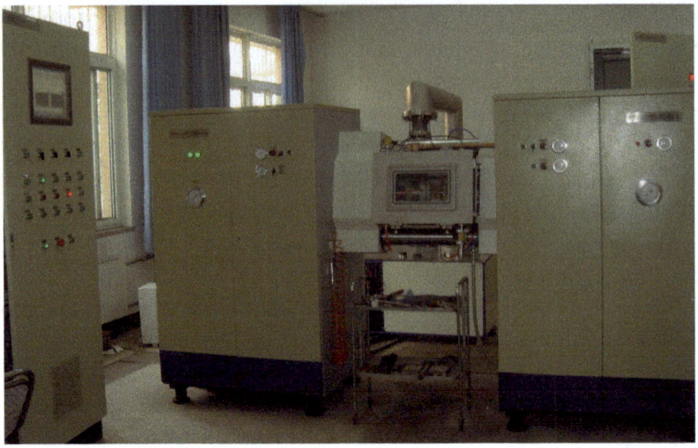

Fig. 7.3 Multifunction thermal–mechanical simulator

control system, and measurement system. Each system has its unique functions and coordinates with other systems.

The experimental specimen size is shown in Fig. 7.4. The typical 22MnB5 is selected as material. Sheet metal thickness is 1.6 mm. The specimen is heated up to 950 °C, held for 5 min, then cooled to test temperature with a cooling rate of over 50 °C/s and maintained constant. Multifunction thermomechanical simulator is utilized for experiment with a loading speed of 0.05/s. The force–displacement curve is recorded.

The stress–strain curve at high temperature is shown in Fig. 7.5.

Figure 7.5 shows that the yield strength of 22MnB5 at 650 °C is 3 times more than that at 950 °C. Therefore, the mechanical property of austenite is strongly dependent on temperature in hot stamping process. Mechanical properties experiment of austenite organization at different temperature ranges is necessary for accurate

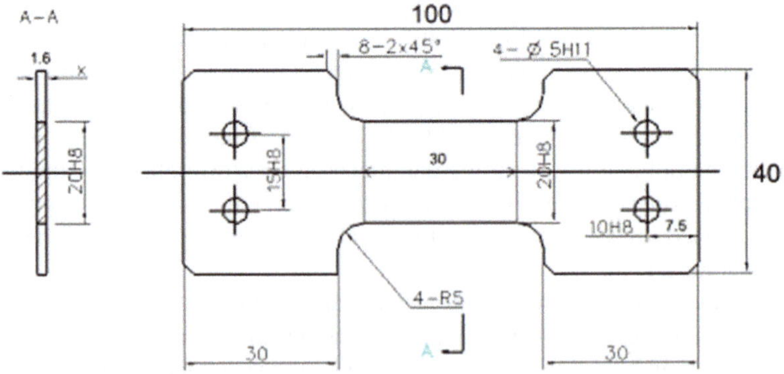

Fig. 7.4 The shape and size of specimen (*unit* mm)

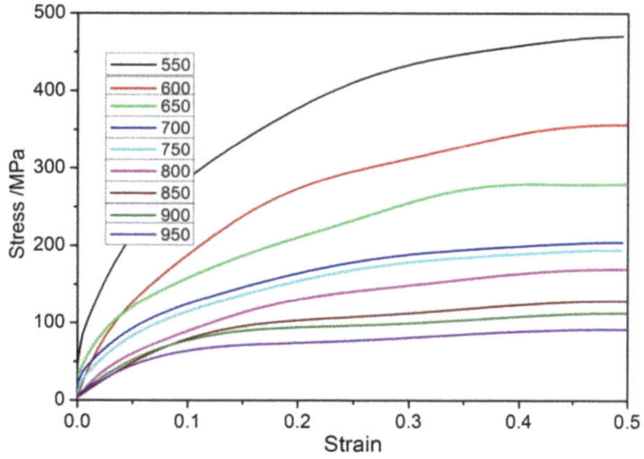

Fig. 7.5 Stress–strain curves at different evaluated temperature ranging from 550 to 950 °C

analysis and simulation of hot stamping process. Finally, the precise expression of mechanical properties depending on the changes of temperature is presented.

7.3.2 *Hardening Model of High-Strength Steel at Elevated Temperature*

The tensile test shows that the flow stress of austenite metal is the function of strain, strain rate, and temperature in hot stamping process. Relevant foreign scholars have studied the flow rule of steel at high temperature from macro and micro perspectives.

Nemat-Nasser [12] established a temperature, strain, and strain rate dependent flow stress model as shown in Eq. (4.3). Where σ_0 is the equivalent yield stress; k is the Boltzmann constant; p, q is strain hardening exponent; ε_0 is initial reference strain; G_0 is free energy of the initial reference; T is temperature parameter.

$$\sigma(\varepsilon,\dot{\varepsilon},T)=\sigma_0\left\{1-\left[-\frac{kT}{G_0}\left(\ln\frac{\dot{\varepsilon}}{\varepsilon_0}+\ln f(\varepsilon,T)\right)\right]^{1/q}\right\}^{1/p}\times f(\varepsilon,T)+\sigma_a^0\times\varepsilon^n$$
$$f(\varepsilon,T)=1+a_0\left[1-\left(\frac{T}{T_m}\right)^2\right]\varepsilon^{1/2} \tag{7.1}$$

Johnson et al. [13] developed a high-temperature material flow model that has an exponential relationship with temperature as shown in Eq. (7.2).

$$\sigma(\varepsilon,\dot{\varepsilon},T)=(A+B\varepsilon^n)\left(1+C\ln\frac{\dot{\varepsilon}}{\varepsilon_0}\right)\left[1-\left(\frac{T-T_0}{T_f-T_0}\right)^m\right],\quad T\geq T_0 \tag{7.2}$$

where A, B, C, n, m are material constants; $\dot{\varepsilon}_0$ is reference strain rate; T is temperature; T_0 is reference temperature; T_f is melting temperature.

Tong et al. [14–16] built up a high temperature flow model which introduced material activation energy. Where A, m_1, m_2, N, β, etc., are material constants; R is molar gas constant; $\dot{\varepsilon}_p$ is stain rate; T is temperature parameters; Q is activation energy of thermal deformation reflecting the difficulty of thermal deformation. Its value is decided by the material organization condition.

$$\sigma(\varepsilon, \dot{\varepsilon}, T) = A\left[\dot{\varepsilon}^{m_1(T-T_0)} \times -\exp\left(\frac{mQ}{RT}\right)\right]\left[1 - \beta\exp(-N\varepsilon_p^n)\right], \quad T \geq T_0 \tag{7.3}$$

Ghost et al. established a high temperature flow model by dislocations and other microscopic material parameters as shown in Eq. (7.4). Where M is Taylor vector; α is dislocation parameters; μ is shear modulus; b is Burgers vector; ρ is dislocation density; k is Boltzmann constant; R is molar gas constant; σ_0 is the initial stress; τ_{va} is shear stress; $\dot{\varepsilon}_0$ is reference strain rate; arsh is the inverse hyperbolic sine function.

$$\sigma(\varepsilon, \dot{\varepsilon}, T) = \left(\sigma_0 + M\alpha\mu b\sqrt{\rho}\right)\left\{1 + \frac{kT}{b^3\tau_{va}}\operatorname{arsh}\left[\frac{\dot{\varepsilon}}{\dot{\varepsilon}_0} \times \exp\left(\frac{Q}{RT}\right)\right]\right\} \tag{7.4}$$

Molinari et al. developed a high-temperature flow model by introducing grain size as shown in Eq. (7.5). Where $\dot{\varepsilon}$ is strain rate; $\dot{\varepsilon}_0$ is reference strain rate; $\hat{\sigma}_0$ is reference initial internal stress; m is material hardening exponent; d is grain size; T is temperature parameters.

$$\sigma(\varepsilon, \dot{\varepsilon}, T) = \hat{\sigma}_0\left(\frac{\dot{\varepsilon}}{\dot{\varepsilon}_0}\right)^{1/m}, \quad \hat{\sigma}_0 = \hat{\sigma}(d)\left(\frac{\delta_0}{\delta}\right) \tag{7.5}$$

Each high temperature flow model mentioned above has its merits and demerits respectively. Based on the simulation accuracy and efficiency the modified Norton–Hoff equation [17–19] is used to expresses the high temperature macroscopic flow stress in austenization stage, which is shown as Eq. (7.6).

$$\begin{aligned}
&\sigma(\varepsilon, \dot{\varepsilon}, T) = K(b+\varepsilon)^{n(T)}\dot{\varepsilon}^{m(T)}\exp(\beta/T)\\
&n(T) = n_0\exp(-c_n(T_i - T_0)),\\
&m(T) = m_0\exp(-c_m(T_i - T_0))
\end{aligned} \tag{7.6}$$

where K is strength factor; b is train correction; β is temperature coefficient; n_0, c_n, m_0, c_m are hardening material parameters. All of them are determined by multi-function thermal–mechanical simulator mentioned above, as shown in Table 7.2.

Table 7.2 Material parameters of flowing stress

K	β	b	n_0	c_n	m_0	c_m
41.54	28.93	0.011	0.2167	0.0033	0.012	0.0031

7.3.3 Effects of Hardening Capacity on Formability

The hardening exponent *n* is used to describe the material hardening strength in plastic deformation. Under the same deformation degree, the true stress of deformed material will increase larger with a larger hardening exponent. Meanwhile, a material with large hardening exponent will produce more uniform deformation in the process of tensile deformation, which can reduce the local thinning and increase the limit strain of blank. The influence of hardening exponent is more significant for sheet formability when the bugling proportion is large and the distribution of deformation is nonuniform, especially for deep drawing process with complex shape.

The high-strength boron steel 22MnB5 has a good plasticity whose elongation can reach 50 % [20] at elevated temperature. Thus the formability of sheet 22MnB5 at elevated temperature is better than other steel sheets for cold stamping when the forming conditions are simple like U shape drawing, etc. For more complex stretch forming part as shown in Fig. 7.6, a reinforced beam of inside door produced by hot stamping, the fracture occurred during the hot stamping process. This shows that even though the plastic flow of the material is good, the distribution of deformation can also be quite nonuniform if hardening ability is bad, which will cause local thinning and fracture. So the hardening exponent *n* is one of the important parameters to determine the formability of blank during hot stamping.

The variation law of the material hardening exponent at 650–900 °C temperature range is obtained according to the high temperature tensile curve of hot forming material [21, 22], which is shown in Fig. 7.7.

Figure 7.7 shows that the material hardening exponent values first increase and then decrease with temperature increasing at 600–900 °C, and the largest value is present to the range of 650–750 °C which means the boron steel has good formability in this range.

The hot stamping simulations of the reinforced beam with material properties at different temperatures were conducted respectively. The thickness of four typical locations *A*, *B*, *C*, and *D* whose stress state were tensile were extracted to analyze (Fig. 7.8). The compared results were shown in Fig. 7.9. It shows that the variation tendency of the thickness of the four points is the same. All of them first increase and then decrease with the temperature increasing at 600–900 °C, and the variation is small at the temperature range of 650–750 °C. The results show that the formability of blank is better at the temperature range of 650–750 °C which is

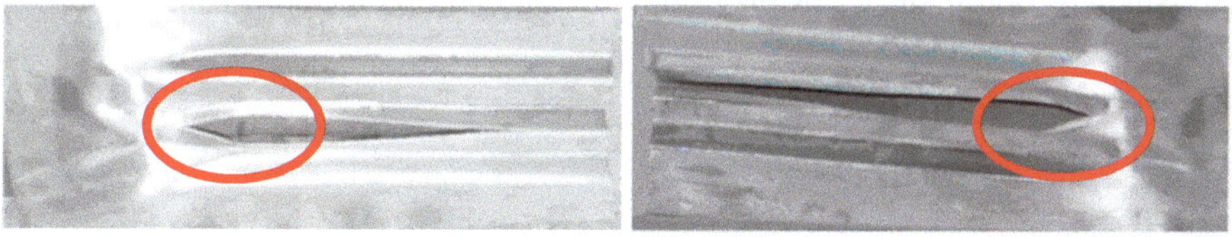

Fig. 7.6 Fracture of the reinforced beam in hot stamping

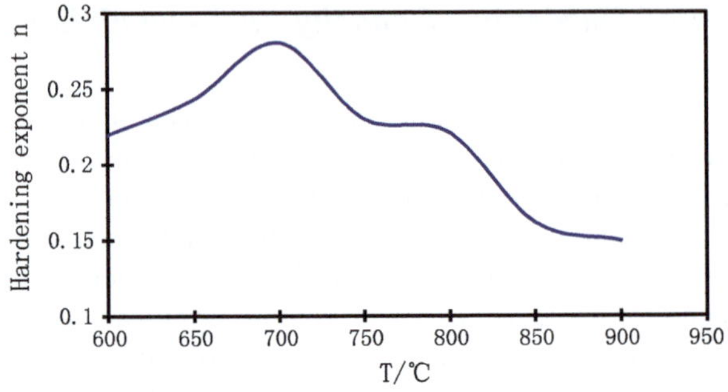

Fig. 7.7 *n* value at different temperatures

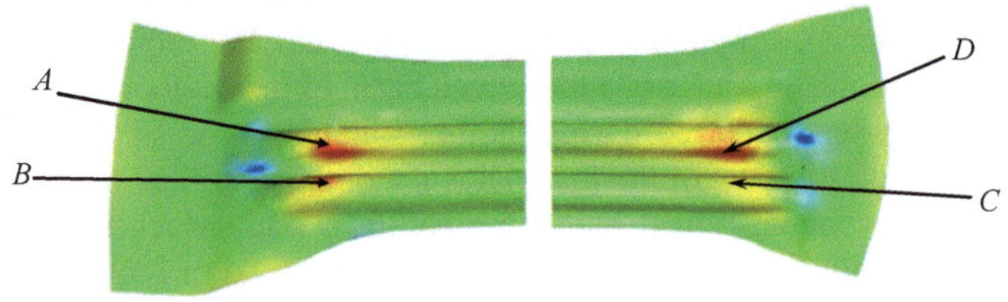

Fig. 7.8 Four locations chosen for analyzing thickness variation at different evaluated temperatures

Fig. 7.9 Thickness distributions of four chosen points at different evaluated temperatures

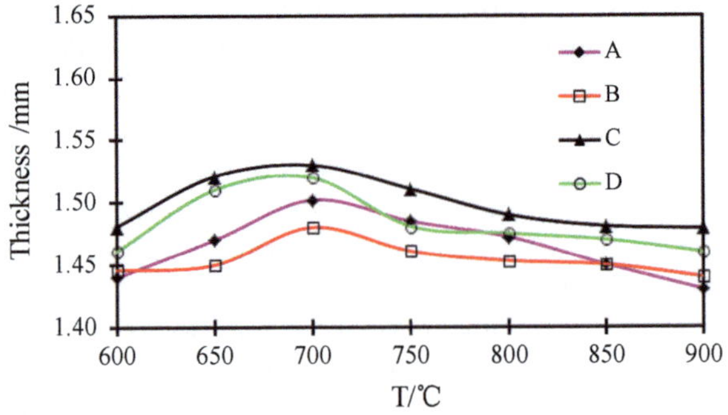

consistent with the above analysis results of hardening exponent *n*. Therefore, for high-strength boron steel 22MnB5, the hot stamping temperature should be controlled at 650–750 °C.

7.3.4 Effects of Directional Anisotropy on Formability

The performance of high-strength steel is quite different in different rolling directions at room temperature. Δr is used to express the planar anisotropy. Large value of Δr reflects great difference in performance with each direction, which results in nonuniform deformation during stamping such as the increasing of local deformation and nonuniform thickness of components.

The planar anisotropy coefficient Δr is a measure of the variation of r with the angle from the rolling direction. This quantity is defined as

$$\Delta r = \frac{r_{0^\circ} - 2r_{45^\circ} + r_{90^\circ}}{2} \tag{7.7}$$

where r_{0°, r_{45°, r_{90° represents the coefficient of normal anisotropy r in 0°, 45°, 90°, respectively.

The value of Δr represents the performance difference in different directions. The difference of performance gets greater when the value of Δr gets higher. For example, when drawing cylinder components, the higher the value of Δr is, the larger earing at the end face of components appears for materials. Therefore, we have to increase material consumption for the increase of the cutting edge margin. A higher value of Δr also increases nonuniform blank deformation in the stamping process, which causes large local distortion and nonuniform wall thickness. A high Δr value has adverse impacts on the quality of almost all stamping components. So the material with low Δr value should be chosen.

At room temperature, the r value is usually tested by the tensile test and defined as the ratio of width strain and thickness strain when the relative elongation is in range of 15–20 % [23, 24], which is also usually used in principle of plastic deformation of volume. The value measured in this way is more accurate, while the test error can be improved, too.

In this study, the 22MnB5 steel is chosen as the object. Considering the anisotropy of sheet metal at room temperature, hot stamping steel with three different rolling directions of 0°, 45°, 90° and different thickness of 1, 2, 2.5 mm respectively at room temperature are investigated. Basic mechanical properties of hot forming materials and anisotropy coefficient Δr through tensile test are studied. The results are shown in Table 7.3. It can be seen that the planar anisotropy coefficient is large for three different directions which means rolling planar anisotropy is obvious for hot forming sheet at room temperature.

But, how is the anisotropy of high-strength boron steel in hot stamping process? Then, we will focus on the study of hot stamping sheet anisotropy and put forward a new method used to study the problem of anisotropy in high temperature.

However, it would be much complicated to take the above traditional tensile method to measure the r value, especial since the interrupt of tensile process would affect test result a lot. Meanwhile, the effect of heat bilges cold shrink of high

Table 7.3 Basic material mechanical properties and planar anisotropy Δr at room temperature

Steel model	Yield strength/MPa	Tensile strengthen/MPa	Hardening exponent n	Δr
TG-1-0	330	500	0.17	−0.31
TG-1-45	340	490	0.16	
TG-1-90	350	520	0.15	
TG-2-0	300	490	0.15	−0.33
TG-2-45	320	495	0.15	
TG-2-90	330	500	0.16	
TG-2.5-0	290	480	0.15	−0.20
TG-2.5-45	300	490	0.14	
TG-2.5-90	320	490	0.15	

Note TG is the name of steel. 1, 2 and 2.5 represents sheet thickness of 1, 2 and 2.5 mm respectively; 0, 45, 90 is the direction along the sheet rolling direction 0°, 45° and 90° respectively

temperature material make it limited for principle of plastic deformation of volume. Thus, to study the anisotropy of hot stamping by method of thermal tensile test is not an option.

Then a new experimental scheme is designed. Experiments are performed to investigate the rolling anisotropic of high-strength steel sheet after thermal tensile test and heat treatment (such as austenization and quenching). We prepare the test sample with different rolling directions of 0°, 45°, 90° and then put them into the furnace, heating to 950 °C by 10 °C/s and maintains constant for 5 min and finally put them into the hot stamping mould with cooling system. After that a tensile test is carried out when the temperature cools to 100 °C to determine the difference of stress–displacement curve and hardening capacity with sample at room temperature. Experimental result is shown in Table 7.4. It can be seen that the anisotropy almost disappeared after high temperature austenitizing and quenching. Meanwhile, the tensile hardening exponent also reduced.

Table 7.4 Basic material mechanical properties and planar anisotropy Δr at room temperature after heat treatment

Steel model	Yield strength/MPa	Tensile strengthen/MPa	Hardening exponent n	Δr
TG-1-0	1098	1608	0.10	−0.04
TG-1-45	1047	1553	0.10	
TG-1-90	1033	1567	0.09	
TG-2-0	1112	1613	0.10	−0.03
TG-2-45	1131	1660	0.10	
TG-2-90	1107	1624	0.09	
TG-2.5-0	1115	1635	0.09	−0.06
TG-2.5-45	1087	1597	0.10	
TG-2.5-90	1052	1591	0.10	

7.4 Prediction of Forming Limit for Hot Stamping

7.4.1 Introduction of Forming Limit and Instability Theory

In the metal stamping process, the moment and location of sheet failure are urgently needed to improve the stamping technology and quality of product. To achieve this, lots of methods to evaluate the sheet formability have been proposed. The forming limit curve (FLC) is an efficient diagnostic tool and has been widely used by the previous scholars.

FLC was first developed by Keeler and Goodwin, which composed of major strain and minor strain in plane. It can predict the ultimate strain levels before the failure occurred during stamping process [25, 26]. Linking the limited strain points under different paths, the FLC can be constructed as shown in Fig. 7.10. In general, FLC is an important evidence for ambient temperature stamping.

The theoretical foundation of forming limit is tensile instability theory. In general, the deformation phase of sheet metal can be divided into five parts, including elastic stage, yield hardening, diffuse necking, localized necking and final fracture, as shown in Fig. 7.11. Therefore, the sheet failure is due to ongoing deformation and the investigation of sheet failure should be focus on the whole process of diffuse and localized necking. The theoretical investigation of sheet instability began in the 1950s, the representative work includes the Swift diffuse necking theory, Hill localized necking theory, Marciniak and Kuczynski model, etc. All these previous works make a tremendous contribution to the prediction of sheet metal forming limit. Next, some brief description of these theories will be given.

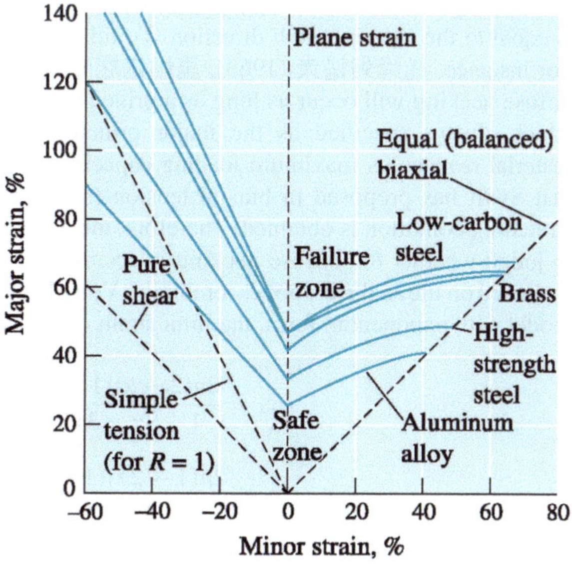

Fig. 7.10 Forming limit curve for different materials

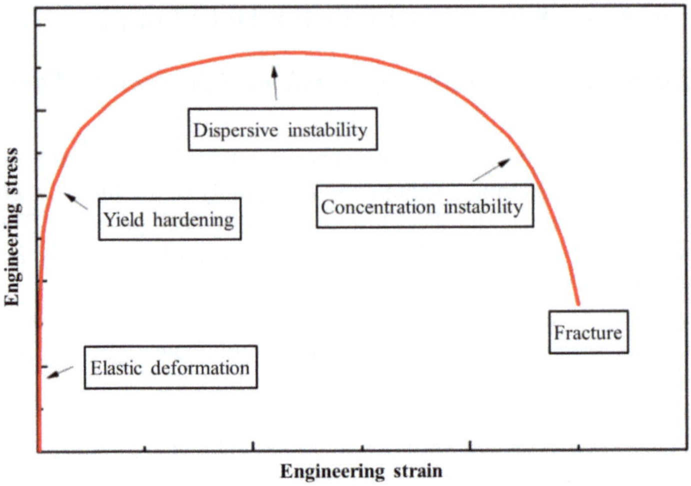

Fig. 7.11 The sheet metal flow stress and strain curve

1. Swift diffuse necking theory

Both the thickness-direction strain and transverse-direction strain are considered in the Swift's diffuse necking theory [27]. According to this theory, sheet failure will occur when the variation of cross-section strain is equal to the current stress state. In uniaxial tension state, for example, the maximum loading capacity P can be used as a criterion of sheet instability. That is, the diffuse necking will happen when the condition $dP = 0$ achieved. After this, the loading capacity of sheet metal decreases and the uniform plastic stage ends. However, in biaxial tension state, Swift thought that diffuse necking will occur only when the load in length direction is equal to the load in width direction. To this, some scholars hold different views. For instance, 托姆列诺夫 (1963), 里格诺尼 and 汾姆逊 (1969), they thought that diffuse necking will occur as long as a principle stress in plane reach the maximum which always specified as the major principle stress direction σ_1, means that material reaches its maximum loading capacity. It is difficult to achieve the status that Swift has proposed in biaxial tension ($dP_1 = 0$, $dP_2 = 0$) unless equibiaxial stretching condition is obtained. Therefore, the evaluation criteria $dP = 0$ is suitable to judge whether the diffuse necking is occurred in plane deformation state.

Based on the analysis above, combining Von Mises yield criterion and constitutive model with exponential form, the limit strain of diffuse necking can be obtained

$$\varepsilon_1 = \frac{2n(2-\alpha)(1-\alpha+\alpha^2)}{4-3\alpha-3\alpha^2+4\alpha^3} \tag{7.8}$$

$$\varepsilon_2 = \frac{2n(1-2\alpha)(1-\alpha+\alpha^2)}{4-3\alpha-3\alpha^2+4\alpha^3} \tag{7.9}$$

2. Hill localized necking theory

The localized necking theory [28] proposed by Hill confirms that sheet fracture will occur when material diffuse instability and localized necking caused by further deformation take place. So, the necking criterion of localized necking is defined as the reduction ratio is equal to the rate of hardening. The deformation rate will increase sharply once the localized necking happened, meanwhile, the deformation will stop beyond the necking zone as the external load doesn't increase anymore. According to Hill localized instability criterion, the localized instability condition can be formulated as following:

$$\frac{d\sigma_1}{\sigma_1} = \frac{d\sigma_2}{\sigma_2} = -\frac{dt}{t} = -d\varepsilon_3 \tag{7.10}$$

Therefore, the limit strain can be derived from Hill localized instability criterion and Von Mises yield criterion:

$$\varepsilon_1 = \frac{n}{1+\beta}, \quad \varepsilon_2 = \frac{n\beta}{1+\beta} \tag{7.11}$$

3. Marciniak-Kuczynski model

Marciniak and Kuczynski developed a groove instability theory [29], which remedies the limitation of Hill localized instability theory in explanation of biaxial tension state. What's more, the M-K model makes it possible to explain the instability of rate-sensitive materials. It has become an important model in predicting forming limitation after the development of many researchers.

The geometrical model of M-K model is showed in Fig. 7.12, where *a* represents the zone with homogeneous deformation while *b* represents inhomogeneous deformation as it is thinner than zone *a*. In tension process, the deformation concentrates in zone *b* due to the thinner thickness. Therefore, fracture will eventually occur in zone *b* because of the quickly decreasing of thickness comparing with zone *a*.

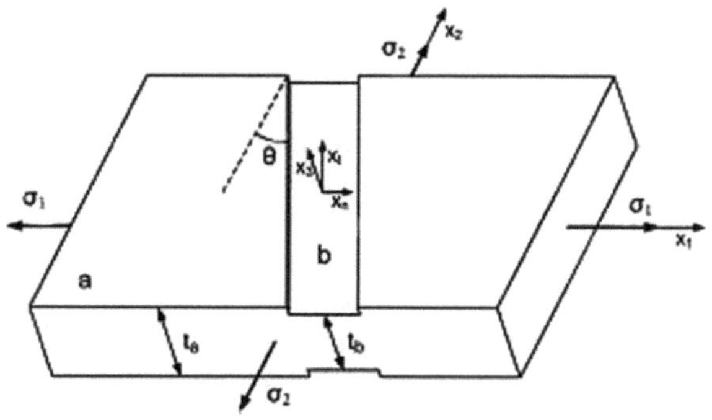

Fig. 7.12 The M-K groove instability theory model

The main equation of the M-K model is as following:

(1) According to the constant volume principle:

$$d\varepsilon_1 + d\varepsilon_2 + d\varepsilon_3 = 0 \tag{7.12}$$

(2) Assuming that the zone a under simple tension condition, where principal stress increases with the increase of major strain, while the ratio keeps constant.

$$\frac{d\sigma_{1a}}{\sigma_{1a}} = \frac{d\sigma_{2a}}{\sigma_{2a}} \tag{7.13}$$

$$\frac{d\varepsilon_{1a}}{\varepsilon_{1a}} = \frac{d\varepsilon_{2a}}{\varepsilon_{2a}} = \frac{d\varepsilon_{3a}}{\varepsilon_{3a}} \tag{7.14}$$

(3) Minor strains in zone a and b are the same based on strain coordination:

$$d\varepsilon_{2a} = d\varepsilon_{2b} = d\varepsilon_2 \tag{7.15}$$

(4) Force equilibrium requirement is contented according to the boundary condition while deforming:

$$\sigma_{1a}t_a = \sigma_{1b}t_b \tag{7.16}$$

where t_a and t_b represent the thicknesses of zone a and b separately.

7.4.2 *Test Principle of Forming Limit at Elevated Temperature*

Although the equipment and test approaches are much more complicated under high temperature condition, for instance, keeping temperature constant, measuring the friction and strain, the principle is the same with which under traditional condition. According to the Chinese standard GB/T15825.8.1995, the Nakazima test with a hemisphere punch is adopted. Figure 7.13 shows the principle for the forming limit test.

In order to obtain the forming limit under different loading path, nine different specimens are prepared previously. As shown in Fig. 7.14, the width of the specimens have decreased from 180 to 20 mm, so that the state of strain varies from biaxial tension to uniaxial tension. Round grids were printed onto the surface of specimens. After stretched by the punch, the round grids become elliptical. The following method is applied to measure the major axis and minor axis of the grids

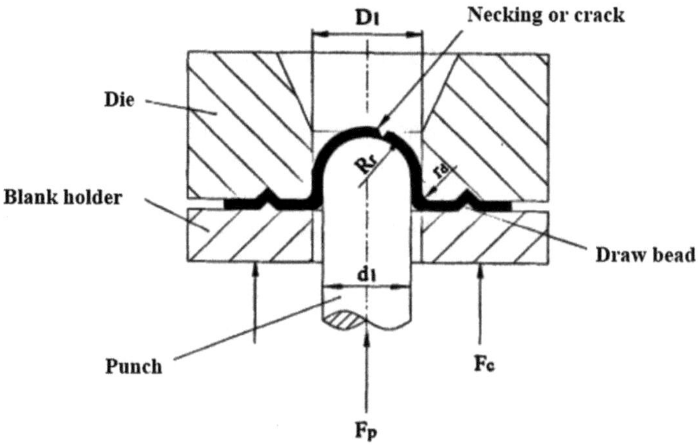

Fig. 7.13 The principle diagram for the forming limit test

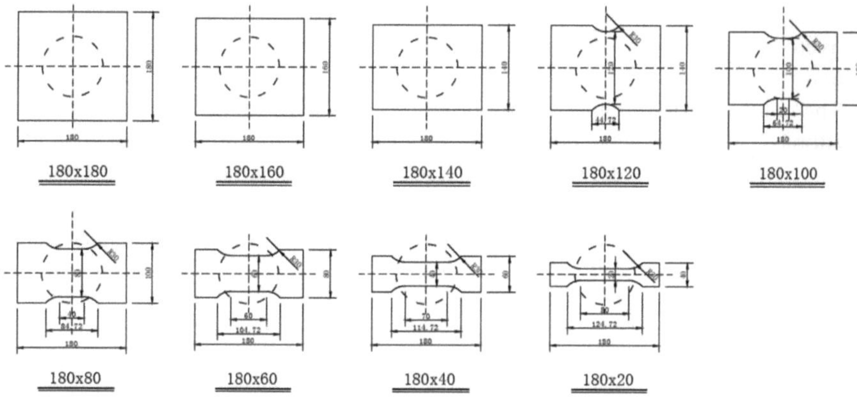

Fig. 7.14 The different specimens for TFLD test

next to the necking area and calculate its major and minor strain in order to obtain forming limit under different loading path.

$$\varepsilon_1 = \ln\left(\frac{d_1}{d_0}\right) \tag{7.17}$$

$$\varepsilon_2 = \ln\left(\frac{d_2}{d_0}\right) \tag{7.18}$$

where ε_1 is the major strain, ε_2 is the minor strain, d_1 is the length of the major axis of measured ellipse, d_2 is the length of the minor axis of measured ellipse, d_0 is the original diameter of the round grid.

7.4.3 Test Equipment and Test Procedure of Forming Limit at Elevated Temperature

Comparing with forming limit test equipment of traditional cold stamping, the difficult points for developing test equipment of high temperature forming limit are the realization of sheet heating and heat preservation as well as the high temperature lubrication. According to Standard GB/T15825.8.1995 *Sheet metal formability and test methods—Forming limit diagram (FLD) test*, the test equipment TFLD-300 for high temperature forming limit which based on the principle of Nakajima bugling experiment is designed and produced, as shown in Fig. 7.15.

Main structures and features of TFLD-300 include the following aspects [30]:

1. The body of TFLD-300 is a vertical structure which can achieve isothermal and antioxidation stamping. A vacuum furnace combined with Nitrogen charging device is adopted to prevent the oxidation of high temperature sheet and heating coil during hot experiment process. Forming mould is placed into a bell shaped furnace and the resistance heating mode is adopted which can provide an isothermal environment during experiment process. Above characteristics overcome the defect of some apparatuses that must use asbestos to reduce the loss of heat.
2. The control system composed by proportional valves in the hydraulic system is adopted. It can exactly control the mould stroke, forming force and holding force which ensures the precision of each experimental parameter. Furthermore, the bugling experiment can be stopped immediately as the control system is adopted which is convenient to acquire the necking or fracture point of the metal sheet.

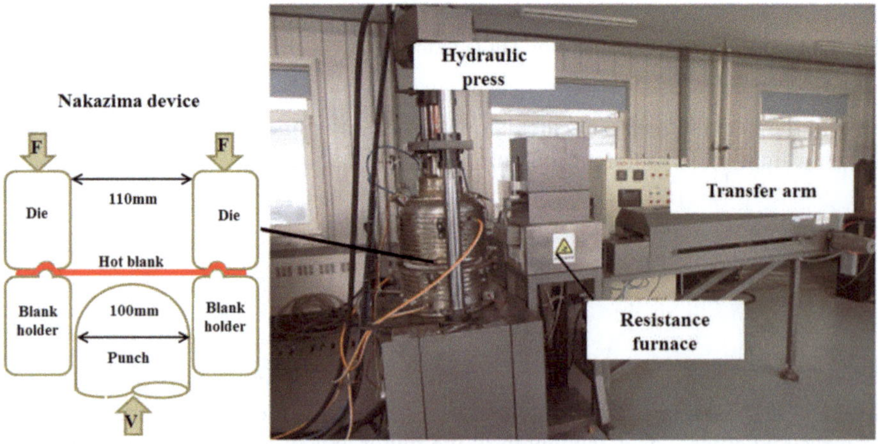

Fig. 7.15 The test equipment for high temperature forming limit

3. The experiment process can be monitored and controlled by computer program in real time. Experiment parameters can be acquired by the measuring control system which is consisted of pressure and displacement sensor, autonomous experiment control system and PCI board that has been integrated in the computer. In concrete terms, monitoring and control of stamping force and velocity can be achieved by this computer flexible test technology and closed-loop control technology.
4. The design method of cooling channel which around the furnace is adopted. This design method can improve the cooling rate effectively; accordingly, the experiment progress for different conditions can be accelerated.

Test process of high temperature forming limit can be summarized as follows:

1. Prepare 9 different size specimens to achieve different stress state as mentioned in Sect. 7.4.2.
2. Use laser marking machine to make round grid on the specimen surface. The diameter of the round grid is suggested from 1.5 to 2.5 mm by national standard and the value 2.5 mm was used in our research. The grid by laser marking can be measured easily even though the metal sheet is heated to a high temperature.
3. Paint high-temperature lubricant on punch and specimen surface. Once the temperature of mould reaches the set value, the austenitic specimen should be cooled down at a cooling rate larger than 50 °C/s and then put into the bell shaped furnace. Then, the bugling experiment is executed by the following steps. The specified holding force is provided by the process of die down firstly. And then the punch draws at a speed of 50 mm/min. The bugling experiment will be stopped when necking or fracture occurs. The specimens of different sizes after bugling are shown in Fig. 7.16.

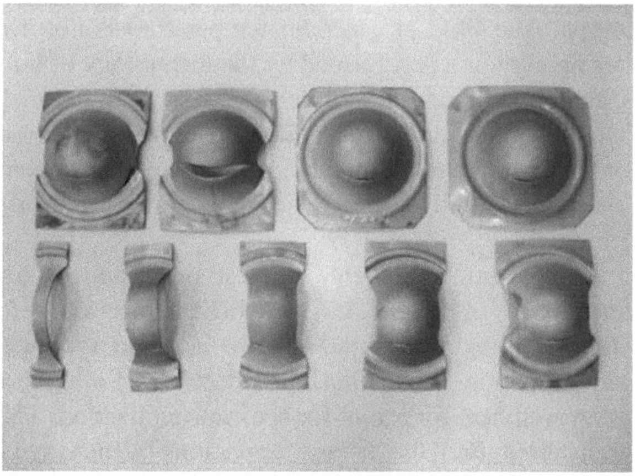

Fig. 7.16 High-strength boron steel specimens after forming limit test

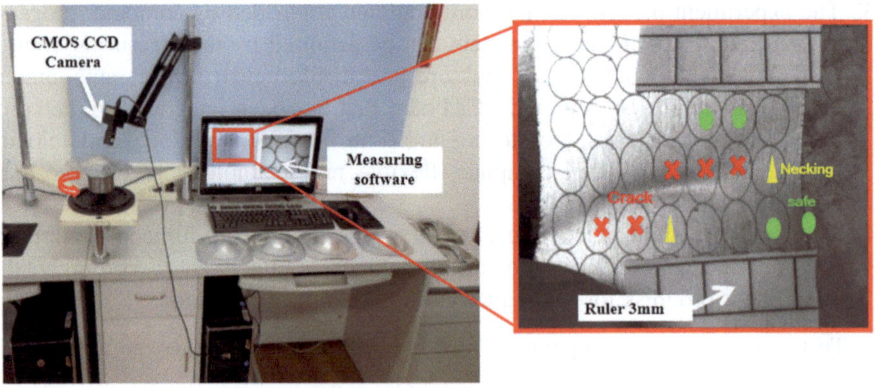

Fig. 7.17 The strain measure system of forming limit test

4. The strain measured system which equipped with CMOS CCD camera is used to measure the major axis and minor axis of specified ellipse. Then, the major and minor strain can be calculated for the safety grid, necking grid and fracture grid as shown in Fig. 7.17.

7.4.4 Three-Dimension Thermal Forming Limit Diagram and Its Application

It's known that hot stamping operation is a non-isothermal process in which the temperature of sheet metal varies according to contact state and deformation condition. The FLC at a certain temperature is not accurate enough to evaluate the formability of a part formed by the technology of hot stamping. And there is no any criterion in literatures to describe the formability of sheet metal during hot stamping which taking into account mutative temperature process. Therefore, it is necessary to establish forming limit criterion at elevated temperature to determine whether the sheet metal will be torn when it is formed into a product shape in hot forming process.

Using test equipment and test procedure of high temperature forming limit mentioned in Sect. 7.4.3, the high temperature forming limit experiments for 22MnB5 ranging from 600 to 800 °C were performed with the experiment process shown in Fig. 7.18. First, the sheet metal was heated to 900 °C in furnace and preserved heat for 5 min for soak austenitization. Then the sheet metal were rapidly transferred into the testing apparatus by auto-feeder and cooled to the testing temperature with an average cooling rate of 50 °C/s and deformed at tested

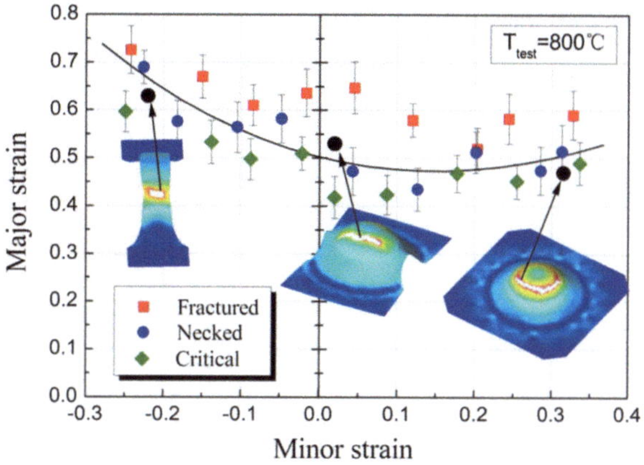

Fig. 7.18 Experiment process of thermal forming limit tests

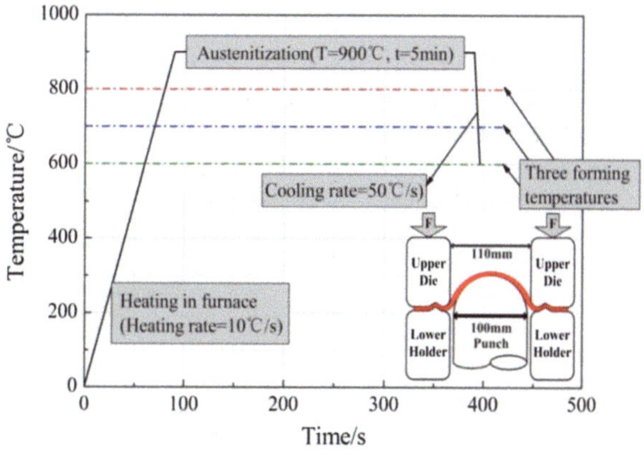

Fig. 7.19 FLC of 22MnB5 obtained from experiment at 800 °C

temperature until failure. And the mould was heated to testing temperature and remained during whole experiment process.

Figure 7.19 shows the forming limit obtained at 800 °C by varying the strain path from equal biaxial stretching to uniaxial stretching with constant punch speed of 500 mm/min. The red dots represent fractured grids, the blue diamonds are necked grids and the green squares are critical grids. And the blank line stands for forming limit curve at 800 °C.

As mentioned above, hot forming is a non-isothermal process (600–800 °C), therefore, a FLC at one temperature is not accurate enough to estimate the formability of hot formed parts. Based on forming limit points obtained at different temperature, a 3-dimensional thermal forming limit diagram (3D-TFLD) can be constructed as shown in Fig. 7.20. Where *X*-axis stands for temperature, *Y*-axis stands for minor strain, and *Z*-axis stands for major strain.

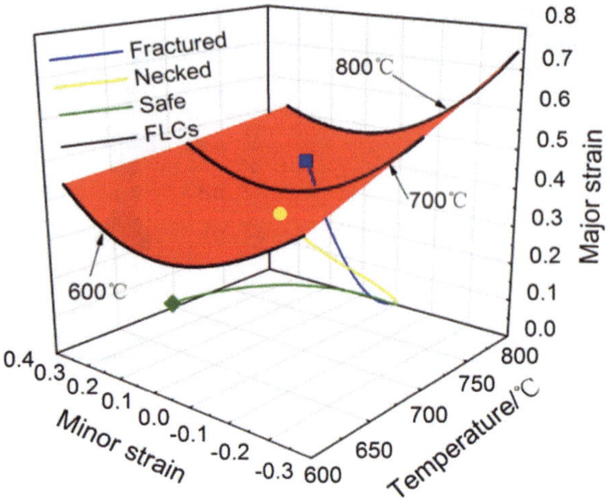

Fig. 7.20 Three-dimensional TFLD for 22MnB5 high stress steel

Fig. 7.21 CAD model of the automobile B-pillar

To prove the validity of 3D thermal forming limit diagram (3D-TFLD), an independently developed hot stamping module, King Mesh Analysis System of Hot Forming (KAMS_HF), was derived to simulate the hot stamping process of a certain B-pillar [31, 32]. Big blank holder force (100 KN) was adopted in order to avoid the fracture during hot stamping. The initial thickness and temperature of boron steel is 2 mm and 800 °C, respectively. The Young modulus and Poisson ratio, which vary with temperature, are obtained from reference [33], and the specific heat capacity and thermal conductivity are obtained from reference [34]. The friction is considered as Coulomb friction, friction coefficient $\mu = 0.4$. Figure 7.21 shows the CAD model.

Figure 7.22a shows the distribution of major strain on the B-pillar after hot stamping. The points include major strain, minor strain, and temperature are printed in the 3D-TFLD, as shown in Fig. 7.22b. It is shown that some points are above the surface, which means failure occurs at the side wall at the top of the B-pillar.

The B-pillar component employing the same process of simulation is shown in Fig. 7.23. It can be seen that the fracture position is the same. Therefore, the validity of 3D-TFLD is proved.

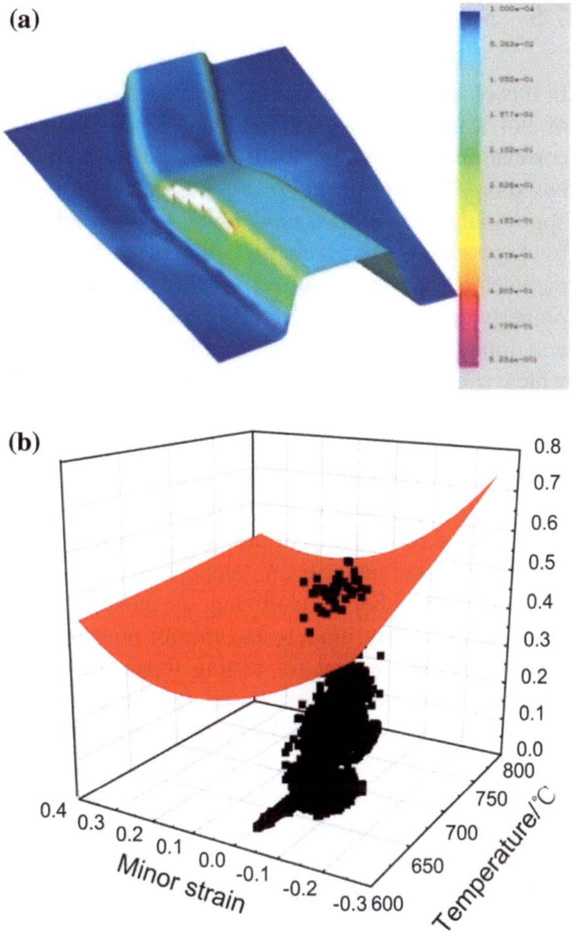

Fig. 7.22 **a** Numerical simulation results (fractured) of B-pillar. **b** The three-dimensional TFLD of hot stamping B-pillar (fractured)

Fig. 7.23 Fractured automotive B-pillar

7.5 Summary

In this chapter, the conception of metal plastic deformation and the effective factor of hot stamping plasticity and resistance of deformation were introduced. The correlational studies under high temperature by other researchers were analyzed based on the experimental researches on the properties of hot stamping material. An appropriate thermal flow model was derived and essential parameters were obtained by experiments. Experimental researches were adopted to study the hardening parameters, which are effective on the material formability and the anisotropy property during rolling. According to the analysis of material hardening property, it was found that the formability is better at the temperature range of 650–750 °C, which is consistent with the simulation result. After introducing the experimental method to test anisotropy of hot stamping material after rolling in room temperature, new experimental scheme to test the anisotropy by rolling under high temperature was proposed, which is more convenient and accurate. It was shown as follows: the specimen was first heated to the required temperature, and then cooled down to the room temperature by the same cooling condition. The rolling anisotropy was tested under room temperature to represent that under high temperature. It is shown that after austenitizing, the anisotropy property almost vanished. Several classical instability models and the forming limit diagram were introduced in this chapter. An equipment for testing forming limit (TFLD-300) was independently researched and developed based on the practical manufacturing technique and international standard. The test philosophy of this equipment was also introduced in this chapter. The experiment to study forming limit of 22MnB5 under different temperatures was carried out and 3D-TFLD was proposed, which can be used in hot stamping simulation. To verify 3D-TFLD, a simulation of a certain B-pillar was derived. It can be seen that 3D-TFLD can be used as a criterion to evaluate the material formability under high temperature.

References

1. Jenner F, Walter ME, Mohan Iyengar R et al (2010) Evolution of phases, microstructure, and surface roughness during heat treatment of aluminized low carbon steel. Metall Mater Trans A 41:1554–1563
2. Ning M, Hu P, Zhai S et al (2009) Hot forming technology of high strength steel and its engineering implementation. Automob Technol Mater 12:12–14 (in Chinese)
3. Kapadia BM, Brown RM, Murphy WJ (1968) The influence of nitrogen, titanium, and zirconium on the boron hardenability effect in constructional alloy steels. US Steel Corp., Monroeville, Pa
4. Eriksson M, Oldenburg M, Somani MC et al (2002) Testing and evaluation of material data for analysis of forming and hardening of boron steel components. Modell Simul Mater Sci Eng 10:277

5. Karbasian H, Brosius A, Tekkaya AE et al (2008) Numerical process design of hot stamping processes based on verified thermo-mechanical characteristics. In: Materials science and technology (MS&T), Pittsburgh, pp 1733–1743
6. Merklein M, Lechler J, Geiger M (2006) Characterisation of the flow properties of the quenchenable ultra high strength steel 22MnB5. CIRP Ann Manuf Technol 55:229–232
7. Grange RA, Kiefer JM (1941) Transformation of austenite on continuous cooling and its relation to transformation at constant temperature. Trans ASM 29:85
8. Liu Z, Wu Z, Wu J (1996) The numerical simulation of heat treatment process. Science Press (in Chinese)
9. Garcia Aranda L, Chastel Y, Fernández Pascual J et al (2002) Experiments and simulation of hot stamping of quenchable steels. Adv Technol Plast 2:1135–1140
10. Mori K, Maki S, Tanaka Y (2005) Warm and hot stamping of ultra high tensile strength steel sheets using resistance heating. CIRP Ann Manuf Technol 54:209–212
11. Ma N (2011) Research on Hot forming of high strength steel. Dalian University of Technology, p 224 (in Chinese)
12. Nemat-Nasser S (2002) Experimentally-based micromechanical modeling of metal plasticity with homogenization from micro-to macro-scale properties. In: IUTAM symposium on micro-and macrostructural aspects of thermoplasticity. Springer, pp 101–113
13. Johnson GR, Cook WH (1983) A constitutive model and data for metals subjected to large strains, high strain rates and high temperatures. In: Proceedings of 7th international symposium on ballistics. The Hague, The Netherlands, pp 541–547
14. Tong L, Stahel S, Hora P (2005) Modeling for the FE-simulation of warm metal forming processes. Numisheet
15. Ghosh S, Noboru K (1988) Finite element formulation for the simulation of hot sheet metal forming processes. Int J Eng Sci 26:143–161
16. Molinari A, Ravichandran G (2005) Constitutive modeling of high-strain-rate deformation in metals based on the evolution of an effective microstructural length. Mech Mater 37:737–752
17. Hoffmann H, So H, Steinbeiss H (2007) Design of hot stamping tools with cooling system. Ann CIRP
18. Merklein M, Lechler J (2008) Determination of material and process characteristics for hot stamping processes of quenchenable ultra high strength steels with respect to a FE-based process design. In: SAE technical paper
19. Turetta A, Bruschi S, Ghiotti A (2006) Investigation of 22MnB5 formability in hot stamping operations. J Mater Process Technol
20. Ma N, Hu P, Yan K et al (2010) Research on boron steel for hot forming and its application. Chin J Mech Eng 14:68–72 (In Chinese)
21. Ma N, Hu P, Shen GZ (2010) Modeling, testing and numerical simulation on hot forming. In: AIP conference proceedings
22. Ma N, Hu P, Guo W et al (2009) Feasible methods applied to the design and manufacturing process of hot forming. In: IDDRG2009 conference, Golden, CO USA
23. China NSOTPSRO (2006) Metallic materials-sheet and strip-determination of plastic strain ratio (r-values). In: China national standardization management committee (in Chinese)
24. Lin Z (2005) Stamping simulation for car body covering parts. Mechanical Industry Press (in Chinese)
25. Goodwin GM (1968) Application of strain analysis to sheet metal forming problems in the press shop. In: SAE international, pp 380–387
26. Keeler SP, Backofen WA (1963) Plastic instability and fracture in sheets stretched over rigid punches. Trans Am Soc Metals 56:25–48
27. Swift HW (1952) Plastic instability under plane stress. J Mech Phys Solids 1:1–18
28. Hill R (1952) On discontinuous plastic states, with special reference to localized necking in thin sheets. J Mech Phys Solids 1:19–30
29. Marciniak Z, Kuczyński K (1967) Limit strains in the processes of stretch-forming sheet metal. Int J Mech Sci 9:609–620

30. Dai M (2014) Research of warm forming technology for 6000 serious aluminum alloy sheet used for automotive body. In: Automobile engineering. Dalian University of Technology, p 82 (in Chinese)
31. Hu P, Shi D, Ying L et al (2015) The finite element analysis of ductile damage during hot stamping of 22MnB5 steel. Mater Des 69:141–152
32. Shi DY, Ying L, Hu P et al (2013) Experimental and numerical determination of thermal forming limit diagrams (TFLD) of high strength steel 22MnB5. In: AIP conference proceedings, pp 406–413
33. Pehlke RD, Jeyarajan A, Wada H (1982) Summary of thermal properties for casting alloys and mold materials. NASA STI/Recon Tech Rep N 83:36293
34. Shapiro AB (2009) Using LS-Dyna for hot stamping. In: Seventh European LS-Dyna conference

Chapter 8
Hot Stamping Simulation Algorithms of High-Strength Steels

8.1 Basic Descriptions of the Hot Stamping Simulation

Hot stamping of high-strength steels is based on the large plastic deformation of metal, with high nonlinear mechanical characteristics of geometry, physics, and boundary friction [1, 2]. As with the traditional stamping process, it is necessary for hot stamping to design and manufacture prototypes, with many trial productions and commissioning, for the design and manufacture of stamping tools with cooling channels, the determination of hot metal sheet's shape and size, and the planning of stamping processes and parameters. It is a process of trial and error, resulting in large consumption of manpower, material and financial resources, high production costs, and long cycle.

Based on thermal–mechanical transformation coupled theory of the large elastic–plastic deformation mechanics, automotive hot stamping computer simulation technology has been successfully used in the automotive mold industry, as a very important solution to the defects of traditional methods to design and manufacture stamping tools. In the application of the technology and its commercial software, automotive engineers in the mold industry can simulate the forming in the real process on a computer and then manufacture molds according to the best results of simulations, which can greatly reduce the mold design and commissioning cycle, lower manufacturing costs, and improve product quality. It is an irreplaceable powerful auxiliary tool to solve the "bottleneck" problem of stamping mold manufacturing industry.

Hot stamping itself has the characteristics of material heat treatment technology and material forming technology, therefore, the author thinks that hot stamping simulation technology is the integration of mechanics, metallurgy, and design technique of auto-body parts and digital car set structure, as shown in Fig. 8.1.

We can imagine such a hot stamping simulation "black box" (closed-loop simulation system)—We enter the metallurgy program (the selection of microalloying elements; parameters of casting, rolling process, etc.), forming process

P. Hu et al., *Hot Stamping Advanced Manufacturing Technology of Lightweight Car Body*, DOI 10.1007/978-981-10-2401-6_8

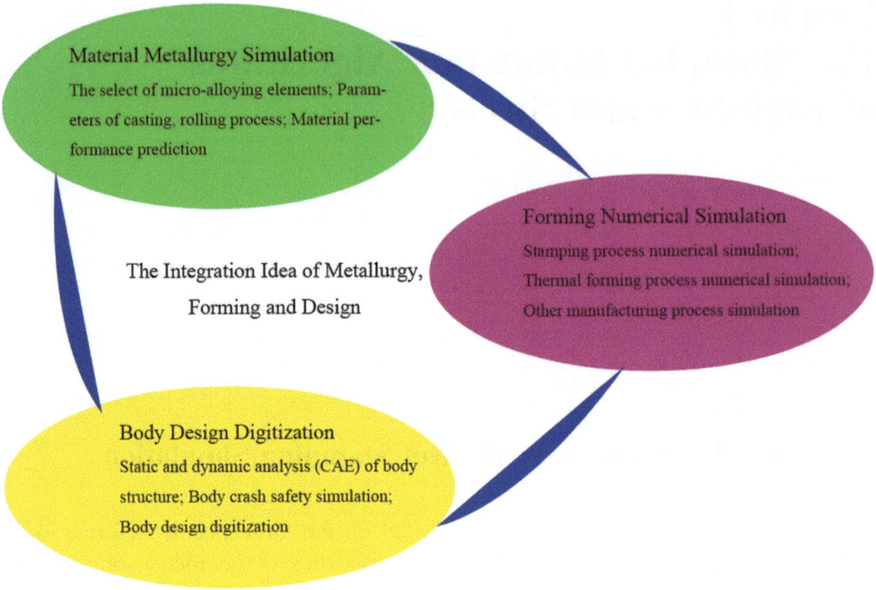

Fig. 8.1 Integrated numerical simulation of metallurgy, forming, and automobile design

program and body design requirements, and then start the simulation system. After running, the system will output the basic thermodynamic properties of new materials, mold design scheme, components and parts forming process parameters, static and dynamic structural properties of the body, etc.; then the closed-loop simulation system will optimize the shape and body design with the feedback of the process parameters, and finally it will output the ideal metallurgy, forming process and body design program for the actual project implementation.

To achieve these goals, a key factor is to establish an accurate and efficient numerical simulation method for hot stamping (including quenching) process of high-strength steels, which can be used to link and guide the design process of mechanics, metallurgy, and body structure. The numerical simulation of hot stamping process involves the sheet temperature changes during heating, thermal expansion, austenite transformation, and homogenization of austenite phase; heat conduction, convection, radiation of the red hot sheets during the sheet delivery to the mold; the complex problems like the thermal conductivity, high temperature contact, heat friction, cooling contraction, phase transition and forming simultaneously during the sheet contact with the mold, and so on. For high-strength steel hot stamping simulation, not only temperature field and phase transition field are added, but also these two physical fields, coupled with stress field of the traditional cold stamping make the numerical simulation of hot stamping a more complex problem.

On the basis of the study on the constitutive theory and experimental analysis of the hot stamping process, hot stamping simulation technology of high-strength steels is also studied in-depth by the author and his team. First of all, this chapter makes a brief explanation of the most important problems in hot stamping simulation. Then, a detailed analysis for the finite element simulation technology of temperature field and phase transition is carried out. Finally, hot stamping static explicit and dynamic explicit coupled fields finite element formulations are established; the related theories of critical time step, contact, and friction model are given.

8.2 Several Key Points in Numerical Simulation of Hot Stamping

8.2.1 Key technology of Multi-field Coupled Problem

As mentioned above, the technology of numerical simulation of hot stamping, apart from the problems of large deformation and nonlinear material in cold stamping, has a critical feature that many components, such as force-heat field and phase transformation, are coupled in the process of hot stamping(Fig. 8.2) [3, 4], just as the multi-field coupled constitutive relation of hot stamping established in Act 4. In this simulation, it should be reflected that how the temperature field and the process of hot stamping affect the phase transformation points, variable, and the evolution of the multi-phases mixed microstructures. First, the microstructure of plate metals, the thermodynamic property (modulus of elasticity, flow rules, heat exchange coefficient, etc.) of every phase should vary with the temperature field [5, 6] which changes continuously in the process of hot stamping. Second, the temperature field

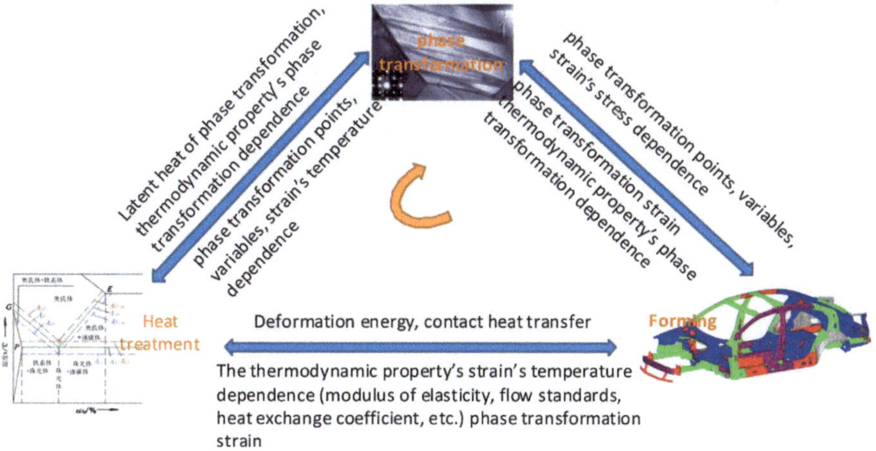

Fig. 8.2 Numerical simulation of hot stamping based on the interaction among heat, stress, and phase transformation

is affected by the latent heat of phase transformation, the plastic work path and the contact between water-cooled mold and hot blank. Third, the thermodynamic property of material is affected by the changing of the microstructure when phase transforms. All above thermal strain, bulk strain and plastic strain of phase transformation affected by the changing of the temperature field are playing a major role in the process of hot stamping, and simultaneously, the strain rates of hot stamping also affect the thermodynamic property of sheets and the changing of the phase transformation field. So, to successfully couple all the components above is one of the key technologies of numerical simulation of hot stamping.

8.2.2 Problems of High Temperature Contact Friction

The contact friction model has great effects on the simulation of plate metals' forming in traditional cold stamping [7, 8]. The mechanism of contact friction, which is changing continuously and synthetically affected by many components (material of molds and sheets, blank holding force, temperature, condition of the surfaces, stamping speed and deformation), is a complicated problem [9, 10] in the process of metal stamping. In the process of hot stamping of HSS, plate metals contact the mold and form in the cavity at over 500 °C, at which plate metals' surface will become very soft, and the oxidation of those without antioxidant coating and other tough components such as unsatisfactory lubrication in high temperature make the contact friction problem highly complicated. Figure 8.3 shows the related methods to test the coefficient of friction in high-temperature stamping.

Stöhr [11] performed an experiment in the way of experiment–analysis–calculation to confirm the coefficient of friction in relative conditions (Fig. 8.3a). Ghiotti and Hardell [9, 12] performed a rotary friction experiment to research how interface parameters (such as temperature, pressure, sliding velocity, and surface roughness) affect the friction between plate metal and mold (Fig. 8.3b). Their analysis showed that the interaction of temperature and pressure become the most effective parameter to the coefficient of friction that reduced when the pressure rose, the reason of which was perhaps Fe–Al, an intermetallic compound, reduced the coefficient of friction when the pressure rose. Dessain [10] performed a convexity friction

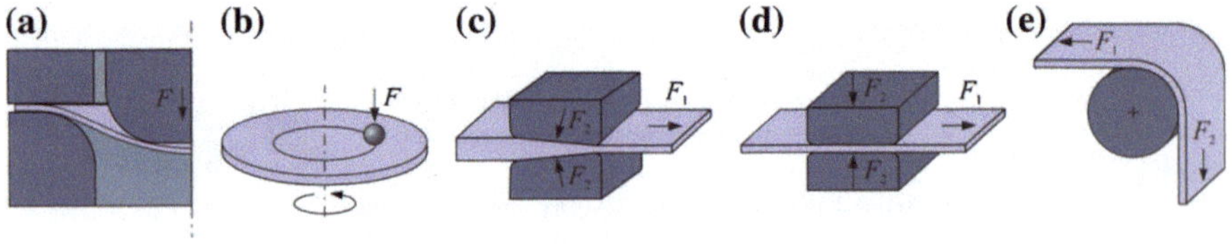

Fig. 8.3 Testing method of friction coefficient at evaluated temperature. **a** Cup drawing test. **b** Rotary friction experiment. **c** Convexity friction experiment. **d** Stretching-friction experiment. **e** Bending-friction method

experiment to research the coefficient of friction in the process of hot stamping (Fig. 8.3c). Through simultaneous measurements of forces of different directions and the contact area that is the convexity of the heated experimental work piece, the test got the data that could be used to calculate the coefficient of friction. Yanagida [13] performed a stretching-friction experiment to research the coefficient of friction of hot stamping (Fig. 8.3d). In addition, the bending-friction method (Fig. 8.3e), which has been used extensively to study the tribology performance of plate metals, also suit the research of the coefficient of contact friction in the process of hot stamping of HSS.

8.2.3 The Simulation Technology of Temperature Field

In the process of hot stamping of HSS, temperature field has a major effect on plate metals' homogenized austenization, performance of hot stamping, and phase transformation. Therefore, an effective and accurate simulation of temperature field is one of the key points during a good numerical simulation of hot stamping [14, 15].

In the hot stamping process, for the start, when the temperature of heated alloy plate reaches the phase point of austenite, thermal insulation measure would be taken to make the microstructure of complete austenitization. And then, plate is transferred to the hot stamping mold quickly, which is a transient heat transfer process, including the transmission of sheet-to-fixture and mold and thermal convection and radiation of the system, which should be simulated accurately in the simulation of temperature field (Fig. 8.4). After the transmission, it is time that the

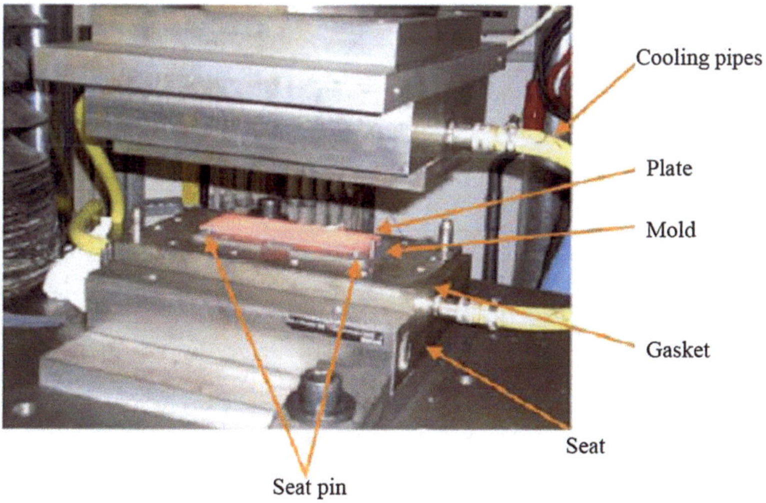

Fig. 8.4 Scheme of thermo transmission and temperature boundary condition in hot forming process

press start to work and make the plate quenched and formed, and during the process a key component involved is the simulation of heat conduction between plate and mold. Different from the traditional cold stamping, in which the mold is always supposed to be a rigid body to increase the efficiency of calculation, in hot stamping process, the mold, as a result of its heat conduction with the plate, must be discretized with shell elements or body elements in the simulation of temperature field of hot stamping and at the same time the layout of cooling line inside the mold and the velocity of cooling water also influence its heat transfer effect. The simulation of heat conduction between plate and mold therefore becomes even more difficult. Additionally, the changing temperature in the thickness direction of plate also should be considered. All the problems together make the simulation even more complicated.

8.2.4 The Simulation Technology of Phase Field

As mentioned in Sect. 8.2.3, for hot stamping, the high-strength steel are heated up to the Ac3-temperature for a fully austenitic microstructure, subsequently hot formed and quenched within the closed tools. The microstructure of the steel sheet will change during the whole process of forming and quenching and the transformation from austenite to martensite can improve mechanical properties of steel sheet markedly. Finally, the tensile strength of the parts can reach over than 1500 MPa.

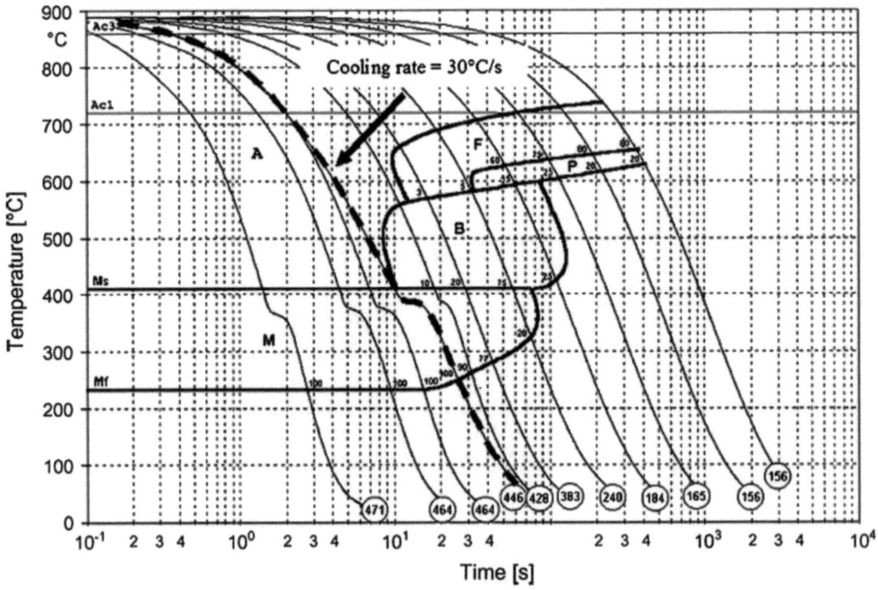

Fig. 8.5 CCT curves of 22MnB5 steel

Considering the complexity of multi-field coupling of hot stamping, the establishment of constitutive relation for microstructure evolution is very important for predicting the microstructure and mechanical properties of the steel sheet during or after hot stamping. When cooling from austenitic temperature, austenite may decompose to different microstructures as ferrite, pearlite, bainite, and martensite depending on the temperature history, as shown in Fig. 8.5. Among these phase transformation, the transformation from austenite to martensite is nondiffusive transformation and the others are diffusive transformation.

8.3 The Model Building and Simulation of Temperature Field in Hot Stamping

The hot stamping of HSS is a technology with forming and quenching combined, in which, metal's temperature field and deformation field both exist inside the plate and interact and couple strongly. In the process of hot stamping, the plate metals cooled down quickly when they started to contact the mold. Then the changing of the temperature field made further changes to the plastic and flow stress of the metal. At the same time, part of the deformation energy in the plastic forming process and part of frictional work were transferred to the metal, becoming an inside heat source leading to temperature rise, by the effect of which, temperature field changed prominently. The interaction and influence between heat and force are called thermal mechanically coupling. The temperature field grows out of the heat conduction between the surface of HSS and mold. So, the design of hot stamping mold and the calculation of temperature field are a major factor in the technology of hot stamping. Building a model that includes the heat conduction between mold and plate and analyzing the temperature field by simulation not only have great significance for the performance of the formed and quenched plate metals, but also provide basis for the optimization design of cooling water ways of molds.

In this section, the author established a hot stamping model and a finite element formulation of transient temperature field based on the Galerkin method and the method of heat-force coupled, performed a 3D numerical simulation of the temperature field of HSS's hot stamping process and obtained the reasonable temperature distribution of the upper and lower surfaces of the mold and plate metals in the hot stamping process [16–18].

8.3.1 Summary of Temperature Field FEA in Hot Stamping Process

In the process of hot stamping, changes of the temperature on the steel are determined by the condition of heat transfer between the steel plate and the outside world, and heat balance. According to the characteristics of heat transfer, the heat transfer process between the steel and the outside is divided into three stages: heat transfer to the air, heat transfer to the mold and mixed heat transfer to both air and mold. The heat flow of the steel plate mainly include: heat convection and radiation happen between the steel and the air during removing the steel out of the heating furnace and putting it to the mold; When stamping begins, the steel plate's surface conduct heat to the mold, or heat convection and radiation happen between the steel and the air; When stamping finishes, the steel plate conduct heat with mold during the holding pressure time, and quenching process happens.

In this section of model derivation, we suppose that the heat conductive high-strength steel is isotropic material. When we analyze the temperature field, we simplify the plate and mold as temperature shell element. The literatures [19] put forward the finite element analysis of the overall program which is applied to axial symmetrical shell temperature field and shell temperature field. This section references literatures' formulation thought, establish the four nodes shell heat transfer analysis column type which is suitable for the program implementation, and use for the transient heat transfer analysis units.

The basic equation of the three-dimensional transient temperature field problem is the Fourier heat conduction differential equation

$$\rho c\frac{\partial T}{\partial t}=\frac{\partial}{\partial x}\left(K_x\frac{\partial T}{\partial x}\right)+\frac{\partial}{\partial y}\left(K_y\frac{\partial T}{\partial y}\right)+\frac{\partial}{\partial z}\left(K_z\frac{\partial T}{\partial z}\right)+\rho\dot{q} \tag{8.1}$$

In the above equation, T is the internal temperature of the materials, it is time variable t and coordinate (x, y, z)'s function; ρ, c are the density and specific heat of material, respectively; K_x, K_y, K_z are the coefficient of heat conduction in the directions of x, y, z respectively, For heat conduction isotropic material, $K_x = K_y = K_z$. $\dot{q}$ is the density of interior heat source of the material. The deduction of this section is only considering heat conduction isotropic material.

Equation (8.1) is a heat balance equation, Its meanings are as follows: on the left is the heat which is needed for the warming of the interior microbody of the material. The first three items on the right are the heat which is conducted from x, y, z directions to the microbody, and the last item is the heat that comes from the heat source of the interior microbody. The equation describes general rules of thermal conductivity in three-dimensional transient temperature field. Initial conditions and boundary conditions must be given when we analysis a certain issue concretely in order to deduce the unique solution. Initial condition refers to the distribution of the temperature field on the blank when $t = 0$. The initial condition is given which means when forming process starts, the temperature of every location on the blank is known. The equation is expressed as

$$\left.\begin{matrix} T = T_0 \\ T = \varphi(x, y, z) \end{matrix}\right\} t = 0 \tag{8.2}$$

In the equation, T_0 is a known constant, which means the initial temperature of object is homogeneous. $\varphi(x, y, z)$ is a known function, which means the initial temperature of object is heterogeneous.

Boundary conditions are the mathematical description that happen in the hot stamping process that the heat exchange between the blank border (i.e., outside surface) and the surrounding environment. The following kinds of common thermal boundary conditions [20] have:

1. The first kind of boundary conditions

The temperature function on the S_1 boundary is known, the equation is expressed as

$$T = \bar{T} \text{ on the } S_1 \tag{8.3}$$

In the equation, $\bar{T}$ is the surface temperature that is given (it is constant or the function of time or location).

2. The second kind of boundary conditions

The Heat flux $\dot{\boldsymbol{q}}$ on the S_2 boundary is known. The direction of $\dot{\boldsymbol{q}}$ is the same as the normal of the bounding surface's. The equation is expressed as

$$K\left(\frac{\partial T}{\partial x}n_x + \frac{\partial T}{\partial y}n_y + \frac{\partial T}{\partial z}n_z\right) = \dot{\bar{q}} \text{ on the } S_2 \tag{8.4}$$

In the equation, n_x, n_y, n_z are orientation cosines which are, respectively, in the direction of x, y, z from the outside boundary normal, $\dot{\boldsymbol{q}}$ is the heat flux which is given on the boundary S_2. We should pay attention, in Eq. (8.4), the direction of the heat flux is the direction of the boundary's surface normal, that means the heat flux outflows from object to the outside. According to the (8.4) type, when using the finite element method to calculate the second kind of boundary conditions of the original input data, all data took positive value if the heat flux $\boldsymbol{q}$ outflows from object to the outside, and a negative value if the heat flux $\boldsymbol{q}$ inflows from outside to the object.

3. The third kind of boundary conditions

In S_3 plane the heat convection on the boundary conditions is given, which means the object's temperature T_α of the contact with liquid or solid medium and heat coefficient h are known. In the process of analyzing the sheet metal's heat conduction, the main boundary conditions are heat transfer processes between mold and sheet metal. Therefore, the numerical simulation process applies the following convection boundary conditions:

$$K\left(\frac{\partial T}{\partial x}n_x + \frac{\partial T}{\partial y}n_y + \frac{\partial T}{\partial z}n_z\right) = h(T_a - T) \tag{8.5}$$

In the equation T_α is mold's temperature, h is heat convection conductivity coefficient between the mold and the sheet, which could be the constant or the function which changes along with the time and the position.

In heat or cooling process, the metal will absorb (heat process) or release (cooling process) latent heat q^{tr} when the change of organization occurs. The latent heat in solid organizational changes cannot be ignored even if it is smaller than that of the melting or solidification process. From the mathematical point of view, the governing equation became highly nonlinear because of the latent heat release, which caused some obstacles to solve the problem. In the simulation computation, three methods are used to solve latent heat problem: Equivalent heat method (Temperature rise method); Equivalent specific heat method, and Specific heat enthalpy method.

Equivalent specific heat method is used in this article which can meet the requirement of engineering simulation for hot stamping. The blank releases latent heat during phase change. Latent heat source is shown as equation [21].

$$\dot{q} = \rho L \frac{\partial \xi}{\partial t} \tag{8.6}$$

where ρ is density, L is latent heat released in the process of austenite transfer into martensite. ξ is the value of martensite transformation. Add the latent heat in temperature field governing equation.

$$\rho c_p \frac{\partial T}{\partial t} = \left(\lambda_x \frac{\partial^2 T}{\partial x^2} + \lambda_y \frac{\partial^2 T}{\partial y^2} + \lambda_z \frac{\partial^2 T}{\partial z^2} \right) + \dot{q} \tag{8.7}$$

And

$$\rho \left(c_p - L \frac{\partial \xi}{\partial T} \right) \frac{\partial T}{\partial t} = \left(\lambda_x \frac{\partial^2 T}{\partial x^2} + \lambda_y \frac{\partial^2 T}{\partial y^2} + \lambda_z \frac{\partial^2 T}{\partial z^2} \right) \tag{8.8}$$

Use the equivalent specific heat method to transform governing equation into

$$\rho c_{\text{eff}} \frac{\partial T}{\partial t} = \left(\lambda_x \frac{\partial^2 T}{\partial x^2} + \lambda_y \frac{\partial^2 T}{\partial y^2} + \lambda_z \frac{\partial^2 T}{\partial z^2} \right) \tag{8.9}$$

$$c_{\text{eff}} = c_p - L \frac{\partial \xi}{\partial T} \tag{8.10}$$

c_{eff} is equivalent specific heat (J/kg K).

In the quench process, the relationship between the amount of nondiffusive martensite transformation and temperature is given in Sect. 4.1. Partial derivatives to temperature on both sides of the equation

$$\frac{\partial \xi}{\partial T} = \theta \cdot \exp[-\theta(M_s - T)] \tag{8.11}$$

Substitute Eq. (8.11) into (8.10), we can get the equivalent specific heat in martensite transformation. Add phase change related material constant L, θ, M_s, etc. in simulation model, introduce Eq. (8.10) when the temperature drops to M_s, then replace c_p by c_{eff}. Thus, the influence of latent heat to hot stamping temperature field can be achieved.

8.3.2 Variational Equation of Temperature Field

Assuming that the approximate function of temperature field T is a $\tilde{T}$:

$$T(x, y, z, t) \approx \tilde{T}(x, y, z, t) \tag{8.12}$$

where $\tilde{T}$ must satisfy the initial conditions (8.2) and the first kind of boundary conditions (8.3). Substitute Eq. (8.12) into (8.1) and boundary conditions (8.4) and (8.5), the residual will be generated as $\tilde{T}$ cannot satisfy these equations accurately. Using the Galerkin method to assume weighted integration of the generated residual at zero, we can obtain

$$\begin{aligned}
&\int\limits_V \left[\rho c \frac{\partial \tilde{T}}{\partial t} - \frac{\partial}{\partial x}\left(K\frac{\partial \tilde{T}}{\partial x}\right) - \frac{\partial}{\partial y}\left(K\frac{\partial \tilde{T}}{\partial y}\right) - \frac{\partial}{\partial z}\left(K\frac{\partial \tilde{T}}{\partial z}\right) - \rho Q\right]\delta\tilde{T}\mathrm{d}V \\
&+ \int\limits_{S_2} \left(K\frac{\partial \tilde{T}}{\partial x}n_x + K\frac{\partial \tilde{T}}{\partial y}n_y + K\frac{\partial \tilde{T}}{\partial z}n_z - \bar{q}\right)\delta\tilde{T}\mathrm{d}S \\
&+ \int\limits_{s_3} \left[K\frac{\partial \tilde{T}}{\partial x}n_x + K\frac{\partial \tilde{T}}{\partial y}n_y + K\frac{\partial \tilde{T}}{\partial z}n_z - h\left(Ta - \tilde{T}\right)\right]\delta\tilde{T}\mathrm{d}S = 0
\end{aligned} \tag{8.13}$$

Partial integrate is used on first item of Eq. (8.13) and made some simplified operation, one has

$$\begin{aligned}
&\int\limits_V \left(\rho c \frac{\partial \tilde{T}}{\partial t} - \rho Q\right)\delta\tilde{T}\mathrm{d}V + \int\limits_V \left(K\frac{\partial \tilde{T}}{\partial x}\frac{\partial(\delta\tilde{T})}{\partial x} + K\frac{\partial \tilde{T}}{\partial y}\frac{\partial(\delta\tilde{T})}{\partial y} + K\frac{\partial \tilde{T}}{\partial z}\frac{\partial(\delta\tilde{T})}{\partial z}\right)\mathrm{d}V \\
&- \int\limits_{s_2} \bar{q}\delta\tilde{T}\mathrm{d}S - \int\limits_{S_3} h\left(Ta - \tilde{T}\right)\delta\tilde{T}\mathrm{d}S = 0
\end{aligned} \tag{8.14}$$

Equation (8.14) is the weak form of equivalent integral for weighted residual method for three-dimensional transient temperature field problem.

The derivational process of finite element equation for three-dimensional transient heat conduction problem which considering the thermo–mechanical coupling will be given below. Supposing that element volume is V; surface area is A; density is ρ; specific heat is c; temperature is T; thermal flux is q_i; the energy equation can be expressed as

$$\rho c\dot{T} = -q_{i,i} + \sigma \tag{8.15}$$

where σ is the internal heat source density.

The internal heat caused by plastic work is

$$\sigma = a\sigma'_{ij}\dot{\varepsilon}_{ij} \tag{8.16}$$

where a is the ratio of plastic work converts into heat; σ'_{ij} is the deviatoric tensor of stress; $\dot{\varepsilon}_{ij}$ is the strain rate tensor. Using Galerkin method, rewrite Eq. (8.15) into weak form:

$$\int\limits_V \delta T\left(-q_{i,i} + \sigma - \rho c\dot{T}\right)\mathrm{d}V = 0 \tag{8.17}$$

Partial integration is used on first item of Eq. (8.17), one has

$$\int\limits_V \left(-\delta Tq_i\right)_{,i}\mathrm{d}V + \int\limits_V \delta T_{,i}q_i\mathrm{d}V + \int\limits_V \delta T\left(\sigma - \rho c\dot{T}\right)\mathrm{d}V = 0 \tag{8.18}$$

According to the divergence theorem, the first item of Eq. (8.18) can be rewritten as

$$-\int\limits_{\partial V} \delta Tq_in_i\mathrm{d}A + \int\limits_V \delta T_{,i}q_i\mathrm{d}V + \int\limits_V \delta T\left(\sigma - \rho c\dot{T}\right)\mathrm{d}V = 0 \tag{8.19}$$

where ∂V is the whole outer boundary of the object; n_i is the normal vector of outer boundary.

According to Fourier Law

$$q_i = -K_{ij}T_{,j} \tag{8.20}$$

with

$$K_{ij} = K\delta_{ij} \tag{8.21}$$

where K is the heat conductivity coefficient. Substitute Eqs. (8.20) and (8.21) into (8.19), one has

$$\int_{\partial V} \delta T \bar{q} \mathrm{d}A - \int_{V} K \delta T_{,i} T_{,i} \mathrm{d}V + \int_{V} \delta T \left(\sigma - \rho c \dot{T}\right) \mathrm{d}V = 0 \tag{8.22}$$

where $\bar{q}$ is the heat flux from object's surface to interior and

$$\bar{q} = -q_i n_i = K T_{,i} n_i \tag{8.23}$$

Supposing that the heat flux produced by the objects in contact is $\bar{q}_1$; the heat flux produced by thermal radiation is $\bar{q}_2$; the heat flux produced by friction is $\bar{q}_3$, one has

$$\bar{q}_1 = H_1 (T_{1w} - T) + H_2 (T_{2w} - T) + \cdots \tag{8.24}$$

$$\bar{q}_2 = \varepsilon_s \sigma_B \left(T_{\infty}^4 - T^4\right) \tag{8.25}$$

$$\bar{q}_3 = \mu \sigma_f v_f \tag{8.26}$$

where H_i is the heat transfer coefficient, T_{iw} is the temperature of the solid or fluid contacted with the studied object, T_{∞} is the temperature of the heat radiation source, T is the temperature of the studied object, ε_s is the radiance, σ_B is the Boltzmann constant, σ_f is the friction stress, v_f is the relative sliding speed; μ is the convert ratio of friction work to heat.

Then the finite element discretization will be given. Supposing the temperature of any point of the studied element is T, one has

$$T = \sum_{i=1}^{n_e} N_i(x, y, z) \theta_i(t) = \mathbf{N} \boldsymbol{\theta}^e \tag{8.27}$$

with

$$\mathbf{N} = \{N_1, N_2, \ldots, N_{n_e}\} \tag{8.28}$$

$$\boldsymbol{\theta}^e = \{\theta_1, \theta_2, \ldots, \theta_{n_e}\}^T \tag{8.29}$$

where n_e is the number of element node, N_i is the shape function, θ_i is the temperature of element node. According to Eq. (8.27), one has:

$$\{T_j\} = \begin{bmatrix} \sum_{i=1}^{n_e} \frac{\partial N_i}{\partial x}\theta_i \\ \sum_{i=1}^{n_e} \frac{\partial N_i}{\partial y}\theta_i \\ \sum_{i=1}^{n_e} \frac{\partial N_i}{\partial z}\theta_i \end{bmatrix} = \mathbf{B}\boldsymbol{\theta}^e \quad (j = 1, 2, 3) \tag{8.30}$$

$$\dot{T} = \mathbf{N}\dot{\boldsymbol{\theta}}^e \tag{8.31}$$

Substitution of Eqs. (8.28), (8.30), and (8.31) into (8.22) gives the following expression:

$$(\delta\boldsymbol{\theta}^e)^T\left(\int_{\partial V^e} \mathbf{N}^T\bar{q}\mathrm{d}A - \int_{V^e} K\mathbf{B}^T\mathbf{B}\mathrm{d}V\boldsymbol{\theta}^e + \int_{V^e} \mathbf{N}^T\sigma\mathrm{d}V - \int_{V^e} \rho c\mathbf{N}^T\mathbf{N}\mathrm{d}V\dot{\boldsymbol{\theta}}^e\right) = 0 \tag{8.32}$$

because of the randomicity of $(\delta\boldsymbol{\theta}^e)^T$, one has:

$$\int_{\partial V^e} \mathbf{N}^T\bar{q}\mathrm{d}A - \int_{V^e} K\mathbf{B}^T\mathbf{B}\mathrm{d}V\boldsymbol{\theta}^e + \int_{V^e} \mathbf{N}^T\sigma\mathrm{d}V - \int_{V^e} \rho c\mathbf{N}^T\mathbf{N}\mathrm{d}V\dot{\boldsymbol{\theta}}^e = 0 \tag{8.33}$$

Rewriting Eq. (8.33), the finite element formulation of the transient temperature field for each element can be written as

$$\mathbf{C}^e\dot{\boldsymbol{\theta}}^e + \mathbf{K}^e\boldsymbol{\theta}^e = \mathbf{F}^e \tag{8.34}$$

where $\mathbf{C}^e$ is the matrix of element heat capacity, $\mathbf{K}^e$ is matrix of element heat conductivity, $\dot{\boldsymbol{\theta}}^e$ and $\boldsymbol{\theta}^e$ are the time derivative tensor of node temperature and the tensor of node temperature, respectively, $\mathbf{F}^e$ is the temperature load tensor of element node. The expressions are given by

$$C_{ij}^e = \int_{V^e} \rho c N_i N_j \mathrm{d}V \tag{8.35}$$

$$K_{ij}^e = \int_{V^e} K\left(\frac{\partial N_i}{\partial x}\frac{\partial N_j}{\partial x} + \frac{\partial N_i}{\partial y}\frac{\partial N_j}{\partial y} + \frac{\partial N_i}{\partial z}\frac{\partial N_j}{\partial z}\right)\mathrm{d}V \tag{8.36}$$

$$F_i^e = \int_{V^e} N_i\sigma\mathrm{d}V + \int_{\partial V^e} N_i\bar{q}\mathrm{d}A \tag{8.37}$$

The finite element formulation of the transient temperature field for the whole object can be obtained by assembling Eq. (8.34) for each element

$$C\dot{\boldsymbol{\theta}} + \mathbf{K}\boldsymbol{\theta} = \mathbf{F} \tag{8.38}$$

8.3.3 The Basic Equation of Temperature Shell Elements

This temperature shell elements adopt $\xi_1\xi_2\xi_3$ curvilinear coordinates, with ξ_1, ξ_2 in the neutral plane, and ξ_3 perpendicular to the neutral plane. On the top surface $\xi_3 = -\frac{1}{2}$ and on the bottom surface $\xi_3 = \frac{1}{2}$. The coordinates x_{10}, x_{20}, x_{30} of the neutral plane in Cartesian coordinate can be expressed as the function of ξ_1 and ξ_2.

$$x_i = x_{i0}(\xi_1, \xi_2) \quad (i = 1, 2, 3) \tag{8.39}$$

The Cartesian coordinate x_1, x_2, x_3 of arbitrary point in the element can be expressed as

$$x_i = x_{i0}(\xi_1, \xi_2) + l_{3i}\bar{t}\xi_3 \quad (i = 1, 2, 3) \tag{8.40}$$

where l_{3i} is normal line, ξ_3 is cosine of three directions in Cartesian coordinate, $\bar{t}$ is the thickness of the element. Seen by the differential geometry

$$l_{3i} = \frac{1}{A_{30}} e_{ijk} \frac{\partial x_{j0}}{\partial \xi_1} \frac{\partial x_{k0}}{\partial \xi_2} \tag{8.41}$$

$$A_{30} = \left(e_{ijk} \frac{\partial x_{j0}}{\partial \xi_1} \frac{\partial x_{k0}}{\partial \xi_2} e_{imn} \frac{\partial x_{m0}}{\partial \xi_1} \frac{\partial x_{n0}}{\partial \xi_2} \right)^{\frac{1}{2}} \tag{8.42}$$

where e_{ijk} is permutation symbol.

The differential relationship between tow coordinates can be deduced by following equation:

$$\frac{\partial}{\partial \xi_i} = \frac{\partial x_j}{\partial \xi_i} \frac{\partial}{\partial x_j} \tag{8.43}$$

where

$$\frac{\partial x_j}{\partial \xi_i} = J_{ij} = \overset{0}{J_{ij}} + \overset{1}{J_{ij}} \xi_3 \tag{8.44}$$

$$\overset{0}{J_{ij}} = \begin{cases} \frac{\partial x_{j0}}{\partial \xi_i} & (i = 1, 2) \\ l_{3j}\bar{t} & (i = 3) \end{cases} \tag{8.45}$$

$$\overset{1}{J}_{ij} = \frac{\partial l_{3j}}{\partial \xi_i}\bar{t} \tag{8.46}$$

$\frac{\partial}{\partial x_i}$ can be expressed as

$$\frac{\partial}{\partial x_j} = J_{ij}^{-1}\frac{\partial}{\partial \xi_j} \tag{8.47}$$

The volume element in coordinate ξ_1, ξ_2, ξ_3 can be expressed as

$$dV = |J_{ij}|d\xi_1 d\xi_2 d\xi_3 \tag{8.48}$$

where $|J_{ij}|$ is Jacobian determinant, which can be calculated by

$$|J_{ij}| = G_0 + G_1\xi_3 + G_2\xi_3^2 \tag{8.49}$$

And

$$\begin{aligned} G_0 &= |\overset{0}{J}_{ij}| = e_{ijk}\overset{0}{J}_{i1}\overset{0}{J}_{j2}\overset{0}{J}_{k3}, \quad G_1 = e_{ijk}\overset{1}{J}_{i1}\overset{0}{J}_{j2}\overset{0}{J}_{k3} + e_{ijk}\overset{0}{J}_{i1}\overset{1}{J}_{j2}\overset{0}{J}_{k3} \\ G_2 &= e_{ijk}\overset{1}{J}_{i1}\overset{1}{J}_{j2}\overset{0}{J}_{k3} \end{aligned} \tag{8.50}$$

The area differential element in coordinate ξ_1, ξ_2, ξ_3 can be calculated by

$$dS = \begin{cases} A_1 d\xi_2 d\xi_3 & \xi_1 = \text{constant} \\ A_2 d\xi_1 d\xi_3 & \xi_2 = \text{constant} \\ A_3 d\xi_1 d\xi_2 & \xi_3 = \text{constant} \end{cases} \tag{8.51}$$

where

$$A_1 = \bar{t}\left(\frac{\partial x_i}{\partial \xi_2}\frac{\partial x_i}{\partial \xi_2}\right)^{\frac{1}{2}} \tag{8.52}$$

$$A_2 = \bar{t}\left(\frac{\partial x_i}{\partial \xi_1}\frac{\partial x_i}{\partial \xi_1}\right)^{\frac{1}{2}} \tag{8.53}$$

$$A_3 = e_{ijk}\frac{\partial x_j}{\partial \xi_1}\frac{\partial x_k}{\partial \xi_2}e_{imn}\frac{\partial x_m}{\partial \xi_1}\frac{\partial x_n}{\partial \xi_2} \tag{8.54}$$

Because ξ_3 is the normal direction of neutral plane, $\overset{0}{J}_{ij}, \overset{1}{J}_{ij}, J_{ij}, J_{ij}^{-1}$, etc., have the following property

$$\overset{0}{J}_{ji}\overset{0}{J}_{3i} = \delta_{3j}\bar{t}^2, \quad \overset{1}{J}_{ji}\overset{0}{J}_{3i} = 0, \quad J_{ji}J_{3i} = \delta_{3j}\bar{t}^2, \quad J_{ij}^{-1}J_{i3}^{-1} = \delta_{3j}\frac{1}{\bar{t}^2} \tag{8.55}$$

where δ is symbol Kronecker.

Use curvilinear coordinate transformation Eq. (8.47). The field Eq. (8.1) can be expressed as

$$J_{ij}^{-1}\frac{\partial}{\partial \xi_j}\left(KJ_{ik}^{-1}\frac{\partial T}{\partial \xi_k}\right) + \rho\dot{q} = \rho c\frac{\partial T}{\partial t} \text{ in } V^{\varepsilon} \tag{8.56}$$

For the element surface $\xi_3 = \pm\frac{1}{2}$, boundary condition Eqs. (8.4)–(8.6) can be rewritten as

$$T = \bar{T} \text{ on } S_1 \tag{8.57}$$

$$\pm\frac{K}{\bar{t}}\frac{\partial T}{\partial \xi_3} = \dot{\bar{q}} \text{ on } S_2 \tag{8.58}$$

$$\pm\frac{K}{\bar{t}}\frac{\partial T}{\partial \xi_3} = h(T_a - T) \text{ on } S_3 \tag{8.59}$$

For the last two equations above, "+" used on surface $\xi_3 = \frac{1}{2}$, "−" used on surface $\xi_3 = -\frac{1}{2}$. Initial condition is

$$T = T_0 \text{ when } t = t_0 \tag{8.60}$$

Introduce Eqs. (8.47) and (8.48) into the weak form of equivalent integral for weighted residual method in three-dimensional shell transient temperature field problem, where

$$\begin{aligned}
&\int_V K\frac{\partial \tilde{T}}{\partial x_i}\frac{\partial \delta\tilde{T}}{\partial x_i}\mathrm{d}V \\
&= \iiint_V {}_{-1/2}^{1/2} KJ_{ij}^{-1}J_{ik}^{-1}\frac{\partial \tilde{T}}{\partial \xi_j}\frac{\partial \delta\tilde{T}}{\partial \xi_k}|J_{pq}|d\xi_3 d\xi_1 d\xi_2 \\
&= \iiint_V {}_{-1/2}^{1/2} K\left\{\left[J_{i1}^{-1}J_{i1}^{-1}\frac{\partial \tilde{T}}{\partial \xi_1} + J_{i1}^{-1}J_{i2}^{-1}\frac{\partial \tilde{T}}{\partial \xi_2}\right]\frac{\partial \delta\tilde{T}}{\partial \xi_1}\right. \\
&+ \left[J_{i1}^{-1}J_{i2}^{-1}\frac{\partial \tilde{T}}{\partial \xi_1} + J_{i2}^{-1}J_{i2}^{-1}\frac{\partial \tilde{T}}{\partial \xi_2}\right]\frac{\partial \delta\tilde{T}}{\partial \xi_2} \\
&\left.+\frac{1}{\bar{t}^2}\frac{\partial \tilde{T}}{\partial \xi_3}\frac{\partial \delta\tilde{T}}{\partial \xi_3}\right\}|J_{pq}|d\xi_3 d\xi_1 d\xi_2
\end{aligned} \tag{8.61}$$

Equation (8.55) is used in the deduction above. Partial integrate is used on last item of Eq. (8.61) and one has

$$
\begin{aligned}
&= \iiint\limits_V {}_{-1/2}^{1/2} K \frac{1}{\bar{t}^2} \frac{\partial \tilde{T}}{\partial \xi_3} \frac{\partial \delta \tilde{T}}{\partial \xi_3} |J_{pq}| d\xi_3 d\xi_1 d\xi_2 \\
&= \iiint\limits_V {}_{-1/2}^{1/2} K \frac{1}{\bar{t}^2} \left\{ \frac{\partial}{\partial \xi_3} \left(\frac{\partial \tilde{T}}{\partial \xi_3} |J_{pq}| \delta \tilde{T} \right) - \frac{\partial^2 \tilde{T}}{\partial \xi_3^2} |J_{pq}| \delta \tilde{T} - \frac{\partial \tilde{T}}{\partial \xi_3} \frac{\partial |J_{pq}|}{\partial \xi_3} \delta \tilde{T} \right\} d\xi_3 d\xi_1 d\xi_2
\end{aligned}
\tag{8.62}
$$

Using Eq. (8.49) to simplify the first item of Eq. (8.62)

$$
\begin{aligned}
&\iiint\limits_V {}_{-1/2}^{1/2} \frac{K}{\bar{t}^2} \frac{\partial}{\partial \xi_3} \left(\frac{\partial \tilde{T}}{\partial \xi_3} |J_{pq}| \delta \tilde{T} \right) d\xi_3 d\xi_1 d\xi_2 \\
&= \iint\limits_S |\xi_3 = \frac{1}{2} A_3 \frac{K}{\bar{t}} \frac{\partial \tilde{T}}{\partial \xi_3} \delta \tilde{T} d\xi_1 d\xi_2 - \iint\limits_S |\xi_3 = -\frac{1}{2} A_3 \frac{K}{\bar{t}} \frac{\partial \tilde{T}}{\partial \xi_3} \delta \tilde{T} d\xi_1 d\xi_2
\end{aligned}
\tag{8.63}
$$

where $S|\xi_3 = \frac{1}{2}$ and $S|\xi_3 = -\frac{1}{2}$ represent the shell surface of $\xi_3 = \frac{1}{2}$ and $\xi_3 = -\frac{1}{2}$, respectively. The following equation is used to deduce the equation above.

$$
|J_{pq}| = \bar{t} A_3 \quad \text{if} \quad \xi_3 = \pm \frac{1}{2} \tag{8.64}
$$

Substitute Eqs. (8.62) and (8.63) into Eq. (8.61), considering the fact that the trial function of T should fulfill the boundary condition not only on S_1, but also on $\xi_3 = \pm\frac{1}{2}$ which belongs to S_2 and S_3. And notice Eq. (8.64), one gets

$$
\begin{aligned}
&\iiint\limits_V {}_{-1/2}^{1/2} \left(\rho c \frac{\partial \tilde{T}}{\partial t} - \rho \dot{q} \right) \delta \tilde{T} |J_{pq}| d\xi_3 d\xi_1 d\xi_2 \\
&\quad + \iiint\limits_V {}_{-1/2}^{1/2} K \left\{ \left[J_{i1}^{-1} J_{i1}^{-1} \frac{\partial \tilde{T}}{\partial \xi_1} + J_{i1}^{-1} J_{i2}^{-1} \frac{\partial \tilde{T}}{\partial \xi_2} \right] \frac{\partial \delta \tilde{T}}{\partial \xi_1} \right. \\
&\quad + \left[J_{i1}^{-1} J_{i2}^{-1} \frac{\partial \tilde{T}}{\partial \xi_1} + J_{i2}^{-1} J_{i2}^{-1} \frac{\partial \tilde{T}}{\partial \xi_2} \right] \frac{\partial \delta \tilde{T}}{\partial \xi_2} \\
&\quad \left. - \frac{1}{\bar{t}^2} \frac{\partial^2 \tilde{T}}{\partial \xi_3^2} \delta \tilde{T} - \frac{1}{\bar{t}^2} \frac{\partial \tilde{T}}{\partial \xi_3} \frac{\partial |J_{mn}|}{\partial \xi_3} \delta \tilde{T} \frac{1}{|J_{ij}|} \right\} |J_{pq}| d\xi_3 d\xi_1 d\xi_2 \\
&\quad - \iint\limits_{S_2} {}_{-1/2}^{1/2} \bar{q} \delta \tilde{T} A_c d\xi_3 dC - \iint\limits_{S_3} {}_{-1/2}^{1/2} h(T_a - \tilde{T}) \delta \tilde{T} A_c d\xi_3 dC \quad = 0
\end{aligned}
\tag{8.65}
$$

where C is the shell boundary surrounded by normal coordinate line ξ_3. If C belongs to the boundary where ξ_1 is constant, then $A_c = A_1$, $dC = d\xi_2$; If C belongs to the boundary where ξ_2 is constant, then $A_c = A_2$, $dC = d\xi_1$.

Assume that the inner element temperature field changes in second-order along ξ_3 direction.

$$T(\xi_1, \xi_2, \xi_3, t) = T_0(\xi_1, \xi_2, t) + T_1(\xi_1, \xi_2, t)\xi_3 + T_2(\xi_1, \xi_2, t)\xi_3^2 \tag{8.66}$$

For the above equation, $T_0(\xi_1, \xi_2, t)$ represents neutral surface temperature, $T_1(\xi_1, \xi_2, t)$ and $T_2(\xi_1, \xi_2, t)$ can be determined by the boundary condition on $\xi_3 = \pm\frac{1}{2}$. For general purpose, assuming there is convection boundary condition on $\xi_3 = \pm\frac{1}{2}$.

$$\frac{K}{\bar{t}}\frac{\partial T}{\partial \xi_3} = h_1(T_{a1} - T) \quad \xi_3 = \frac{1}{2} \tag{8.67}$$

$$-\frac{K}{\bar{t}}\frac{\partial T}{\partial \xi_3} = h_2(T_{a2} - T) \quad \xi_3 = -\frac{1}{2} \tag{8.68}$$

T_1 and T_2 can be determined by Eqs. (8.66)–(8.68). Then Eq. (8.66) becomes

$$T(\xi_1, \xi_2, \xi_3, t) = \left(1 + C_1\xi_3 + C_2\xi_3^2\right)T_0(\xi_1, \xi_2, t) + C_3(\xi_1, \xi_2, t)\xi_3 + C_4(\xi_1, \xi_2, t)\xi_3^2 \tag{8.69}$$

where

$$C_1 = (\bar{h}_2 - \bar{h}_1)/\Delta \tag{8.70}$$

$$C_2 = -(\bar{h}_1 + \bar{h}_2 + \bar{h}_1\bar{h}_2)/\Delta \tag{8.71}$$

$$C_3(\xi_1, \xi_2, t) = \left(1 + \frac{1}{4}\bar{h}_2\right)\bar{h}_1 T_{a1}(\xi_1, \xi_2, t)/\Delta - \left(1 + \frac{1}{4}\bar{h}_1\right)\bar{h}_2 T_{a2}(\xi_1, \xi_2, t)/\Delta \tag{8.72}$$

$$C_4(\xi_1, \xi_2, t) = \left(1 + \frac{1}{2}\bar{h}_2\right)\bar{h}_1 T_{a1}(\xi_1, \xi_2, t)/\Delta + \left(1 + \frac{1}{2}\bar{h}_1\right)\bar{h}_2 T_{a2}(\xi_1, \xi_2, t)/\Delta \tag{8.73}$$

$$\Delta = 2 + \frac{3}{4}\bar{h}_1 + \frac{3}{4}\bar{h}_2 + \frac{1}{4}\bar{h}_1\bar{h}_2 \tag{8.74}$$

$$\bar{h}_1 = \frac{\bar{t}h_1}{K} \quad \bar{h}_2 = \frac{\bar{t}h_2}{K} \tag{8.75}$$

Even though Eq. (8.69) is obtained under convention boundary condition, it actually includes given temperature and given flux boundary conditions. What can be seen from Eqs. (8.67) and (8.68) is if flux boundary condition is given on $\xi_3 = \frac{1}{2}$, just assuming $\bar{h}_1 = 0$.

And

$$\bar{h}_1 T_{a1}(\xi_1, \xi_2, t) = \frac{\bar{t}}{K} \bar{q}(\xi_1, \xi_2, t) \tag{8.76}$$

If temperature is given on the surface of $\xi_3 = \frac{1}{2}$, just assume $\bar{h}_1 = H$ (H is arbitrary but $H \gg 1$) in equations above. At the same time, $T_{a1}(\xi_1, \xi_2, t) = \bar{T}(\xi_1, \xi_2, t)$. For surface $\xi_3 = -\frac{1}{2}$, one can use the same method.

Use interpolating function to express undetermined function $T_0(\xi_1, \xi_2, t)$ in Eq. (8.64)

$$T(\xi_1, \xi_2, \xi_3, t) = \left(1 + C_1\xi_3 + C_2\xi_3^2\right) \sum_{i=1}^{n_e} N_i(\xi_1, \xi_2) T_i(t) + C_3(\xi_1, \xi_2, t)\xi_3 + C_4(\xi_1, \xi_2, t)\xi_3^2 \tag{8.77}$$

where $N_i(\xi_1, \xi_2)$ is two-dimensional interpolating function. $T_i(t)$ is the temperature on element middle surface node, n_e is node amount. On adding Eq. (8.69) into Eq. (8.65) one can get shell temperature element finite element formulation

$$\boldsymbol{C}\dot{\boldsymbol{T}} + \boldsymbol{K}\boldsymbol{T} = \boldsymbol{F} \tag{8.78}$$

where

$$\mathbf{C} = \sum_e \mathbf{C}^e \quad \mathbf{K} = \sum_e \mathbf{K}^e \quad \mathbf{F} = \sum_e \mathbf{F}^e \tag{8.79}$$

Corresponding elements are

$$C_{ij}^e = \iiint_e {}_{-1/2}^{1/2} \rho c D_1^2 N_i N_j |J_{pq}| d\xi_3 d\xi_1 d\xi_2 \tag{8.80}$$

$$\begin{aligned} K_{ij}^e = \iiint_e {}_{-1/2}^{1/2} K \Bigg\{ & D_1^2 \left[E_1 \frac{\partial N_i}{\partial \xi_1}\frac{\partial N_j}{\partial \xi_1} + E_2 \frac{\partial N_i}{\partial \xi_2}\frac{\partial N_j}{\partial \xi_2} + E_3 \left(\frac{\partial N_i}{\partial \xi_1}\frac{\partial N_j}{\partial \xi_2} + \frac{\partial N_i}{\partial \xi_2}\frac{\partial N_j}{\partial \xi_1} \right) \right] \\ & - \frac{D_1}{\bar{t}^2} \left[\frac{\partial}{\partial \xi_3}(|J_{mn}|) \frac{1}{|J_{\alpha\beta}|} \frac{\partial D_1}{\partial \xi_3} + \frac{\partial^2 D_1}{\partial \xi_3^2} \right] N_i N_j \Bigg\} |J_{pq}| d\xi_3 d\xi_1 d\xi_2 \\ & + \iint_{S_3^e} {}_{-1/2}^{1/2} h D_1^2 N_i N_j A_c d\xi_3 dC \end{aligned} \tag{8.81}$$

$$F_i^e = -\iiint_e \int_{-1/2}^{1/2}\Bigg\{KD_1\left[E_1\frac{\partial D_2}{\partial\xi_1}\frac{\partial N_i}{\partial\xi_1} + E_2\frac{\partial D_2}{\partial\xi_2}\frac{\partial N_i}{\partial\xi_2} + E_3\left(\frac{\partial D_2}{\partial\xi_1}\frac{\partial N_i}{\partial\xi_2} + \frac{\partial D_2}{\partial\xi_2}\frac{\partial N_i}{\partial\xi_1}\right)\right]$$
$$-\frac{1}{t^2}\left[\frac{\partial|J_{mn}|}{\partial\xi_3}\frac{1}{|J_{\alpha\beta}|}\frac{\partial D_2}{\partial\xi_3} + \frac{\partial^2 D_2}{\partial\xi_3^2}\right] - \rho\dot{q}D_1N_i\Bigg\}|J_{pq}|d\xi_3 d\xi_1 d\xi_2$$
$$+\iint_{S_3^e}\int_{-1/2}^{1/2}\bar{q}D_1N_iA_c d\xi_3 dC + \iint_{S_3^e}\int_{-1/2}^{1/2}h(T_a - D_2)D_1N_iA_c d\xi_3 dC \tag{8.82}$$

where

$$\begin{aligned} &E_1 = J_{i1}^{-1}J_{i1}^{-1} \quad E_2 = J_{i2}^{-1}J_{i2}^{-1} \quad E_3 = J_{i1}^{-1}J_{i2}^{-1} \\ &D_1 = 1 + C_1\xi_3 + C_2\xi_3^2 \quad D_2 = C_3(\xi_1,\xi_2,t)\xi_3 + C_4(\xi_1,\xi_2,t)\xi_3^2 \end{aligned} \tag{8.83}$$

After the middle surface node temperature T_i is obtained by Eq. (8.78), the inner element temperature $T(\xi_1,\xi_2,\xi_3,t)$ is obtained by substituting T_i back into Eq. (8.77). And the average temperature and temperature difference in thickness direction can be obtained by two following equations:

$$\begin{aligned} T_{av} &= \int_{-\frac{1}{2}}^{\frac{1}{2}} T(\xi_1,\xi_2,\xi_3,t)d\xi_3 = \left(1 + \frac{C_2}{12}\right)T_0(\xi_1,\xi_2,t) + \frac{C_4(\xi_1,\xi_2,t)}{12} \\ \Delta T &= C_1T_0(\xi_1,\xi_2,t) + C_3(\xi_1,\xi_2,t) \end{aligned} \tag{8.84}$$

where T_{av} is the average temperature in thickness direction, ΔT is temperature difference in thickness direction.

It should be noted that the inner element temperature change in thickness direction (8.66) can be assumed as linear. Typically one can meet the accuracy requirements by linear change assumption. If there is a large gradient change in thickness direction, taking the inner heat producer as an example, then second-order change should be adopted. Because there is internal heat source in hot stamping process (i.e., plastic work), second-order change is applied in this act.

8.3.4 Discreteness of Space Domain and Time Domain in Shell Transient Temperature Field

The finite element equation in transient temperature field not only involves the discreteness of space domain but also time domain. But the space domain and time domain are not coupled. So the two domains can be discreted, respectively.

In this section, four-node quadrilateral shell element (DKT or BT shell element) is adapted to make deformation analysis for hot stamping blank. Use the same mesh (i.e., four-node quadrilateral element) which is used to analyze deformation to carry out temperature field analysis.

For first-order time derivatives equations, step-difference algorithm is usually used. That is for every time step starting from initial temperature field to calculate the end time temperature from the initial time temperature. This method is a single-step method. In this method, two loop formulae are commonly used. This book uses a multi-step method—Houbolt difference method. The temperature value at t, $t-\Delta t$ and $t-2\Delta t$, should be used to obtain the temperature at $t+\Delta t$. Generally speaking, multi-step method is more accurate than single-step method. The formula of this difference method is

$$\dot{T}_{t+\Delta t}=\frac{1}{6\Delta t}(11T_{t+\Delta t}-18T_t+9T_{t-\Delta t}-2T_{t-2\Delta t}) \tag{8.85}$$

where $\dot{T}_{t+\Delta t}$ is the derivative of T at $t+\Delta t$. $T_{t+\Delta t}$, T_t, $T_{t-\Delta t}$ and $T_{t-2\Delta t}$ are the corresponding temperature at $t+\Delta t$, t, $t-\Delta t$ and $t-2\Delta t$.

Substitute Eq. (8.85) for transient temperature field finite element equation $\boldsymbol{C}\dot{\boldsymbol{T}}+\boldsymbol{K}\boldsymbol{T}=\boldsymbol{F}$ one obtains

$$\left(\frac{11}{6\Delta t}\mathbf{C}^e+\mathbf{K}^e\right)\mathbf{T}_{t+\Delta t}=\mathbf{F}^e+\frac{1}{\Delta t}\mathbf{C}^e\left(3\mathbf{T}_t-1.5\mathbf{T}_{t-\Delta t}+\frac{1}{3}\mathbf{T}_{t-2\Delta t}\right) \tag{8.86}$$

Equation (8.86) is the finite element equation used in reality temperature field analysis. $\boldsymbol{T}_{t-\Delta t}$ and $\boldsymbol{T}_{t-2\Delta t}$ are initial values at $t=0$. Because T at $t+\Delta t$ is used in Eq. (8.86) when the derivative of T divided for time at $t+\Delta t$ is calculated, this difference method is an implicit algorithm because it is unconditionally stable.

8.4 The Modeling and Simulation of Phase Field in Hot Stamping

8.4.1 Summary of Phase Field

The microstructure such as the volume fraction of each phase and the mechanical property of the final part are determined by the phase transformation that austenite may decompose to martensite, ferrite, pearlite, and bainite during the stage of quenching after stamping forming. To obtain optimized technological parameter and mold design, which will be used to realize ideal distribution of product mechanical property, it is necessary to further study the law of microstructure evolution during hot stamping [22, 23].

With the wide and deepening application of computer simulation technology in the field of materials science, a fair amount of the microstructure control work is gradually transformed from the traditional empirical research to computer simulation technology to realize the prediction of microstructure evolution. The model of microstructure evolution established by computer numerical simulation can predict the grain size, grain shape, grain distribution etc, which will be used to optimize process parameters and schemes, and make the formulation of a forming process more scientific. Then the material development method, which depends on the accumulation of experience and trial and error method, can be replaced by the computer numerical simulation, which will realize the prediction of structure property and make the experiments to be carried out within the valid range, shorten the test cycle, reduce development costs.

8.4.2 The Basic Equation of Phase Field

The predicted model of microstructure evolution is used to simulate the continuous cooling transformation of austenite during the quenching process after hot stamping forming. Many dynamic models of phase transformation have been established to investigate the microstructure evolution of hot steel during cooling process. The generally used models, which can predict the diffusive transformation (austenite to ferrite, pearlite and bainite), include the Avrami model, Kirkaldy, and Venugopalan model et al.

The model proposed by Kirkaldy and Venugopalan [21] has the form

$$\frac{dX}{dt} = f(G)f(T)f(C)f(X) \tag{8.87}$$

The left side of Eq. (8.86) is phase rate, $f(G)$ represents the effect of the austenite grain size, $f(C)$ is the effect of alloy composition, $f(X)$ is the effect of current fraction formed. To improve the prediction accuracy of K–V model under the condition of continuous cooling and extend the predicted scope of K–V model for more alloying elements, Li et al. [24] modified the K–V model as following:

When $B_s < T < A_{e3}$, austenite decomposes to ferrite

$$\begin{cases} f_F(G) = 2^{0.41G} \\ f_F(T) = (A_{e3} - T)^3 \exp(-27,500/RT) \\ f_F(C) = [\exp(1.00 + 6.31\text{C} + 1.78\text{Mn} + 0.31\text{Si} + 1.12\text{Ni} + 2.70\text{Cr} + 4.06\text{Mo})]^{-1} \end{cases} \tag{8.88}$$

When $B_s < T < A_{e1}$, austenite decomposes to pearlite

$$\begin{cases} f_P(G) = 2^{0.32G} \\ f_P(T) = (A_{e1} - T)^3 \exp(-27,500/RT) \\ f_P(C) = [\exp(-4.25 + 4.12\text{C} + 4.36\text{Mn} + 0.44Si + 1.71\text{Ni} + 3.33\text{Cr} + 5.19\sqrt{\text{Mo}})]^{-1} \end{cases} \tag{8.89}$$

When $M_s < T < B_s$, austenite decomposes to bainite

$$\begin{cases} f_B(G) = 2^{0.29G} \\ f_B(T) = (B_s - T)^2 \exp(-27,500/RT) \\ f_B(C) = [\exp(-10.23 + 10.18\text{C} + 0.85\text{Mn} + 0.55\text{Ni} + 0.90\text{Cr} + 0.36\text{Mo})]^{-1} \end{cases} \tag{8.90}$$

where G is the ASTM grain size which will be kept constant during cooling process, A_{e3} and A_{e1} are phase balance temperature, B_s is onset temperature of bainite transformation, R is universal gas constant, Q is the activation energy of the

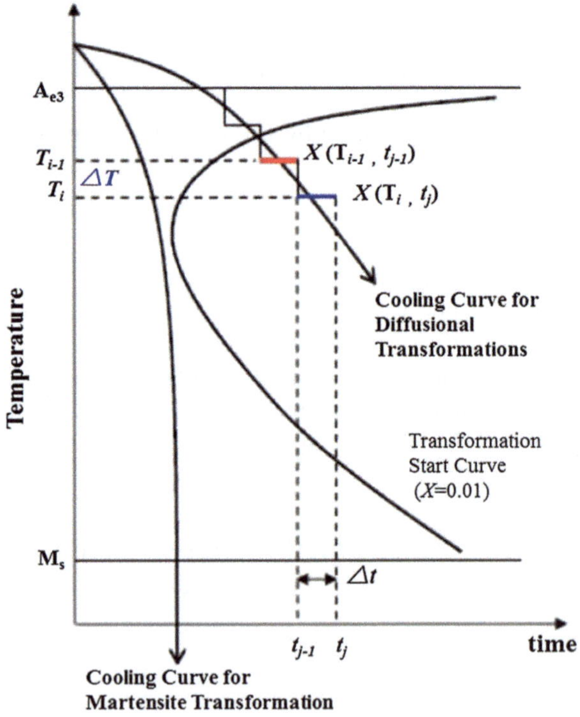

Fig. 8.6 Schematic for the application of additivity rule

diffusional reaction (−27,500 J/mol/K), T is the temperature. C, Mn, Si are the mass fraction of alloying elements.

When $T < M_s$, austenite decomposes to martensite. This nondiffusive transformation only depends on the temperature and has nothing to do with the time. It can be described by Koistinen-Marburger equation [25], a modified type can be found in Chap. 4.

$$X_M = X_A(1 - e^{-\alpha(M_s - T)}) \tag{8.91}$$

where X_M is the martensite quantity, X_A is the volume fraction of residual austenite, $\alpha = 0.011$ is the constant of martensite transformation rate, M_s is the onset temperature of martensite transformation, T is temperature.

The hardness is calculated by use of the mixtures rule

$$Hv = (X_F + X_P)Hv_{F+P} + X_B Hv_B + X_M Hv_M \tag{8.92}$$

where Hv is the hardness of multiphase mixture. And Hv_{F+P}, Hv_B and Hv_M are the empirical hardness value of the mixture of ferrite and pearlite, bainite, and martensite, respectively. The detailed equations are shown in literature [26].

8.4.3 The Solving Method of Phase Field

For diffusive transformation, according to Scheil rule as shown in Fig. 8.6, the continuous cooling can be converted into step cooling and each step can be treated as constant temperature. Then, the isothermal kinetics model mentioned above can be used to calculate the transformation quantity of each phase

$$X(T_i, t_j) = X(T_i, t_{j-1}) + \Delta X \tag{8.93}$$

where $X(T_i, t_j)$ is the volume fraction at time t_j and temperature T_i, $X(T_i, t_{j-1})$ is the volume fraction at time t_{j-1} and temperature T_i; ΔX is the transformation quantity at a time period Δt and temperature T_i.

The dynamic models of phase transformation can be solved by classical Runge–Kutta method. The cooling curve of hot blank can be fitted by polynomial, then set it as the initial data together with the volume fraction of alloying element. The initial temperature of the input temperature curve must be higher than phase balance temperature A_{e3}. The predicted procedure is shown in Fig. 8.7.

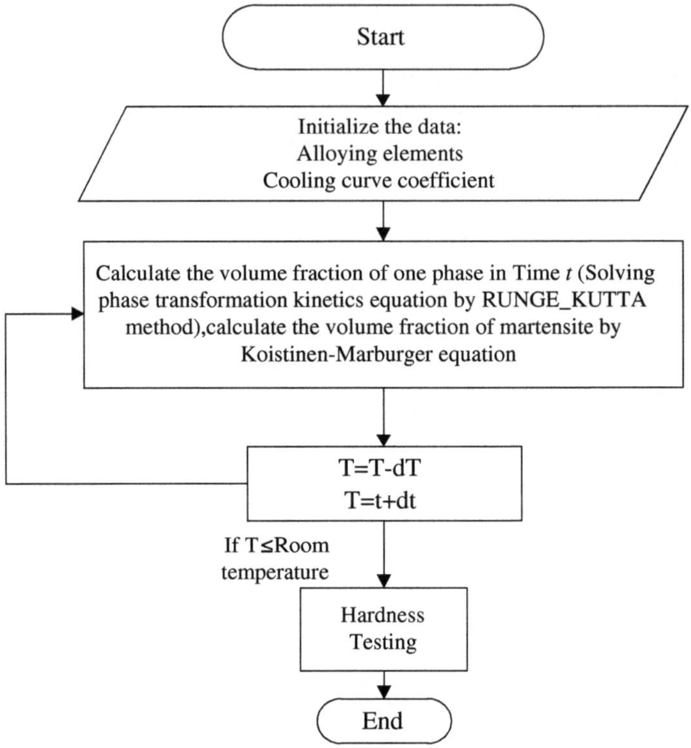

Fig. 8.7 Schematic structure for the prediction program

8.5 Hot Stamping Multi-Field Coupled Numerical Simulation

8.5.1 Static Explicit Algorithm for Hot Stamping Multi-Field Coupled Numerical Simulation

1. **Hot stamping multi-field coupled constitutive equation based on sustained equilibrium equations**

High-strength steel hot stamping process is a process of thermal–mechanical phase change coupled with viscoelastic-plastic large strain deformation contacting non-linear large strain deformation, so incremental theory (rate form) constitutive equation [27, 28] and finite element formula [29] should be adopted.

Suppose the plasmodium virtual power equation for two neighboring configuration at current time t and adjacent time $\tau = t + dt$ is

$$\int_{\Omega} t_{Ij}\delta\left(\frac{\partial v_j}{\partial x_I}\right)\mathrm{d}\Omega = \int_{\Omega} P_I\delta v_i \mathrm{d}\Omega + \int_{A_T} \bar{P}_I\delta v_i \mathrm{d}A \tag{8.94}$$

$$\int_{\Omega} (t_{Ij}+dt_{Ij})\delta\left(\frac{\partial v_j}{\partial x_I}\right)\mathrm{d}\Omega = \int_{\Omega} (P_I+dP_I)\delta v_i \mathrm{d}\Omega + \int_{A_T} (\bar{P}_I+d\bar{P}_I)\delta v_i \mathrm{d}A \tag{8.95}$$

Subtracting the two equations, one can obtain

$$\int_{\Omega} dt_{Ij}\delta\left(\frac{\partial v_j}{\partial x_I}\right)\mathrm{d}\Omega = \int_{\Omega} dP_I\delta v_i \mathrm{d}\Omega + \int_{A_T} d\bar{P}_I\delta v_i \mathrm{d}A \tag{8.96}$$

If express dt_{Ij}, dP_I and $d\bar{P}_I$ as material deviation, one obtains continuous equilibrium equation

$$\int_{\Omega} \dot{t}_{Ij}\delta\left(\frac{\partial v_j}{\partial x_I}\right)\mathrm{d}\Omega = \int_{\Omega} \dot{P}_I\delta v_i \mathrm{d}\Omega + \int_{A_T} \dot{\bar{P}}_I\delta v_i \mathrm{d}A \tag{8.97}$$

The configuration at t is reference configuration. If $\tau \to t$, one can obtain $\frac{\partial v_j}{\partial x_I} \to l_{ji}$, $\dot{P}_I(\tau) \to \dot{p}_i(t)$, $\dot{\bar{P}}_I(\tau) \to \dot{\bar{p}}_i(t)$, so continuous equilibrium equations are used to element e, at the same time considering the change with respect to time for element nodal forces $\dot{f}_i^{(k)}(t)$ $(k = 1, 2, \ldots, n;\ i = 1, 2, 3)$ (n is the amount of element nodal) at t, one obtains element continuous equilibrium equation.

$$\int_{e} \dot{t}_{ij}(t)\delta l_{ji}\mathrm{d}v = \sum_{k=1}^{n} \dot{f}_i^{(k)}(t)\delta v_i^{(k)} + \int_{e} \dot{p}_i(t)\delta v_i \mathrm{d}v + \int_{a_\sigma} \dot{\bar{p}}_i(t)\delta v_i \mathrm{d}a \tag{8.98}$$

By analyzing the hot stamping stress in Chap. 4, total strain rate tensor $\dot{\varepsilon}_{ij}$ in hot stamping process is constituted by the five following parts:

$$\dot{\varepsilon}_{ij} = \dot{\varepsilon}_{ij}^{e} + \dot{\varepsilon}_{ij}^{p} + \dot{\varepsilon}_{ij}^{th} + \dot{\varepsilon}_{ij}^{tr} + \dot{\varepsilon}_{ij}^{tp} \tag{8.99}$$

where $\dot{\varepsilon}_{ij}^{e}$ is elastic strain rate tensor caused by external force, $\dot{\varepsilon}_{ij}^{p}$ is plastic strain rate tensor caused by external force, $\dot{\varepsilon}_{ij}^{th}$ is thermal strain rate tensor caused by temperature, $\dot{\varepsilon}_{ij}^{tr}$ is phase change volume strain rate tensor, and $\dot{\varepsilon}_{ij}^{tp}$ is phase change plastic strain rate tensor. In the assumption of isotropy, set the linear expansion coefficient as α, Modulus as E, Poisson's ratio as μ. Set $\alpha_{ij} = \alpha^{\text{effective}}\delta_{ij}$. E and $\alpha^{\text{effective}}$ are function of T, μ do not relate with T. If the temperature changes from T_0 to T, the elastic strain ${\varepsilon^{e}}_{ij}$ and thermal strain caused by force can be expressed, respectively, as

$$\varepsilon_{ij}^{e} = B_{ijkl}^{e}\left(T, \dot{\bar{\varepsilon}}\right)\sigma_{kl} \tag{8.100}$$

$$\varepsilon_{ij}^{th} = \alpha_{ij}(T - T_0) \tag{8.101}$$

So

$$\dot{\varepsilon}_{ij}^{e} = B_{ijkl}^{e}\dot{\sigma}_{kl} + \sigma_{kl}\left(\frac{\partial B_{ijkl}^{e}}{\partial T}\dot{T} + \frac{\partial B_{ijkl}^{e}}{\partial \dot{\bar{\varepsilon}}}\ddot{\bar{\varepsilon}}\right) \tag{8.102}$$

$$B_{ijkl}^{e} = \frac{1+\mu}{E}\left\{\frac{1}{2}\left(\delta_{ik}\delta_{jl} + \delta_{jk}\delta_{il}\right) - \frac{\mu}{1+\mu}\delta_{ij}\delta_{kl}\right\} \tag{8.103}$$

where

$$B_{ijkl}^{e} = \frac{1+\mu}{E}\left\{\frac{1}{2}\left(\delta_{ik}\delta_{jl} + \delta_{jk}\delta_{il}\right) - \frac{\mu}{1+\mu}\delta_{ij}\delta_{kl}\right\} \tag{8.104}$$

And

$$\frac{\partial B_{ijkl}^{e}}{\partial T} = -\frac{1}{E^2}\frac{\partial E}{\partial T}\tilde{B}_{ijkl}^{e} \tag{8.105}$$

$$\frac{\partial B_{ijkl}^{e}}{\partial \dot{\bar{\varepsilon}}} = -\frac{1}{E^2}\frac{\partial E}{\partial \dot{\bar{\varepsilon}}}\tilde{B}_{ijkl}^{e} \tag{8.106}$$

$$\tilde{B}_{ijkl}^{e} = \frac{1+\mu}{2}\left(\delta_{ik}\delta_{jl} + \delta_{jk}\delta_{il}\right) - \mu\delta_{ij}\delta_{kl} \tag{8.107}$$

So Eqs. (8.102) and (8.103) can be combined as

$$\begin{aligned}\dot{\varepsilon}_{ij}^{e} + \dot{\varepsilon}_{ij}^{th} = {} & B_{ijkl}^{e}\dot{\sigma}_{kl} + \left\{\alpha_{ij} + \frac{\partial \alpha_{ij}}{\partial T}(T - T_0)\right\}\dot{T} \\ & - \frac{1}{E^2}\left(\frac{\partial E}{\partial T}\dot{T} + \frac{\partial E}{\partial \dot{\bar{\varepsilon}}}\ddot{\bar{\varepsilon}}\right)\left\{(1+\mu)\sigma_{ij} - \mu\delta_{ij}\sigma_{kk}\right\}\end{aligned} \tag{8.108}$$

If the change of α with temperature T and the change of E with equilibrium are not considered, the equation above can be simplified as

$$\dot{\varepsilon}_{ij}^{e} + \dot{\varepsilon}_{ij}^{th} = B_{ijkl}^{e}\dot{\sigma}_{kl} + \alpha_{ij}\dot{T} - \frac{1}{E^2}\frac{\partial E}{\partial T}\dot{T}\left\{(1+\mu)\sigma_{ij} - \mu\delta_{ij}\sigma_{kk}\right\} \tag{8.109}$$

Plastic strain rate caused by external force adopting the rate form of deformation theory can be expressed as.

$$\dot{\varepsilon}_{ij}^{p} = \dot{\varepsilon}_{ij}^{vp} = \frac{3}{2H'}\frac{\dot{\bar{\sigma}}}{\bar{\sigma}}\sigma'_{ij} + \frac{3}{2H_s}\left(\dot{\sigma}'_{ij} - \frac{\dot{\bar{\sigma}}}{\bar{\sigma}}\sigma'_{ij}\right) \tag{8.110}$$

where

$$H' = \frac{\dot{\bar{\sigma}}}{\dot{\bar{\varepsilon}}^{vp}}, \quad H_s = \bar{\sigma}/\bar{\varepsilon}^{vp} \tag{8.111}$$

and

$$\beta_{ij}^{k} = \beta^{k}\delta_{ij} \tag{8.112}$$

The phase change volume strain rate is

$$\dot{\varepsilon}_{ij}^{tr} = \sum_{k=F}^{M} \beta_{ij}^{k}\frac{\partial \Delta V_k}{\partial T}\dot{T}. \tag{8.113}$$

Assume that the hot deformation is subject to isotropic orthogonal flow rule and misses yield principle. The phase change plastic strain rate is

$$\dot{\varepsilon}_{ij}^{tp} = 3kS_{ij}(1-\xi)\frac{\partial \xi}{\partial T}\dot{T} \tag{8.114}$$

where S_{ij} is Cauchy deviatoric stress coefficient. One can obtain the constitutive law between inner true stress rate $\dot{\sigma}_{ij}$, total strain rate $\dot{\varepsilon}_{ij}$, and temperature change rate $\dot{T}$ by Eqs. (8.99), (8.102), (8.103), (8.110), (8.113), and (8.114).

$$\dot{\sigma}_{ij} = D_{ijkl}^{ep}\dot{\varepsilon}_{kl} - \dot{\bar{\varepsilon}}^{vp}P_{ij} - \dot{T}\beta'_{ij} \tag{8.115}$$

where

$$D_{ijkl}^{ep} = \frac{H_s}{H_s+3G}\left\{D_{ijkl}^{e} + \frac{3G}{H_s}\left[\frac{1}{3}\delta_{ij}\delta_{kl}\frac{2G(1+\mu)}{1-2\mu} + \frac{3G}{\bar{\sigma}^2}\sigma'_{ij}\sigma'_{kl}\right]\right\} \tag{8.116}$$

$$\beta'_{ij} = \frac{H_s}{H_s+3G}\left\{\frac{3G}{H_s}\beta_{kl}\left(\frac{1}{3}\delta_{ij}\delta_{kl} + \frac{3}{2\bar{\sigma}^2}\sigma'_{ij}\sigma'_{kl}\right) + \beta_{ij}\right\} \tag{8.117}$$

$$\beta_{ij} = \frac{E}{1-2\mu}\left[\left(\alpha^{\text{affective}} + \sum_{k=F}^{M}\beta^{k}\frac{\partial \Delta V_k}{\partial T}\right)\delta_{ij} + 3kS_{ij}(1-\xi)\frac{\partial \xi}{\partial T}\right] - \frac{1}{E}\frac{\partial E}{\partial T}\sigma_{ij} \tag{8.118}$$

where

$$P_{ij} = D^e_{ijkl} p_{kl}, \quad p_{kl} = \frac{3\sigma'_{kl}}{2\bar{\sigma}} \tag{8.119}$$

2. **Finite element formula of large deformation**

Based on constitutive equations of hot stamping rate form, continuous equilibrium equation and the research about hot stamping material properties mentioned in previous section, this section is to establish finite element formula of hot stamping large deformation.

According to Mises's definition of equivalent stress

$$\bar{\sigma}^2 = \frac{3}{2}\sigma'_{ij}\sigma'_{ij} \tag{8.120}$$

One obtains

$$\dot{\bar{\sigma}} = \frac{3}{2\bar{\sigma}}\sigma'_{ij}\dot{\sigma}_{ij} \tag{8.121}$$

First, assume that at time t and $t+\Delta t$, the equivalent plastic strain rate are $\dot{\bar{\varepsilon}}^{vp}_t$ and $\dot{\bar{\varepsilon}}^{vp}_{t+\Delta t}$, respectively. For tangent coefficient method

$$\Delta\bar{\varepsilon}^{vp} = \Delta t\left\{(1-\varsigma)\dot{\bar{\varepsilon}}^{vp}_t + \varsigma\dot{\bar{\varepsilon}}^{vp}_{t+\Delta t}\right\}, \quad 0 \le \varsigma \le 1 \tag{8.122}$$

Notice that $\dot{\bar{\varepsilon}}$ is related to $\bar{\sigma}$, $\bar{\varepsilon}^{vp}$ and T, given $\dot{\bar{\varepsilon}}^{vp}_{t+\Delta t}$ by Taylor expansion at t time

$$\dot{\bar{\varepsilon}}^{vp}_{t+\Delta t} = \dot{\bar{\varepsilon}}^{vp}_t + \frac{\partial\dot{\bar{\varepsilon}}^{vp}_t}{\partial\bar{\sigma}}\Delta\bar{\sigma} + \frac{\partial\dot{\bar{\varepsilon}}^{vp}_t}{\partial\bar{\varepsilon}^{vp}}\Delta\bar{\varepsilon}^{vp} + \frac{\partial\dot{\bar{\varepsilon}}^{vp}_t}{\partial T}\Delta T \tag{8.123}$$

$\Delta\bar{\sigma}$ is obtained by multiplying Δt at both sides of Eq. (8.121)

$$\Delta\bar{\sigma} = \Delta t P_{ij}\dot{\varepsilon}_{ij} - \Delta\bar{\varepsilon}^{vp} p_{ij} P_{ij} + \Delta t\frac{1}{E}\frac{\partial E}{\partial T}\bar{\sigma}\dot{T} \tag{8.124}$$

Substitute (8.124) into (8.123), then substitute into (8.122), one obtains

$$\Delta\bar{\varepsilon}^{vp} = \Delta t\left\{\frac{\dot{\bar{\varepsilon}}^{vp}_t}{1+\omega} + \frac{1}{h}\frac{\omega}{1+\omega}\left(P_{ij}\dot{\varepsilon}_{ij} + \frac{1}{E}\frac{\partial E}{\partial T}\bar{\sigma}\dot{T}\right) + \frac{\varsigma}{1+\omega}\frac{\partial\dot{\bar{\varepsilon}}^{vp}_t}{\partial T}\Delta T\right\} \tag{8.125}$$

where

$$\omega = (\varsigma\Delta t)h\frac{\partial\dot{\bar{\varepsilon}}^{vp}_t}{\partial\bar{\sigma}} \tag{8.126}$$

$$h = p_{ij}P_{ij} - \left(\frac{\partial\dot{\bar{\varepsilon}}_t^{vp}}{\partial\bar{\varepsilon}^{vp}}\right)\left(\frac{\partial\dot{\bar{\varepsilon}}_t^{vp}}{\partial\bar{\sigma}}\right)^{-1} \tag{8.127}$$

Divided both sides of Eq. (8.125) by Δt then substitute in (8.115)

$$\dot{\sigma}_{ij} = \tilde{D}^{ep}_{ijkl}\dot{\varepsilon}_{kl} - \frac{\eta}{1+\omega}P_{ij} - \dot{T}\beta'_{ij} \tag{8.128}$$

where

$$\tilde{D}^{ep}_{ijkl} = D^{ep}_{ijkl} - \frac{1}{h}\frac{\omega}{1+\omega}P_{ij}P_{kl} \tag{8.129}$$

$$\eta = \dot{\bar{\varepsilon}}_t^{vp} + \frac{\omega}{h}\frac{1}{E}\frac{\partial E}{\partial T}\bar{\sigma}\dot{T} + \varsigma\Delta T\frac{\partial\dot{\bar{\varepsilon}}_t^{vp}}{\partial T} \tag{8.130}$$

Introduce Eq. (8.128) into current configuration large deformation elasto-viscoplastic constitutive equation

$$\overset{\nabla}{\sigma}_{ij} = \tilde{D}^{ep}_{ijkl}d_{kl} - \frac{\eta}{1+\omega}P_{ij} - \dot{T}\beta'_{ij} \tag{8.131}$$

In the equation above, $\overset{\nabla}{\sigma}_{ij}$ is Jaumann derivative tensor of Cauchy stress, d_{kl} is deformation rate tensor. Stress rate effect which corresponds to strain rate in Eqs. (8.103), (8.110), and (8.111) are expressed by the second and third term in the equation above.

Introduce Eq. (8.131) in continuous equilibrium equation, namely Eq. (8.98). For Piola stress material deviation $\dot{t}_{ij}$ is not explicit in Eq. (8.98), the relationship establishment between $\overset{\nabla}{\sigma}_{ij}$ and $\dot{t}_{ij}$ is necessary. According to reference, considering incompressibility for plastic deformation caused by force, one obtains

$$\dot{t}_{ij}(t) = \overset{\nabla}{\sigma}_{ij} - \sigma_{ik}d_{ki} - \sigma_{kj}d_{ki} + \sigma_{ik}l_{jk} \tag{8.132}$$

Take the symmetry of $\overset{\nabla}{\sigma}_{ij}$ and the material isotropic, one obtains

$$\dot{t}_{ij}(t) = \overset{\nabla}{\sigma}_{ij} - F_{ijkl}d_{kl} + \sigma_{ik}l_{jk} \tag{8.133}$$

$$F_{ijkl}d_{kl}\delta l_{ji} = F_{ijkl}d_{kl}\delta d_{ij} \tag{8.134}$$

$$\overset{\nabla}{\sigma}_{ij}\,\delta l_{ji} = \overset{\nabla}{\sigma}_{ij}\,\delta d_{ij} \tag{8.135}$$

where

$$F_{ijkl} = \frac{1}{2}\left(\sigma_{lj}\delta_{ki} + \sigma_{kj}\delta_{li} + \sigma_{li}\delta_{kj} + \sigma_{ki}\delta_{lj}\right) \tag{8.136}$$

Substitute (8.133) in (8.98)

$$\int_e \left(\overset{\nabla}{\sigma}_{ij} - F_{ijkl}d_{kl} + \sigma_{ik}l_{jk}\right)\delta l_{ji}\mathrm{d}v = \sum_{k=1}^{n} \dot{f}_i^{(k)}(t)\delta v_i^{(k)} + \int_e \dot{p}_i(t)\delta v_i \mathrm{d}v + \int_{a_\sigma} \dot{p}_i(t)\delta v_i \mathrm{d}a \tag{8.137}$$

Taking Eqs. (8.134) and (8.135) into account, Eq. (8.137) can be transformed into

$$\int_e \left\{\delta d_{ij}\left(\overset{\nabla}{\sigma}_{ij} - F_{ijkl}d_{kl}\right) + \delta l_{ji}\sigma_{ik}l_{jk}\right\}\mathrm{d}v = \sum_{k=1}^{n} \dot{f}_i^{(k)}(t)\delta v_i^{(k)} + \int_e \dot{p}_i(t)\delta v_i \mathrm{d}v + \int_{a_\sigma} \dot{p}_i(t)\delta v_i \mathrm{d}a \tag{8.138}$$

Substitute (8.131) for the equation above

$$\begin{aligned}&\int_e \left\{\delta d_{ij}\left(\tilde{D}^{ep}_{ijkl} - F_{ijkl}\right)d_{kl} + \delta l_{ji}\sigma_{ik}l_{jk}\right\}\mathrm{d}v \\ &= \sum_{k=1}^{n} \dot{f}_i^{(k)}(t)\delta v_i^{(k)} + \int_e \dot{p}_i(t)\delta v_i \mathrm{d}v + \int_{a_\sigma} \dot{\bar{p}}_i(t)\delta v_i \mathrm{d}a + \int_e \frac{\eta}{1+\omega}\delta d_{ij}P_{ij}\mathrm{d}v + \int_e \delta d_{ij}\dot{T}\beta'_{ij}\mathrm{d}v\end{aligned} \tag{8.139}$$

Remark

$$\{v\}^e = \left(\{v^1\}^{\mathrm{T}}, \{v^2\}^{\mathrm{T}}, \ldots, \{v^N\}^{\mathrm{T}}\right)\{v^I\}^{\mathrm{T}} = \left(v_x^I, v_y^I, v_z^I\right)^{\mathrm{T}} \quad (I = 1, 2, \ldots N)$$

$$\{\sigma\} = \left(\sigma_{xx}, \sigma_{yy}, \sigma_{zz}, \sigma_{xy}, \sigma_{yz}, \sigma_{zx}\right)^{\mathrm{T}}$$

$$\{\dot{p}\} = \left(\dot{p}_x, \dot{p}_y, \dot{p}_z\right)^{\mathrm{T}}, \{\dot{\bar{p}}\} = \left(\dot{\bar{p}}_x, \dot{\bar{p}}_y, \dot{\bar{p}}_z\right)^{\mathrm{T}}, \{v\} = \left(v_x, v_y, v_z\right)^{\mathrm{T}}$$

$$\{d\} = \left(d_{xx}, d_{yy}, d_{zz}, 2d_{xy}, 2d_{yz}, 2d_{zx}\right)^{\mathrm{T}}$$

$$\{l\} = \left(v_{x,x}, v_{y,y}, v_{z,z}, v_{x,y}, v_{x,z}, v_{y,x}, v_{y,z}, v_{z,x}, v_{z,y}\right)^{\mathrm{T}}$$

$$\{\dot{f}\}^e = \left[\dot{f}_1^{(1)}, \dot{f}_2^{(1)}, \dot{f}_3^{(1)}, \ldots, \dot{f}_1^{(N)}, \dot{f}_2^{(N)}, \dot{f}_3^{(N)}\right]^{\mathrm{T}} \quad (N \text{ is the number of element node})$$

Then Eq. (8.139) can be written in matrix form as

$$\begin{aligned}
&\int\limits_{e} \left[\delta\{d\}^{\mathrm{T}}\left(\left[\tilde{D}_{ep}\right]-[F]\right)\{d\}+\delta\{l\}^{\mathrm{T}}[Q]\{l\}\right]\mathrm{d}v \\
&\quad=\int\limits_{e}\frac{\eta}{1+\omega}\delta\{d\}^{\mathrm{T}}\{P\}\mathrm{d}v+\int\limits_{e}\dot{T}\delta\{d\}^{\mathrm{T}}\{\beta'\}\mathrm{d}v \\
&\quad+\int\limits_{a_\sigma}\delta\{v\}^{\mathrm{T}}\{\dot{\bar{p}}\}\mathrm{d}a+\int\limits_{e}\delta\{v\}^{\mathrm{T}}\{\dot{p}\}\mathrm{d}v+(\delta\{v\}^{e})^{\mathrm{T}}\{\dot{f}\}^{e}
\end{aligned} \tag{8.140}$$

where

$$\left[\tilde{D}_{ep}\right]=\left[D_{ep}\right]-\frac{1}{h}\frac{\omega}{1+\omega}\frac{9G^2}{\bar{\sigma}^2}[D_2] \tag{8.141}$$

$$\left[D_{ep}\right]=\frac{H_s}{H_s+3G}\left\{[D_e]+\frac{3G}{H_s}\left(\frac{1}{3}\frac{2G(1+\mu)}{1-2\mu}[D_1]+\frac{3G}{\bar{\sigma}^2}[D_2]\right)\right\} \tag{8.142}$$

$$[D_1]=\begin{bmatrix} 1 & 1 & 1 & 0 & 0 & 0 \\ & 1 & 1 & 0 & 0 & 0 \\ & & 1 & 0 & 0 & 0 \\ & \text{sym} & & 0 & 0 & 0 \\ & & & & 0 & 0 \\ & & & & & 0 \end{bmatrix} \tag{8.143}$$

$$[D_2]=\begin{bmatrix} \sigma'^2_{xx} & \sigma'_{xx}\sigma'_{yy} & \sigma'_{xx}\sigma'_{zz} & \sigma'_{xx}\sigma'_{xy} & \sigma'_{xx}\sigma'_{yz} & \sigma'_{xx}\sigma'_{zx} \\ & \sigma'^2_{yy} & \sigma'_{yy}\sigma'_{zz} & \sigma'_{yy}\sigma'_{xy} & \sigma'_{yy}\sigma'_{yz} & \sigma'_{yy}\sigma'_{zx} \\ & & \sigma'^2_{zz} & \sigma'_{zz}\sigma'_{xy} & \sigma'_{zz}\sigma'_{yz} & \sigma'_{zz}\sigma'_{zx} \\ & \text{sym} & & \sigma'_{xy} & \sigma'_{xy}\sigma'_{yz} & \sigma'_{xy}\sigma'_{zx} \\ & & & & \sigma'^2_{yz} & \sigma'_{yz}\sigma'_{zx} \\ & & & & & \sigma'^2_{zx} \end{bmatrix} \tag{8.144}$$

$$[D_e]=\frac{2G}{1-2\mu}\begin{bmatrix} 1-\mu & \mu & \mu & 0 & 0 & 0 \\ & 1-\mu & \mu & 0 & 0 & 0 \\ & & 1-\mu & 0 & 0 & 0 \\ & \text{sym} & & (1-2\mu)/2 & 0 & 0 \\ & & & & (1-2\mu)/2 & 0 \\ & & & & & (1-2\mu)/2 \end{bmatrix} \tag{8.145}$$

Let $\{v\}=[N]\{v\}^e$ (where $[N]$ is shape function matrix, $\{v\}^e$ is element nodal velocity vector). Then $\{d\}=[B]\{v\}^e$, $\{l\}=[E]\{v\}^e$. By substituting them in (8.140), one obtains.

$$\delta(\{v\}^e)^{\mathrm{T}}\left[\int\limits_e \left\{[B]^{\mathrm{T}}\left([\tilde{D}_{ep}] - [F]\right)[B] + [E]^{\mathrm{T}}[Q][E]\right\}\mathrm{d}v\{v\}^e \right.$$
$$\left. - \int\limits_e \frac{\eta}{1+\omega}[B]^{\mathrm{T}}\{P\}\mathrm{d}v - \int\limits_e \dot{T}[B]^{\mathrm{T}}\{\beta'\}\mathrm{d}v - \int\limits_{a_\sigma} [N]^{\mathrm{T}}\{\dot{\bar{p}}\}\mathrm{d}a - \int\limits_e [N]^{\mathrm{T}}\{\dot{p}\}\mathrm{d}v - \{\dot{f}\}^e \right] \tag{8.146}$$

For the arbitrariness of variation $\delta(\{v\}^e)^T$.

$$[K_p]\{v\}^e = \{\dot{f}_p\} + \{\dot{g}_p\} + \{\dot{f}\} \tag{8.147}$$

where the element stiffness matrix is

$$[K_p] = \int\limits_e \left\{[B]^{\mathrm{T}}\left([\tilde{D}_{ep}] - [F]\right)[B] + [E]^{\mathrm{T}}[Q][E]\right\}\mathrm{d}v \tag{8.148}$$

$$\{\dot{g}_p\} = \int\limits_e [B]^{\mathrm{T}}\left(\frac{\eta}{1+\omega}\{P\} + \dot{T}\{\beta'\}\right)\mathrm{d}v \tag{8.149}$$

$$\{\dot{f}_p\} = \int\limits_{a_\sigma} [N]^{\mathrm{T}}\{\dot{\bar{p}}\}\mathrm{d}a + \int\limits_e [N]^{\mathrm{T}}\{\dot{p}\}\mathrm{d}v \tag{8.150}$$

It is worth noting that in the finite element formula mentioned above, the stress related to temperature and phase change of Eqs. (8.103), (8.110), and (8.111) is corresponding to phase change. By the disintegration of strain rate, one obtains the right side of the finite element directly, and action as equivalent load of heat and phase change.

8.5.2 Dynamic Explicit Finite Element Formulation of Multi-Filed Coupled Hot Stamping Simulation

1. **Dynamic equation of single degree of freedom damped system**

For a linear system spring–damper vibration system with single degree of freedom, as shown in Fig. 8.8, the dynamic equilibrium equation is

$$ma_n + cv_n + kx_n = P_n \tag{8.151}$$

where n represents the nth time step.

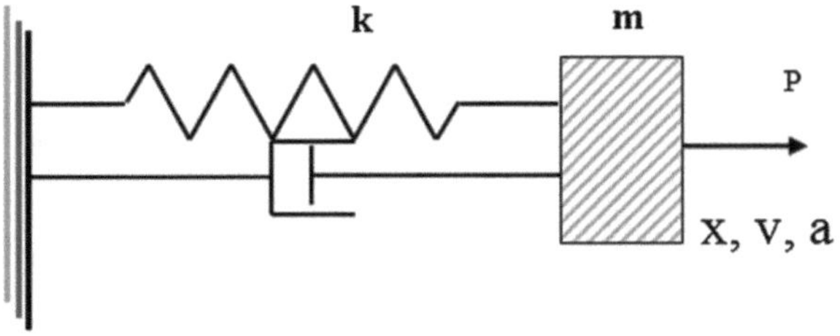

Fig. 8.8 Single freedom degree vibration system containing linear spring and damping

2. **Central difference solving format of discrete dynamic equilibrium equation**

To solve the dynamic equilibrium Eq. (8.151), the time course is divided into numerous discrete point. The period between two points is referred as time step. In this way, we can calculate displacement, velocity, acceleration, stress, strain, and other physical quantities at each time point by central difference method.

Specific central difference method is derived as following: let n corresponds to the state at time t, given the physical quantities before t and at t, we define (Fig. 8.9):

Remark

$t-\Delta t_{n-1}$: the state of $n-1$;
$t-\frac{1}{2}\Delta t_{n-1}$: the state of $n-\frac{1}{2}$;
$t+\frac{1}{2}\Delta t_n$: the state of $n+\frac{1}{2}$;
$t+\Delta t_n$: the state of $n+1$.

And then assume $\beta=\frac{\Delta t_n}{\Delta t_{n-1}}$, the velocity and acceleration in (8.151) are rewritten as the differential form:

$$v_n=\frac{\beta}{\beta+1}v_{n+\frac{1}{2}}+\frac{1}{1+\beta}v_{n-\frac{1}{2}} \tag{8.152}$$

$$a_n=\frac{2}{(1+\beta)\Delta t_{n-1}}\left(v_{n+\frac{1}{2}}-v_{n-\frac{1}{2}}\right) \tag{8.153}$$

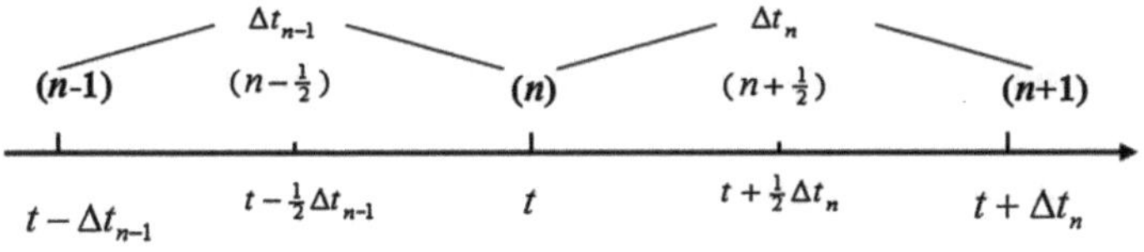

Fig. 8.9 Schematic of discrete points in time

The total displacement at $t+\Delta t_n$ (state of $n + 1$) time can be obtained by cumulation

$$x_{n+1} = x_n + v_{n+\frac{1}{2}}\Delta t_n \tag{8.154}$$

Using (8.152) and (8.153), and let $c = \gamma \cdot m$ (proportional damping), we can write (8.151) as

$$\frac{2m}{(1+\beta)\Delta t_{n-1}}\left(v_{n+\frac{1}{2}} - v_{n-\frac{1}{2}}\right) + \gamma m\left(\frac{\beta}{1+\beta}v_{n+\frac{1}{2}} + \frac{1}{1+\beta}v_{n-\frac{1}{2}}\right) = P_n - kx_n \tag{8.155}$$

We arrive at the solution

$$v_{n+\frac{1}{2}} = \frac{2-\gamma\Delta t_{n-1}}{2+\gamma\beta\Delta t_{n-1}}v_{n-\frac{1}{2}} + \frac{(1+\beta)\Delta t_{n-1}}{(2+\gamma\beta\Delta t_{n-1})m}\cdot(P_n - kx_n) \tag{8.156}$$

For multidegree of freedom system, v, a, x, m, k, p, in (8.152)–(8.156) can be expressed as vector or matrix, respectively, we have

$$\{v\}_n = \frac{\beta}{1+\beta}\{v\}_{n+\frac{1}{2}} + \frac{1}{1+\beta}\{v\}_{n-\frac{1}{2}} \tag{8.157}$$

$$\{a\}_n = \frac{2}{(1+\beta)\Delta t_{n-1}}\left(\{v\}_{n+\frac{1}{2}} - \{v\}_{n-\frac{1}{2}}\right) \tag{8.158}$$

$$\{v\}_{n+1} = \{v\}_{n+\frac{1}{2}} + \{v\}_{n-\frac{1}{2}}\Delta t_n \tag{8.159}$$

$$\begin{aligned}&\frac{2[m]}{(1+\beta)\Delta t_{n-1}}\left(\{v\}_{n+\frac{1}{2}} - \{v\}_{n-\frac{1}{2}}\right) + \gamma[m]\left(\frac{\beta}{1+\beta}\{v\}_{n+\frac{1}{2}} + \frac{1}{1+\beta}\{v\}_{n-\frac{1}{2}}\right)\\ &= \{P\}_n - [k]\{x\}_n\end{aligned} \tag{8.160}$$

If $[m]$ can be written as a diagonal matrix, the above equations can be written in the form of components, so that an explicit velocity expression format can be obtained

$$v^i_{n+\frac{1}{2}} = \frac{2-\lambda\Delta t_{n-1}}{2+\lambda\beta\Delta t_{n-1}}v^i_{n-\frac{1}{2}} + \frac{(1+\beta)\Delta t_{n-1}}{(2+\gamma\beta\Delta t_{n-1})m^i}\cdot\left(P^i_n - kx^i_n\right) \tag{8.161}$$

where i is the number of degree of freedom.

3. **Explicit finite element algorithm of large deformation dynamic in continuum**

(1) Finite element equation of dynamic explicit

For an arbitrary infinitesimal parallelepiped taken from the configuration at time t, dx, and dy are dimensions in the x and y directions, and the dimension in the z direction is unit length, as shown in Fig. 8.10. In the x direction, we have

Fig. 8.10 Dynamic equilibrium schematic diagram of representative elemental volume

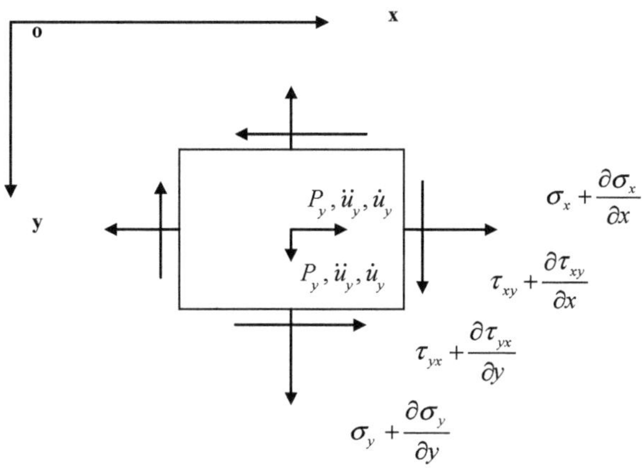

$$\left(\sigma_x + \frac{\partial \sigma_x}{\partial x} dx\right) dy \times 1 - \sigma_x dy \times 1 + \left(\tau_{yx} + \frac{\partial \tau_{yx}}{\partial y} dy\right) dx \times 1 - \tau_{yx} dx \times 1 + (P_x - \rho \ddot{u}_x - c\ddot{u}_x) \times dxdy \times 1 = 0 \tag{8.162}$$

Both sides of the equation divided by $dxdy$

$$\frac{\partial \sigma_x}{\partial x} + \frac{\partial \tau_{yx}}{\partial y} + P_x - \rho \ddot{u}_x - c\dot{u}_x = 0 \tag{8.163}$$

Similarly, in the y direction

$$\frac{\partial \sigma_y}{\partial y} + \frac{\partial \tau_{xy}}{\partial x} + P_y - \rho \ddot{u}_y - c\dot{u}_y = 0 \tag{8.164}$$

Using dynamic explicit algorithm, in terms of the two equations above, we obtain the kinematic differential equation for sheet metal

$$\frac{\partial \sigma_{ij}}{\partial x_j} + p_i - \rho \ddot{u}_i - c\dot{u}_i = 0 \tag{8.165}$$

where ρ is the material density, c is the damping coefficient, $\dot{u}_i$ and $\ddot{u}_i$ are the speed and acceleration at any point within the material, p_i is volume force of the point, and σ_{ij} is Cauchy stress of the point.

According to divergence theorem

$$\int \frac{\partial \sigma_{ij}}{\partial x_j} \delta \dot{u}_i \mathrm{d}V = \int_{\Gamma} q_i \delta \dot{u}_i \mathrm{d}\Gamma - \int_{V} \sigma_{ij} \delta \dot{\varepsilon}_{ij} \mathrm{d}V \tag{8.166}$$

and boundary conditions, from (8.165), we can get the system virtual work equation

$$\int_V \rho\ddot{u}_i\delta\dot{u}_i\mathrm{d}V + \int_V c\dot{u}_i\delta\dot{u}_i\mathrm{d}V = \int_V p_i\delta\dot{u}_i\mathrm{d}V + \int_\Gamma q_i\delta\dot{u}_i\mathrm{d}\Gamma - \int_V \sigma_{ij}\delta\dot{\varepsilon}_{ij}\mathrm{d}V \quad (8.167)$$

where $\delta\dot{u}_i$ is the virtual velocity, $\delta\dot{\varepsilon}_{ij}$ is the virtual strain rate corresponding to Cauchy stress σ_{ij} (virtual deformation rate δd_{ij}).

Discrete the body into m units, and any element in the body has α nodes, taking their shape function as N^α, so that the displacement component of any point within the element u_i, velocity component u_i and acceleration component $\dot{u}_i$ are given by

$$\begin{cases} u_i = N^a u_i^a \\ \dot{u}_i = N^a \dot{u}_i^a \\ \ddot{u}_i = N^a \ddot{u}_i^a \end{cases} \quad (8.168)$$

From geometric equation, we have

$$\dot{\varepsilon}_{ij} = B_j^\alpha \dot{u}_i^\alpha \quad (8.169)$$

where u_i^α, $\dot{u}_i^\alpha$ and $\ddot{u}_i^\alpha$ are the displacement component, velocity component, and acceleration component of node α, B_j^α is strain matrix (tensor) (for specific solution, the time step is about 3–4 orders of magnitude smaller than the previous static explicit algorithm, so linear process can be used according to the case of small deformation strain matrix, but the configuration needs to be updated after each step is solved).

Substitute (8.168) and (8.169) into (8.167), we obtain element dynamic equilibrium equation:

$$\begin{aligned} &\int_{\Omega_e} \rho N^\alpha \ddot{u}_i^\alpha N^\beta \delta\dot{u}_i^\beta \mathrm{d}\Omega + \int_{\Omega_e} cN^\alpha \dot{u}_i^\alpha N^\beta \delta\dot{u}_i^\beta \mathrm{d}\Omega \\ &\quad = \int_{\Omega_e} p_i N^\beta \delta\dot{u}_i^\beta \mathrm{d}\Omega + \int_{\bar{\Omega}_e} \bar{q}_i N^\beta \delta\dot{u}_i^\beta \mathrm{d}\bar{\Omega} - \int_{\Omega_{\dot{e}}} \sigma_{ij} B_j^\beta \delta\dot{u}_i^\beta \mathrm{d}\Omega \end{aligned} \quad (8.170)$$

where $\delta\dot{u}_i^\beta$ is the virtual velocity of node β, the matrix form is given by

$$\begin{aligned} &\int_{\Omega_e} \rho[\boldsymbol{N}]^{\mathrm{T}}[\boldsymbol{N}]\mathrm{d}\Omega\{\ddot{\boldsymbol{u}}\} + \int_{\Omega_e} c[\boldsymbol{N}]^{\mathrm{T}}[\boldsymbol{N}]\mathrm{d}\Omega\{\dot{\boldsymbol{u}}\} \\ &\quad = \int_{\Omega_e} [\boldsymbol{N}]^{\mathrm{T}}\{\boldsymbol{p}\}\mathrm{d}\Omega + \int_{\bar{\Omega}_e} [\boldsymbol{N}]^{\mathrm{T}}\{\bar{\boldsymbol{q}}\}\mathrm{d}\bar{\Omega} - \int_{\Omega_{\dot{e}}} [\boldsymbol{B}]^{\mathrm{T}}\{\sigma\}\mathrm{d}\Omega \end{aligned} \quad (8.171)$$

Assemble element equations, we obtain overall finite element equation, that is

$$\boldsymbol{M}\ddot{u} + \boldsymbol{C}\dot{u} = \boldsymbol{P} - \boldsymbol{F} \tag{8.172}$$

where $\boldsymbol{M}$ is the consistent mass matrix, $\boldsymbol{C}$ is the damping matrix, $\boldsymbol{P}$ is the nodal external force vector, and $\boldsymbol{F}$ is the nodal internal force vector

$$\boldsymbol{F} = \sum \int_{\Omega_e} \boldsymbol{B}^{\mathrm{T}} \boldsymbol{\sigma}^{\text{total}} \, \mathrm{d}\Omega \tag{8.173}$$

where $\boldsymbol{B}$ is the strain matrix, $\boldsymbol{\sigma}$ is the total stress containing the shape induced stress, thermal expansion induced stress and phase change induced stress.

When using the finite element method to analyze dynamic problems, proper description of the mass matrix is the basic requirement for exact dynamic analysis. Although the consistent mass matrix can well reflect the distribution of quality, consistent mass matrix is a full matrix, for sheet metal forming analysis, finite element meshes are very small with a large number of elements and nodes usually, so mass matrix is a large square in this case. As we need to calculate the inverse matrix in the solution process, significant time is required for the solution.

Therefore, for actual calculation, lumped mass matrix is usually used. We can diagonalize the mass matrix and set the damping matrix proportional to the mass matrix, taking $\boldsymbol{C} = \alpha\boldsymbol{M}$ as an example, where α is the proportional damping coefficient. After simplification, the equation set is no longer coupled, which eliminates the need to solve linear equations and greatly reduces the computation time, and the accuracy of this algorithm can still meet the engineering requirements. Then the simultaneous Eq. (8.172) are translated into the number of nodes multiplied by the number of DOFs independent equations

$$m_i\ddot{u}_i + c_i\dot{u}_i = P_i - F_i \tag{8.174}$$

where i is not summed here, i = 1—the total number of degrees of freedom.

(2) Determine the critical time step

As the central difference algorithm is conditionally stable, the time increment step size Δt must be limited in order to ensure calculation stability. Stability conditions are usually determined by the system's highest frequency $\omega_{\max}$, and the time step increment satisfying stability condition is

$$\Delta t \leq \frac{2}{\omega_{\max}}\left(\sqrt{1+\xi^2} - \xi\right) \tag{8.175}$$

where ξ is the model critical damping ratio with the highest frequency. The introduction of the damping system actually reduces the critical stability condition, which violates engineering intuition apparently. The maximum frequency of the system depends on the expansion mode of the largest element.

The time increment step satisfying stability condition can be obtained from minimum traversing time of expansion wave passing through element approximately

$$\Delta t_n \leq \gamma \frac{L_n^e}{c} \tag{8.176}$$

where $\gamma = 0.6$–0.8, the expansion wave velocity in the material c is given by

$$c = \left(\frac{\rho}{E}\right)^{1/2} \tag{8.177}$$

L_n^e is the nominal length of element e at time t_n.

Stability condition can guarantee that the disturbance just spread one element length in a time increment. If the system includes only one material, the length of time increment step satisfying the stability condition is proportional to the smallest element size; if the system is divided into uniform element but a variety of materials, the stable system time step is determined by the smallest element of the material with the maximum expansion velocity.

For a simple truss element, in the case of lumped mass matrix, the stability criterion gives a critical time step $\Delta t \leq (l/c)$, where c is the material velocity of sound, l is the element length, Δt indicates the time expansion wave across the element l. This is the so-called courant–friedrichs–lewy (CFL) stability condition.

For triangular element and quadrilateral plate element, the critical time step depends on the determination of element nominal length, which is generally determined according to Fig. 8.11, for high-grade element, the critical time step is much lower than lower grade element, which means explicit integration algorithm is quite inappropriate for high grade element.

Strictly speaking, the stability criterion given above is just based on linear systems, but it can also supply a useful stability estimation for nonlinear problem. A 80–90 % reduced linear time step is suitable for the majority of nonlinear problems to keep the system stable enough. However, it is very important to check the balance of energy continually in the calculation process, because any gain or loss in the total energy (5 % or more) will lead to instability. It is worth noting that the constant incremental time step cannot always maintain the stability of the solution, even if the highest frequency $\omega_{\max} l$ of the system decreases continuously.

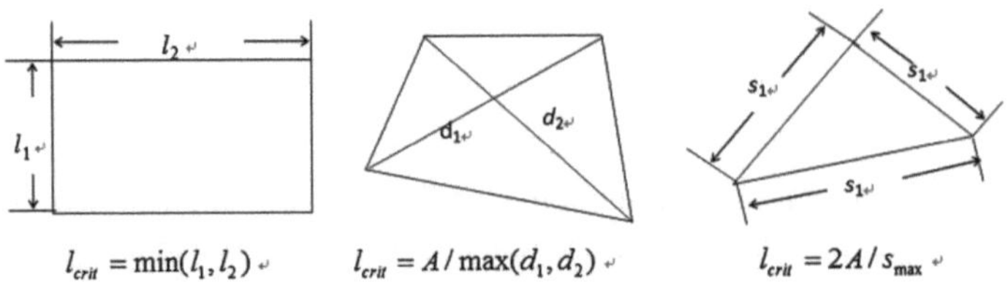

Fig. 8.11 Nominal length of element

(3) Internal stress calculation of hot stamping

Hot stamping is a complex thermal–mechanical transformation coupled process. In order to describe the thermal–mechanical transformation response, the strain rate expressions can be generally written in incremental form, then the total strain incremental can be written as [23, 30, 31]:

$$d\varepsilon_{ij} = d\varepsilon_{ij}^{e} + d\varepsilon_{ij}^{p} + d\varepsilon_{ij}^{th} + d\varepsilon_{ij}^{tr} + d\varepsilon_{ij}^{tp} \tag{8.178}$$

where $d\varepsilon_{ij}^{e}$ is the elastic component, $d\varepsilon_{ij}^{p}$ is the plastic component, $d\varepsilon_{ij}^{th}$ is the thermal component, $d\varepsilon_{ij}^{tr}$ is the transformation volume component and $d\varepsilon_{ij}^{tp}$ is the transformation-induced plasticity component.

The thermal strain can be expressed as

$$d\varepsilon_{ij}^{th} = \alpha dT + (T - T_{ref})d\alpha \tag{8.179}$$

where α is the thermal expansion coefficient, T is temperature, T_{ref} is the reference temperature. In the book, the thermal expansion coefficient of tested boron steel is treated as constant and the value is $1.3 \times 10^{-5}\ \text{K}^{-1}$.

During the continuous cooling process, the additional strain named transformation volume strain which likes the thermal strain will be caused by the different molar volume of austenite and ferrite, pearlite, bainite, and martensite. It can be written as

$$d\varepsilon_{ij}^{tr} = \sum_{I}^{N} \beta_{A\to I}\delta_{ij}dV_I \quad I = F, P, B, M \tag{8.180}$$

where $\beta_{A\to I}$ is the phase transformation expansion coefficient of phase A to phase I. The phase transformation expansion coefficient of austenite to ferrite is 0.0025, austenite to pearlite is 0.003, austenite to bainite is 0.0045, austenite to martensite is 0.008. A represents austenite, F represents ferrite, P represents pearlite, B represents bainite, M represents martensite. Among them, ferrite, pearlite, and bainite transformation, which belong to diffusive transformation, are associated with temperature and time. Martensite transformation, which belong to nondiffusive transformation, is only associated with temperature. V_I is the volume fraction of each phase, N is total number of phase, δ_{ij} is Kronecker symbol.

Even though the stress is lower than yield strength, permanent plastic deformation will be caused by phase transformation. The transformation-induced plasticity strain increment can be written as

$$d\varepsilon_{ij}^{tp} = \sum_{I}^{N} 3K_{A\to I}(1 - V_I)s_{ij}dV_I \quad I = F, P, B, M \tag{8.181}$$

where s_{ij} is the component of deviatoric stress tensor, $K_{A\to I}$ is the phase transformation plasticity coefficient of phase A to phase I. The phase transformation plasticity coefficient of austenite to ferrite, pearlite and bainite is $4.18 \times 10^{-5}\,\text{MPa}^{-1}$, austenite to martensite is $5.08 \times 10^{-5}\,\text{MPa}^{-1}$.

Based on the generalized Hooke's law, the stress can be expressed as

$$\sigma_{ij} = C^e_{ijkl}\varepsilon^e_{kl} \tag{8.182}$$

where C^e_{ijkl} is the component of elastic matrix tensor.

Taking differentiation for the two sides of Eq. (8.182) and considering Eq. (8.178), we can obtain

$$d\sigma_{ij} = C^e_{ijkl}\left(d\varepsilon_{kl} - d\varepsilon^p_{kl} - d\varepsilon^{th}_{kl} - d\varepsilon^{tr}_{kl} - d\varepsilon^{tp}_{kl}\right) + dC^e_{ijkl}\varepsilon^e_{kl} \tag{8.183}$$

Using the Von Mises yield criterion and isotropic hardening model

$$f(\sigma_{ij}, \sigma_y) = J_2^2 - \frac{\sigma_y^2}{3} = 0 \tag{8.184}$$

where $J_2^2 = \frac{1}{2}s_{ij}s_{ij}$ and σ_y are the yield strength, s_{ij} is the deviator tensor of Cauchy stress. In the process of hot stamping, to consider the influence of each microstructure on the yield strength of material, the yield strength can be expressed by the mixture law as

$$\sigma_y = V_I\sigma_{y\_I} \quad I = A, F, P, B, M \tag{8.185}$$

where σ_{y_I} is the yield strength of phase I and can be expressed by linear law as

$$\sigma_{y\_I} = \sigma_{y0\_I} + H_I\bar{\varepsilon}^p_I \tag{8.186}$$

where σ_{y0_I} is the initial yield strength of phase I, H_I is the plastic modulus of phase I which represents the slope of strain hardening part of the stress–strain curve. The assumption that the equivalent plastic strain of each phase is equal to the total equivalent plastic strain $\bar{\varepsilon}^p$ is adopted in this book. Because of the mechanical properties of studied boron steel related to temperature and strain rate, thus σ_{y0_I} and H_I are the function of temperature and strain rate.

According to the consistency condition, one obtains

$$df = \frac{\partial f}{\partial \sigma_{ij}}d\sigma_{ij} + \frac{\partial f}{\partial \sigma_y}d\sigma_y = 0 \tag{8.187}$$

Substitute (8.185) and (8.186) into (8.187), one obtains

$$\begin{aligned} & s_{ij}C^e_{ijkl}\left(d\varepsilon_{kl} - d\varepsilon^p_{kl} - d\varepsilon^{th}_{kl} - d\varepsilon^{tr}_{kl} - d\varepsilon^{tp}_{kl}\right) + s_{ij}dC^e_{ijkl}\varepsilon^e_{kl} - \\ & \tfrac{2}{3}\sigma_y\left\{\sum_{I=A}^{M} V_I H_I d\bar{\varepsilon}^p + \sum_{I=A}^{M} V_I\left[\left(\tfrac{\partial\sigma_{y0\_I}}{\partial T} + \tfrac{\partial H_I}{\partial T}\bar{\varepsilon}^p\right)dT + \left(\tfrac{\partial\sigma_{y0\_I}}{\partial\dot{\varepsilon}} + \tfrac{\partial H_I}{\partial\dot{\varepsilon}}d\dot{\varepsilon}\right)\right]\right\} = 0 \end{aligned} \tag{8.188}$$

where the plastic strain increment can be calculated using the Drucker postulation

$$d\varepsilon^p_{ij} = d\lambda\frac{\partial f}{\partial\sigma_{ij}} = d\lambda s_{ij} \tag{8.189}$$

where $d\lambda$ is plastic scale factor.

The equivalent plastic strain increment can be expressed as

$$d\bar{\varepsilon}^p = \frac{2}{3}d\lambda\sigma_y \tag{8.190}$$

Substitute (8.189) and (8.190) into (8.188), the plastic scale factor can be obtained as

$$\begin{aligned} d\lambda = & \frac{s_{ij}C^e_{ijkl}\left(d\varepsilon_{kl} - d\varepsilon^p_{kl} - d\varepsilon^{th}_{kl} - d\varepsilon^{tr}_{kl} - d\varepsilon^{tp}_{kl}\right) + s_{ij}dC^e_{ijkl}\varepsilon^e_{kl}}{s_{ij}C^e_{ijkl}s_{kl} + \frac{4}{9}\sigma_y^2\sum_{I=A}^{M}V_I H_I} \\ & \frac{-\frac{2}{3}\sigma_y\sum_{I=A}^{M}V_I\left[\left(\frac{\partial\sigma_{y0\_I}}{\partial T} + \frac{\partial H_I}{\partial T}\bar{\varepsilon}^p\right)dT\left(\frac{\partial\sigma_{y0\_I}}{\partial\dot{\varepsilon}} + \frac{\partial H_I}{\partial\dot{\varepsilon}}\bar{\varepsilon}^p\right)d\dot{\varepsilon}\right]}{s_{ij}C^e_{ijkl}s_{kl} + \frac{4}{9}\sigma_y^2\sum_{I=A}^{M}V_I H_I} \end{aligned} \tag{8.191}$$

where

$$dC^e_{ijkl}\varepsilon^e_{kl} = \frac{d\mu}{\mu}s_{ij} + \frac{\sigma_{kk}}{3}\frac{dK}{K}\delta_{ij} \tag{8.192}$$

where $\mu = E/[2(1+v)]$ is shear modulus, $K = E/[3(1-2v)]$ is bulk modulus, is Poisson's ratio.

Substitute (8.191) into (8.189) and then substitute (8.189) and (8.192) into (8.183), we can obtain

$$d\sigma_{ij} = C^{ep}_{ijkl}\left(d\varepsilon_{kl} - d\varepsilon^{th}_{kl} - d\varepsilon^{tr}_{kl} - d\varepsilon^{tp}_{kl}\right) + dM_{ij} \tag{8.193}$$

where

$$C^{ep}_{ijkl} = C^e_{ijkl} - \frac{C^e_{ijkl}s_{kl}s_{ij}C^e_{ijkl}}{W} \tag{8.194}$$

$$dM_{ij} = \left(s_{ij} - \frac{2}{3}\frac{C^e_{ijkl}s_{kl}\sigma_y^2}{W}\right)\frac{d\mu}{\mu} + \left(1 - \frac{C^e_{ijkl}s_{kl}s_{ij}}{W}\right)\frac{\sigma_{kk}}{3}\frac{dK}{K}\delta_{ij} + \\ \frac{2}{3}\frac{C^e_{ijkl}s_{kl}\sigma_y}{W}\sum_{I=A}^{M} V_I\left[\left(\frac{\partial\sigma_{y0\_I}}{\partial T} + \frac{\partial H_I}{\partial T}\bar{\varepsilon}^p\right)dT + \left(\frac{\partial\sigma_{y0\_I}}{\partial\dot{\varepsilon}}\bar{\varepsilon}^p\right)d\dot{\varepsilon}\right] \tag{8.195}$$

The parameter W in the above two equations can be expressed as

$$W = s_{ij}C^e_{ijkl}s_{kl} + \frac{4}{9}\sigma_y^2\sum_{I=a}^{M} V_I H_I \tag{8.196}$$

The Eq. (8.193) is the final thermal–mechanical phase coupled constitutive equation of hot stamping.

For the dynamic explicit algorithm of thermo–mechanical phase coupled large deformation hot stamping, rate-dependent constitutive equation [15] can be adopted. Jaumann derivative $\overset{\nabla}{\sigma}_{ij}$ of objective Cauchy true stress σ_{ij} is given by

$$\overset{\nabla}{\sigma}_{ij} = \dot{\sigma}_{ij} - \sigma_{ij}\omega_{ij} - \sigma_{kj}\omega_{ki} \tag{8.197}$$

where σ_{ij} is the material derivative of Cauchy stress, ω_{ij} is spin tensor.

The rate-dependent constitutive equation of hot stamping large deformation thermo–mechanical phase coupled is similar with the constitutive Eq. (8.193) of small deformation elastic-plastic incremental in form. As long as the stress and strain are replaced by the Jaumann derivative of Cauchy stress and deformation rate, stress deviator and equivalent stress with Cauchy true stress deviator and equivalent Cauchy true stress, respectively, small deformation elastic-plastic incremental constitutive equation is transformed to large deformation thermo–mechanical-metallurgical coupled rate constitutive equation. In thermoplastic large deformation finite element formulation of the dynamic explicit algorithm, $\dot{\varepsilon}_{kl}(t)$, $\dot{\varepsilon}^p_{kl}(t)$, $\dot{\varepsilon}^{th}_{kl}(t)$, $\dot{\varepsilon}^{tr}_{kl}(t)$, $\dot{\varepsilon}^{tp}_{kl}(t)$, are Green strain rate tensor, plastic strain rate tensor caused by internal forces, thermal strain rate tensor caused by temperature change, isotropic transformation strain rate tensor and phase transformation plastic strain rate tensor, respectively, relative to the adjacent configuration at time t as a reference configuration. The internal stress in internal force tensor F of Eq. (8.173) is the Cauchy true stress relative to current configuration.

Then, using the central difference scheme, the Cauchy true stress σ_{ij}^{n+1} corresponding to time $t + \Delta t$ (the $n + 1$ state of quantity to be determined) configuration is expressed as

$$\sigma_{ij}^{n+1} = \sigma_{ij}^{n} + \dot{\sigma}_{ij}^{n+\frac{1}{2}}\Delta t \tag{8.198}$$

where $\dot{\sigma}_{ij}^{n+\frac{1}{2}}$ can be obtained from known mechanical quantities at time $n + \frac{1}{2}$.

4. **Contact and friction models**

Hot stamping of high-strength steel involves not only the above geometric large deformation, material nonlinearity, and complex thermal–mechanical metallurgical coupled mechanical issues, but also includes contact nonlinear boundary conditions under continuously variable temperature conditions [29].

Dynamic equilibrium equation in view of damping is given by [32]

$$\boldsymbol{M}\ddot{\boldsymbol{u}} + \boldsymbol{C}\dot{\boldsymbol{u}} + \boldsymbol{F}^{\text{int}} = \boldsymbol{F}^{\text{ext}} + \boldsymbol{\lambda} \tag{8.199}$$

where $\boldsymbol{F}^{\text{int}}$ and $\boldsymbol{F}^{ext}$ are nodal internal force s and external forces; $\boldsymbol{\lambda}$ is frictional contact reaction force vector, and $\boldsymbol{u}$, $\dot{\boldsymbol{u}}$ and $\ddot{\boldsymbol{u}}$ are the node displacement, velocity, and acceleration vector, respectively. The discrete motion equation can be written as follows [33]

$$\frac{\boldsymbol{M}}{\Delta t_n^2}[\boldsymbol{u}_{n+1} - (1+\beta)\boldsymbol{u}_n + \beta\boldsymbol{u}_{n-1}] + \frac{\boldsymbol{C}}{\Delta t_n}(\boldsymbol{u}_{n+1} - \boldsymbol{u}_n) + \boldsymbol{F}_{n+1}^{\text{int}} = \boldsymbol{F}_{n+1}^{\text{ext}} + \boldsymbol{\lambda}_{n+1} \tag{8.200}$$

where the subscript n represents time t_n, and time step is expressed as

$$\Delta t_n = t_{n+1} - t_n \tag{8.201}$$

$$\rho = \Delta t_n / \Delta t_{n-1} \tag{8.202}$$

For high-strength steel hot stamping, the material was formed at temperature ranging from 950 °C to the martensitic phase transformation. The hardness and strength of high temperature sheet will reduce to almost $1/3 - 1/5$ [29], when the interaction between sheet metal and tool reduces and depends on the temperature change of sheet metal and tool.

From (8.199), equivalent nodal force of Cauchy true stress is $\boldsymbol{F}^{\text{int}}$ in (8.173). For frictional contact reaction force, due to the assumption that the quality of element are concentrated in nodes, the contact between blank and tool can be seen as node-surface contact, which is shown in Fig. 8.12, the point p with mass m penetrates through the mold surface ab and reaches p' point, the normal penetration distance and tangential sliding distance on the mold surface of p point are

$$d_N = \boldsymbol{P'Q} \cdot \boldsymbol{V}_N, \quad d_T = \boldsymbol{P'Q} \cdot \boldsymbol{V}_T \tag{8.203}$$

where $\mathbf{V}_N$ and $\mathbf{V}_T$ are the outward normal direction and tangential direction of point Q at the mold surface; the direction of $\mathbf{V}_T$ depends on the movement direction of the node, which is calculated as

$$\boldsymbol{V} = \frac{\boldsymbol{P'P} \times \boldsymbol{V}_N}{\|\boldsymbol{P'P} \times \boldsymbol{V}_N\|}, \quad \boldsymbol{V}_T = \frac{\boldsymbol{V}_N \times \boldsymbol{V}}{\|\boldsymbol{V}_N \times \boldsymbol{V}\|} \tag{8.204}$$

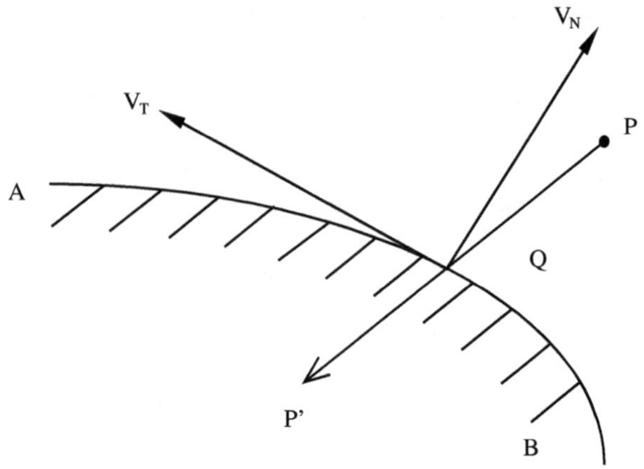

Fig. 8.12 Friction and contact model

Taking into account the temperature-dependent properties, normal contact reaction force, and tangential friction can be expressed as

$$F_N = \varsigma(T)\frac{md_N}{\Delta t_n \Delta t_{n-1}} \tag{8.205}$$

$$F_T = \varsigma(T)\frac{md_T}{\Delta t_n \Delta t_{n-1}} \tag{8.206}$$

respectively, where $\varsigma(T)$ is defined as contact force control coefficient, the range is

$$\frac{\sigma_s^T}{\sigma_s^R} \leq \varsigma(T) \leq 1 \tag{8.207}$$

where σ_s^T, σ_s^R are yield strength of sheet under high temperature and room temperature, respectively.

Using (8.205) and (8.206), λ in (8.200) can be expressed as

$$\boldsymbol{\lambda} = F_N \boldsymbol{V}_N + F_T \boldsymbol{V}_T \tag{8.208}$$

For the dynamic explicit algorithm of thermal–mechanical phase coupled metal material hot stamping, the simulation procedure can be concluded by Fig. 8.13.

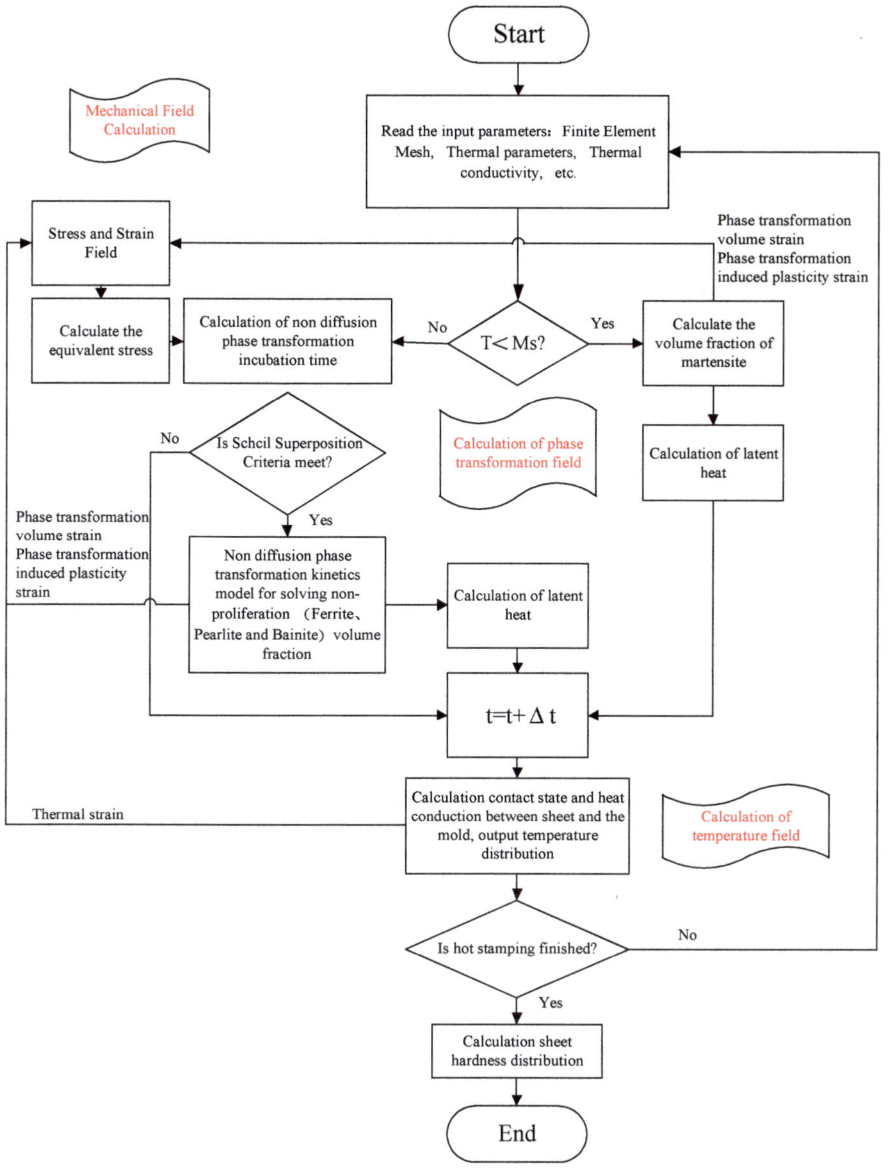

Fig. 8.13 The flow chart for finite element prediction of hot stamping

8.6 Summary

In this chapter, the conception of hot stamping numerical simulation technology was introduced. The concrete content includes

In Sect. 8.2, based on the unique characteristics of hot stamping, the differences between hot stamping numerical simulation and traditional cold stamping numerical simulation were analyzed; the key technology of hot stamping numerical simulation was introduced and the coupled relation schema of heat treatment, phase transformation and deformation were given, which illustrates the interdependent relation of each thermal–mechanical parameters in hot stamping; the contact friction problem was analyzed in hot stamping process and the test method of friction coefficient at elevated temperature was given; the simulation characteristics of temperature field were indicated which include the heat exchange among metal blank, air and mold and also include the heat exchange between mold and cooling channel; according to the CCT curves, the process that austenite decomposes to other phase was indicated which has deep significance to accurately predict the hot stamping process, microstructure and mechanical property of final components.

In Sect. 8.3, based on Fourier heat conduction differential equation and combined the initial condition and boundary condition of transient heat conduction problem, the weak form of equivalent integral for weighted residual method and finite element formulation for three-dimensional transient temperature field problem were derived. A temperature shell element for transient temperature field problem of general shell was introduced to the finite element analysis of temperature field during hot stamping. By using this element, the element for temperature field analysis is not meshed any more and the element for deformation analysis can be used directly, thus the input data and the degree of freedom of the analyzed system are reduced. Furthermore, the pathological problem of system matrix that may occur when the shell is very thin can also be solved. This book uses a multi-step method to discrete time domain. Generally speaking, multi-step method is a high precision and unconditionally stable difference method.

In Sect. 8.4, the dynamic model of phase transformation proposed by Kirkaldy and Venugopalan and modified by Li was introduced in detail. For diffusive transformation, the continuous cooling can be converted into step cooling and each step can be treated as constant temperature. Then, the isothermal kinetics model can be used to calculate the transformation quantity of each phase. The calculation flow of microstructure transformation was also given in this section.

In Sect. 8.5, the thermal–mechanical phase transformation-coupled constitutive equation was established; Then, the static explicit and dynamic explicit nonlinear element finite formation for large deformation were developed; The method to determine the critical time step of dynamic explicit algorithm was introduced; by defining the concept of contact control parameters, the high temperature properties were introduced into contact and friction model; The simulation procedure of dynamic explicit algorithm of thermal–mechanical phase coupled metal material hot stamping was given at the end of this section.

References

1. Liang Y (2013) Research and application on key process experiment of high strength steel for hot forming. Dalian University of Technology, p 179 (in Chinese)
2. Ma N (2011) Research on hot forming of high strength steel. Dalian University of Technology, p 224 (in Chinese)
3. Kerström P (2006) Modelling and simulation of hot stamping. In: Luleå tekniska universitet/Tillämpad fysik, maskin-och materialteknik/Hållfasthetslära
4. Bergman G, Oldenburg M (2004) A finite element model for thermomechanical analysis of sheet metal forming. Int J Numer Meth Eng 59:1167–1186
5. Merklein M, Lechler J (2008) Determination of material and process characteristics for hot stamping processes of quenchenable ultra high strength steels with respect to a FE-based process design. In: SAE technical paper
6. Merklein M, Lechler J, Geiger M (2006) Characterisation of the flow properties of the quenchenable ultra high strength steel 22MnB5. CIRP Ann Manuf Technol 55:229–232
7. Hus TC, Liu CC (1998) Internal variable. Eff Punch Friction Charact 121:510–516
8. Saha PK, Wilson WR (1994) Influence of plastic strain on friction in sheet metal forming. Wear 172:167–173
9. Borsetto F, Ghiotti et al (2009) Investigation of the high strength steel Al-Si coating during hot stamping operations. Key Eng Mater 410–411:289–296
10. Dessain C, Hein P, Wilsius J et al (2008) Experimental investigation of friction and wear in hot stamping of usibor 1500P. 1st international conference on hot sheet metal forming of high-performance steel. Kassel, Germany, pp 217–227
11. Stoehr T, Merklein M, Lechler J (2008) Determination of frictional and thermal characteristics for hot stamping with respect to a numerical process design. In: Proceedings of 1st international conference in hot sheet metal forming of high performance steel, pp 293–300
12. Hardell J, Prakash B (2008) High-temperature friction and wear behaviour of different tool steels during sliding against Al–Si-coated high-strength steel. Tribol Int 41:663–671
13. Yanagida A, Azushima A (2009) Evaluation of coefficients of friction in hot stamping by hot flat drawing test. CIRP Ann Manuf Technol 58:247–250
14. Lenze FJ, Banik J, Sikora S (2008) Applications of hot formed parts for body in white. In: International deep drawing research group proceedings
15. Suehiro M, Maki J, Kusumi K et al (2003) Properties of aluminium-coated steels for hot-forming: Tech. Rep 88:16–21
16. Daxin J, Wu W, Hu P (2013) Research on temperature numerical simulation of hot stamping process of high strength steel. Eng Mech 1:419–424 (in Chinese)
17. Jiang D, Wu W, Hu P et al (2012) Thermo-mechanical-martensitic transformation numerical simulation of high strength steel in hot forming. J Mech Eng 12:18–23 (in Chinese)
18. Shi DY, Liang Y, Ping HU et al (2014) Numerical simulation of 3D temperature field in hot forming of high strength steel. J Jilin Univ (Eng Technol Edn) 44:946–952 (in Chinese)
19. Xucheng W, Youmei P (1985) Finite element analysis of temperature field of shells. In: Transactions of the 8th international conference on structural mechanics in reactor technology, vol B
20. Daxin J (2012) Research on high strength multi-filed coupled relationships in hot forming. Dalian University of Technology, p 94 (in Chinese)
21. Kirkaldy JS, Venugopalan D (1983) Prediction of microstructure and hardenability in low-alloy steels. In: Phase transformations in ferrous alloys, pp 125–148
22. Hu P, Shi DY, Ying L et al (2013) Coupled of thermal-mechanical-transformation numerical simulation on hot stamping with static explicit algorithm. In: AIP conference proceedings, pp 394–405
23. Shi D, Hu P, Ying L et al (2014) Numerical simulation on hot forming based on dynamic explicit algorithm. Eng Mech 12:186–192 (in Chinese)

24. Li MV, Niebuhr DV, Meekisho LL et al (1998) A computational model for the prediction of steel hardenability. Metall Mater Trans B 29:661–672
25. Koistinen DP, Marburger RE (1959) A general equation prescribing the extent of the austenite-martensite transformation in pure iron-carbon alloys and plain carbon steels. Acta Metall 7:59–60
26. Maynier P, Jungmann B, Dollet J (1977) Creusot–Loire system for the prediction of the mechanical properties of low alloy steel products. In: Hardenability concepts with applications to steel, pp 518–545
27. Budiansky B (1958) A reassessment of deformation theories of plasticity. In: DTIC document
28. Hu P (2004) Rate-dependent quasi-flow corner theory for elastic/visco-plastic materials. Int J Solids Struct 41:1263–1284
29. Ma N, Hu P, Guo W et al (2009) Feasible methods applied to the design and manufacturing process of hot forming. In: IDDRG2009 conference, Golden, CO USA
30. Hu P, Liu W, Ying L et al (2014) Theory, experiment and simulation research on High strength steel hot forming and the application in lightweight car body. In: The 11th international symposium on advanced forming and material processing technology. Dalian, pp 52–63 (in Chinese)
31. Ma N, Hu P, Wu W et al (2011) The constitutive theory and experimental analysis for high strength steel hot forming. Chin J Theor Appl Mech 2:346–354 (in Chinese)
32. Schweizerhof K, Hallquist JO (1991) Explicit integration schemes and contact formulations for thin sheet metal forming. In: FE-Simulation of, pp 405–439
33. Mura T (1987) Micromechanics of defects in solids. Springer

Chapter 9
Lightweight of Car Body Structure Applied by Hot Stamping Parts

Hot stamping parts are highly fit for vehicle structures design and manufacture due to their ultra-high strength and hardness. In what aspect the hot stamping parts are reflected when applied in vehicle lightweight? How can we judge the service performance and proceed the optimization design? How does it work when the vehicle's service performance is increased? In this chapter, we will research all the above questions through both simulation method and experiment method.

First, based on, vehicle's door anti-crash beam, side beams, concept of vehicle beam structures, and side beams of school bus, examples [1–5], this chapter researches on hot stamping parts application on vehicle and the differences are compared to the traditional parts, furthermore, tries to find the best design parameters with different optimization methods.

Second, there are two researches on the hardness gradient hot stamping parts, one is for typical thin-wall beam structure focused on energy absorption [6], the other is for B-pillar based on multi-objective optimization design [7].

9.1 Lightweight of Car Body Structure Applied by Hot Stamping Parts

9.1.1 Hot Stamping Door Anti-crash Beam and Its Process Optimization

According to process research from Chap. 3, cooling rate in hot stamping process has important influence on material final properties [8, 9]. When forming temperature is different, cooling speed, and the final parts properties can also be affected [10]. Through 500, 650, and 800 °C hot stamping process, we can get different properties 1.6 mm thickness 22MnB5 hot stamping anti-crash beam, select three points of them to do the strength properties test, see Fig. 9.1. Table 9.1 shows that

P. Hu et al., *Hot Stamping Advanced Manufacturing Technology of Lightweight Car Body*, DOI 10.1007/978-981-10-2401-6_9

Fig. 9.1 Hot stamping door beam and test positions

Table 9.1 Tensile strength of hot stamping door beams

Temperature/°C	Tensile strength/MPa			
	Location 1	Location 2	Location 3	Average
500	1078	1105	1123	1102
650	1550	1546	1562	1553
800	1608	1622	1629	1620

the average tensile strength is 1102, 1553, and 1620 MPa. The 500 °C hot stamping process makes the tensile strength the lowest which cannot meet the basic demands. However the 650 and 800 °C hot stamping process make the tensile strength more than 1500 MPa.

The thickness distribution analysis of door anti-crash beam meddle section is shown in Fig. 9.2. Stamping temperature has a great influence on thickness, the smaller thickness the higher stamping temperature. Besides location 1, 7, and 4 thickness decrease little, location 3 and 5 decreases because of drawing. Stamping temperature 500 °C makes the thickness more than others, each part's thickness can reach to 1.4 mm above due to better fluidity. 800 °C makes the thickness smallest compared with the other two temperatures, the smallest thickness is 1.31 mm which is easy to crack.

In order to study how different hot stamping process affect the door anti-crash beam bending resistance and energy absorption properties, we design the quasi-static bending experiment and dynamic impact simulation. The quasi-static

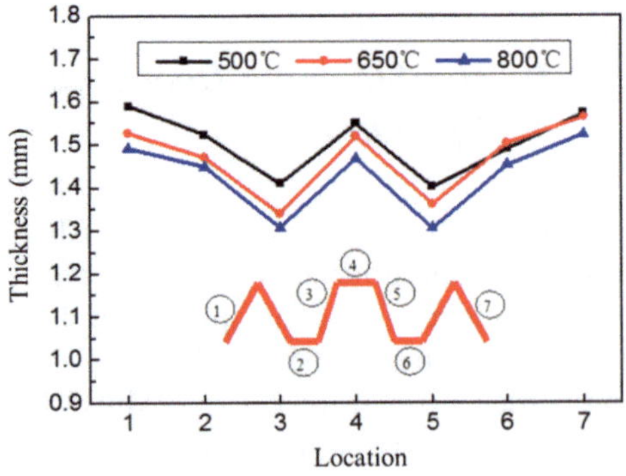

Fig. 9.2 Thickness distribution of hot stamping door beams

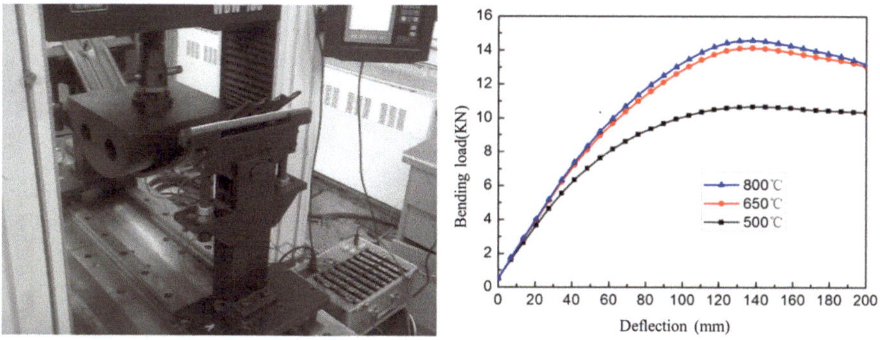

Fig. 9.3 Three-point bending test setup and the bending load curves

bending experiment setup is shown in Fig. 9.3, the supporter can rotate and translate, diameter of press block is 300 mm, and its speed is 500 mm/min.

The hot stamping door anti-collision beam flexural load deflection curve of quasi static bending experiment contrasts as shown in Fig. 9.3. When the bending deflection reaches about 131 mm, bending load is max, the biggest bending forming temperature 500 °C anti-collision beam load is only 10.76 kN, flexural capacity is weak and cannot meet the use requirements. The forming temperature of 800 and 650 °C anti-collision beam maximum bending load were 14.58, 14.13 kN; flexural capacity is strong and can satisfy the use requirement. The local damage is shown in Fig. 9.4. As you can see, hot stamping car anti-collision beam deformation mainly concentrates in the center area, forming the temperature 800 °C; anti-collision beam center first cracks and then fails to fracture. The forming temperature of 500 and 500 °C anti-collision beam does not have obvious cracks. Forming temperature 650 °C hot stamping car anti-collision beam compared to the forming temperature of 800 and 500 °C anti-collision beam has the best comprehensive performance.

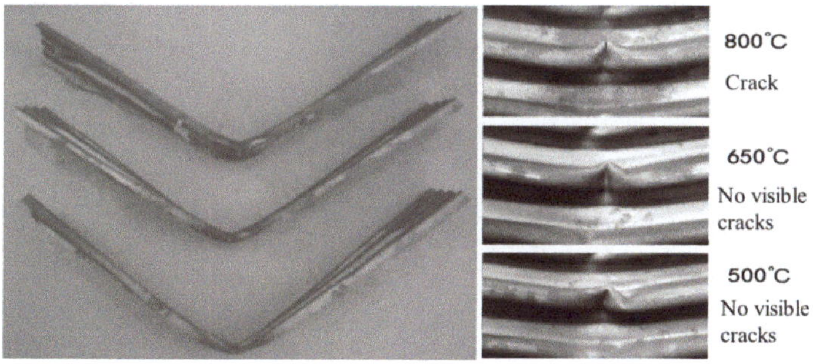

Fig. 9.4 Local failure states of hot stamping door beams

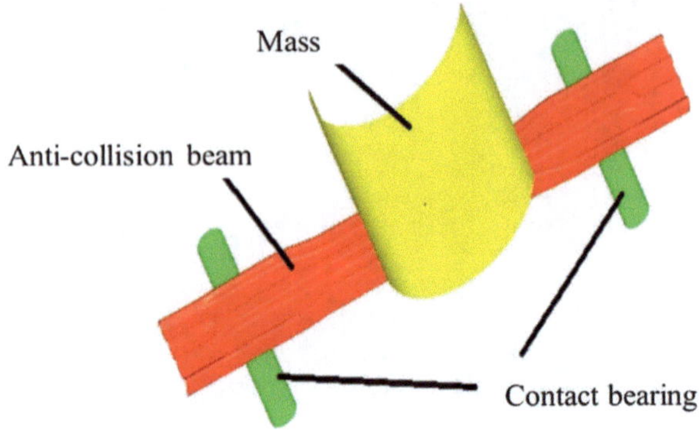

Fig. 9.5 Finite element model of dynamic impact

Based on LS-DYNA module of Hyper Mesh, dynamic impact simulation of the door anti-collision beam was carried out, the finite element model is shown in Fig. 9.5. Using rigid block impacting on the hot stamping door anti-collision beam, to test its energy absorption, the rigid block weighs 25 kg, initial speed is 50 km/h with the central vertical impact anti-collision beam.

The energy absorption—time curve is shown in Fig. 9.6. Forming temperature 500 °C hot stamping door anti-collision beam can be continuously deformed with energy absorption, 2.15 kJ is final energy absorption. Forming temperature 800 °C hot stamping car anti-collision beam energy absorption is 2.19 kJ, local plastic failure happens first, easy to cause harm to the crew. Anti-collision beam forming the temperature 650 °C hot stamping door is 2.25 kJ, its deformation and energy absorption ability are better than anti-collision beam the forming temperature 800 °C and it can effectively protect the security of the crew.

Analysis shows that the forming temperature of 650 °C hot stamping car anti-collision beam with high strength and good plastic deformation energy absorption ability has the best comprehensive performance and is suitable for car anti-collision beam side impact resistance.

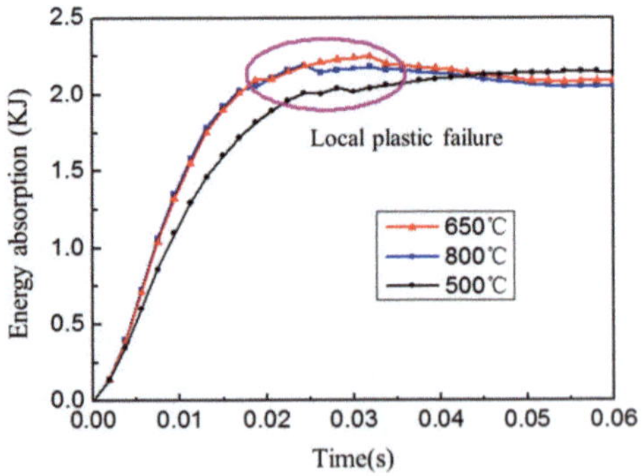

Fig. 9.6 Energy absorption curves of hot stamping door beams

9.1.2 *Application of Hot stamping Parts Based on CAE Crash of Whole Vehicle*

Study the application of hot stamping parts in the automobile body design, not only needs a single independent component analysis, also the analysis in the vehicle model based on CAE collision [3, 11]. The finite element model of the vehicle used in the study of this section is shown in Fig. 9.7, in order to analyze the high-strength steel plate hot stamping parts in the vehicle collision mechanism, the body of the B-pillar plate, door strengthened beams, and threshold plate (as shown in Fig. 9.8) materials were replaced with the corresponding hot stamping parts for the vehicle side impact analysis.

Side impact speed in accordance with the relevant laws and regulations is 36 km/h, four kinds of designed working conditions are simulated. First condition is original simulation model, second condition is 1.5 mm B-pillar panel replaced by 1.0 mm hot forming parts, the third is the B-pillar and door anti-collision beam replaced with hot forming parts, the fourth condition is the B-pillar, door anti-collision beam and the threshold reinforced beam replaced by hot forming parts. By tracking the displacement amount of intrusion of point B1, B2, B3, and

Fig. 9.7 Finite element model for side-wall crash analysis of the whole car

Fig. 9.8 Sketch map of the replaced parts and the testing points' locations on B-pillar of intrusion displacement

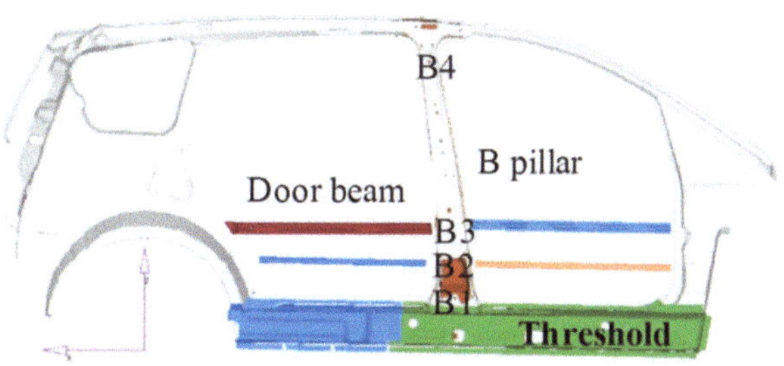

B4 (as shown in Fig. 9.8) on B-pillar in the process of collision compare the vehicle collision performance of four kinds of design condition.

The above four kinds of working conditions, the intrusion displacement are shown in Figs. 9.9, 9.10, 9.11 and 9.12. Amount of intrusion comparison results are shown in Table 9.2.

It can be seen from the Fig. 9.10, for the second condition, B-pillar is replaced by hot forming parts and after the vehicle collision, the intrusion displacement of each test point on B-pillar does not reduce but increase. Because only the B-pillar is strengthened but surrounding components is relatively soft the side impact B-pillar actually plays a reaction on intrusion damage. It shows that using ultra-high strength hot forming parts for automobile body design can not only consider local individual components, but also prevent the damage caused by high-strength hot forming parts in high speed impact.

Fig. 9.9 Intrusion displacement of B-pillar in the first model

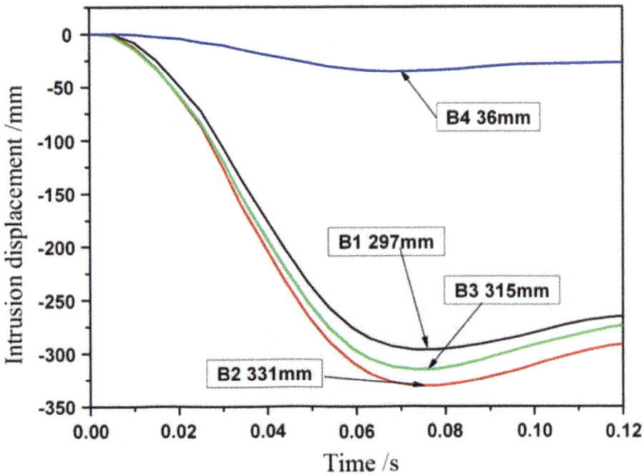

Fig. 9.10 Intrusion displacement of B-pillar in the second model

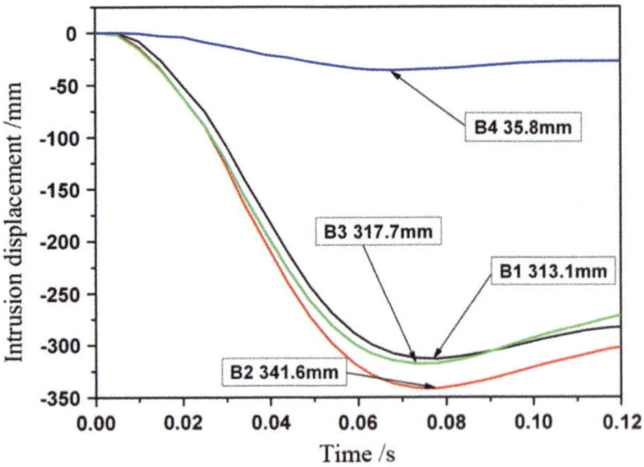

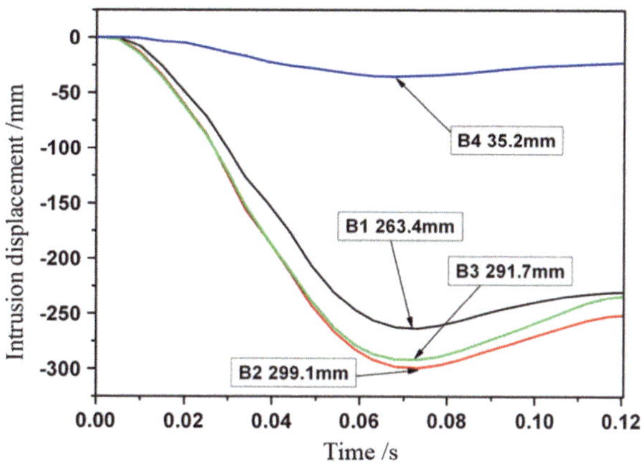

Fig. 9.11 Intrusion displacement of B-pillar in the third model

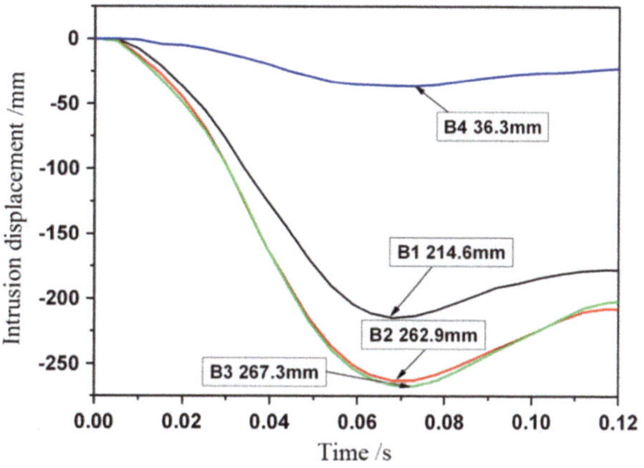

Fig. 9.12 Intrusion displacement of B-pillar in the fourth model

Table 9.2 Comparison of intrusion displacement for B-pillar (unit: mm)

Measure points	B1	B2	B3	B4
Condition 1	297	331	315	36
Condition 2	313.1	341.6	317.7	35.8
Condition 3	263.4	299.1	291.7	35.2
Condition 4	214.6	262.9	267.3	36.3

As you can see from Fig. 9.11, replace the B-pillar and door anti-collision beam at the same time and after side impact the intrusion displacement of each test point on B-pillar decreases; because the B-pillar and door anti-collision beam strengthened together, we can pull the door components deforming together and let ultra-high strength hot stamping parts play a big role on its advantages.

In Fig. 9.12, at the same time, replace the B-pillar, door reinforced beam and the threshold with hot stamping parts; intrusion displacement of each test point on

B-pillar drops significantly, the biggest drop is closer to 30 %, which fully shows that the hot forming parts have the huge advantage in strengthening body resistance to impact. B-pillar, and the threshold of the gate reinforced beam strengthening for hot forming parts, make the whole body side to form a whole solid skeleton and when impacted by high speed, the skeleton plays its protective role and drives the whole body parts to resist the side impact load, thereby significantly reducing the collision during the invasion of the body.

Compared to the side crash dummy damage index of the HFS body, according to C-NCAP is shown in Table 9.3 [12]. It can be seen from Table 9.3 that when the HFS body head acceleration peak value increases, HTC value also increases, but not more than the lower limit of human tolerance value, therefore the head score does not change. Due to the reduction of deformation of car body side, the HFS car body collision chest compression deformation is significantly reduced, so that the chest scores improve. Decrease of belly and pubic forces also makes the score of the abdomen and pelvis higher. HFS body C-NCAP scored 11 points, compared with the original body improving two points. HFS body, therefore, is to improve the crashworthiness of vehicle side impact and occupant safety.

The high peak power problem caused by high-strength hot forming parts can be solved by adding energy absorption material in the right place. For the layout of the energy absorption material, the author and other authors have done many studies [13–15].

Based on the mechanism analysis of the roles, hot forming parts play in body design, the design method of hot forming parts used in automobile body design is put forward, namely:

1. In body design, high-strength hot forming parts should be the overall frame layout, forming ultra-high strength protect storehouse for drivers and passenger;
2. In the body design, avoid simple application of a single high-strength hot forming part, preventing from invading the body when impact;
3. In body design, the application of high-strength hot forming parts, must be in the right location to layout energy absorption material, in order to reduce the peak force of impact.

Table 9.3 Comparison of injury criteria of occupant

Items		Original car		HFS	
		Value	Score	Value	Score
Head	HIC36	313.92	4	325.56	4
Chest	Compressive deformation	30.22 mm	2.4	24.78 mm	3.4
Belly	Belly force	2.23 kN	0.5	2.07 kN	1.1
Pelvis	Pubic forces	4.38 kN	2.1	4.16 kN	2.5
		Total	9	Total	11

9.1.3 Application of Hot stamping Parts in Concept Body Lightweight Design

Hot forming parts in the application of lightweight design concept body needs to be based on the design modeling, the digital model, CAE key performance parameters of the simulation and optimization design at the same time; usually the four functions need to be independent in different software, and then through the design personnel to manually connect each part to the practical application of the final product design to product, no commercial software can realize hot forming parts design and application integration. Professor HuPing's team from School of Automotive Engineering, Dalian University of Technology spent five years independent research and develop automobile body structure Concept Design tool, VCD-ICAE (Vehicle Concept Design-Intelligent CAE) system [16, 17], the system is developed based on the UG OPEN API language from UG NX5. It uses parametric technology, template technology, database technology, and knowledge engineering technology such as advanced design technology. The system can quickly build parameterized concept body geometry model, generate finite element mesh model and the boundary constraint load seamlessly integrate the NX Nastran solver and at the same time the plugins can read other CAE analysis software Nastran data interface (such as MSC Nastran, Ansys and Abaqus, etc.), the strength, stiffness and modal calculation, can automatically generate the postprocessing of view, and product report needed by the designers. The method also suits the concept of beam and plate, parameter optimization, to guide and help users to get a satisfactory design. The system has been successfully implemented in UG NX5.0 commercialization, and has been applied in the domestic auto makers.

This section based on VCD-ICAE system, will replace the typical parts of a white car (such as A, B, C pillar, the anti-collision beam, etc.) with hot forming parts using lightweight body thickness optimization module developed by the author, to assess the lightweight potential and effect of hot forming parts.

Car body structure concept design phase is the important stage to guarantee the performance; in the stage of design flaws are difficult to make up in the back of the detailed design stage [18]. Body concept design which can shorten body design cycle time and reduce the late repeat design has important significance. Frame beam element section design is a key task, because section determines the actual performance of the beam element, and therefore affects the whole vehicle performance. At present, this work relies mainly on the reference to a specific model and experimental simulation, so it is easy to cause too much surplus on cross-section thickness, resulting in material waste and body mass increase. Body concept design phase, therefore, needs a better way to optimize design of section to improve the reliability and rationality of concept car body model.

At present, some scholars have done in-depth research on calculation on the beam section properties and its optimization problem [19]. Aleksandar Prokic gave the open and closed thin-walled beam section properties computer algorithm [20, 21], Banichuke discussed the round rod cross-section optimization problem [22],

Vinot studied thin-walled beam in detail, such as cross-section shape optimization problem [23]. Ragnedda and Serra had done in-depth research about the thin-walled beam cross-section shape optimization problems [24], however, all the researches done in the optimization problem ignore the buckling characteristics of the section; the optimization variables and the optimization constraints are not considered comprehensively, and the practical effects are not combined with the engineering application in this paper, so the optimization goal cannot be directly applied to the beam cross-section optimization design of car body.

To this end, the author has established the arbitrary thin-walled beam cross-section geometric attributes algorithm and cross-section shape ratio vector control, successfully developed cross-section properties, shaped control of beam section optimization module, and integrated the model into the conceptual design system VCD_ICE [25]. Lightweight car body is always the goal of car design; this article uses the above modules, to realize the beam cross-section optimization design of car body and lightweight body quality and improve the efficiency and rationality of the concept car design.

Vehicle carrier beam performance is decided by beam section properties, so through the analysis of section thickness, section area, and structure performance, we can preliminarily assess both the bending and torsion performance of the main carrier beam of the vehicle, and it is also very important for conceptual design guidance. Beam section geometry area, moment of inertial, torsion constant, and warping constant are main parameters during calculation of section stiffness which is key factor of stiffness of BIW [26, 27].

Assume that there are n_c smallest cavities and n section parts, section is shown in Fig. 9.13. Beam section properties can be calculated based on the Vlasov thin-walled beam theory [28].

Fig. 9.13 The schematic diagram of *i*th wall on the section

$$A = \sum_{i=1}^{n} l_i t_i \tag{9.1}$$

$$I_x = \frac{1}{3}\sum_{1}^{n} (y_i^2 + y_{i+1}^2 + y_i y_{i+1}) l_i t_i \tag{9.2}$$

$$I_y = \frac{1}{3}\sum_{1}^{n} (x_i^2 + x_{i+1}^2 + x_i x_{i+1}) l_i t_i \tag{9.3}$$

$$I_{xy} = \frac{1}{6}\sum_{1}^{n} (2x_i y_i + 2x_{i+1} y_{i+1} + x_i y_{i+1} + x_{i+1} y_i) l_i t_i \tag{9.4}$$

$$I_t = \frac{1}{3}\int_{0}^{s} \{t(s)\}^3 \mathrm{d}s + 2\sum_{1}^{n_c} A_i \overline{q_i} \tag{9.5}$$

$$I_\omega = \frac{1}{3}\sum_{1}^{n} (\omega_i^2 + \omega_{i+1}^2 + \omega_i \omega_{i+1}) l_i t_i \tag{9.6}$$

Among them, A_i is the i cavity area, ω is the generalized sectorial coordinate, $\overline{q_i}$ is the i torsion function.

Equation (9.1) calculates the section area, Eqs. (9.2)–(9.4) calculate the moment of inertia of beam section and the product of inertia, Eq. (9.5) calculates the torsion constant, and Eq. (9.6) does the warping constant; through these we can see that the thickness directly affects the section properties.

Based on the method above, we develop a module of section design in VCD_ICAE in order to calculate the section properties and edit section such as create or delete a section and geometry edition. The main interface is shown in Fig. 9.14, and the development code refers to Appendix A.

Section shape which is also called topological structure affects directly beam section properties and beam element mechanical performance. The best section shape should get both high performance and small material consumption. There are several section shape control methods, like offset method, rectangle simulation method, and polar coordinates. Proportion vector method is taken into account in this book which has less control amount, continuous change, and better operation [29] Proportion vector method controls section shape through the vector angle (θ) and vector number (SV), see Fig. 9.15.

θ is proportional vector rotation angle relative to the original coordinate system, which controls the direction of section shape change, the vector SV determines the size of cross-section shape change. Proportional vector method makes the two variables control section shape, greatly reducing the number of variables.

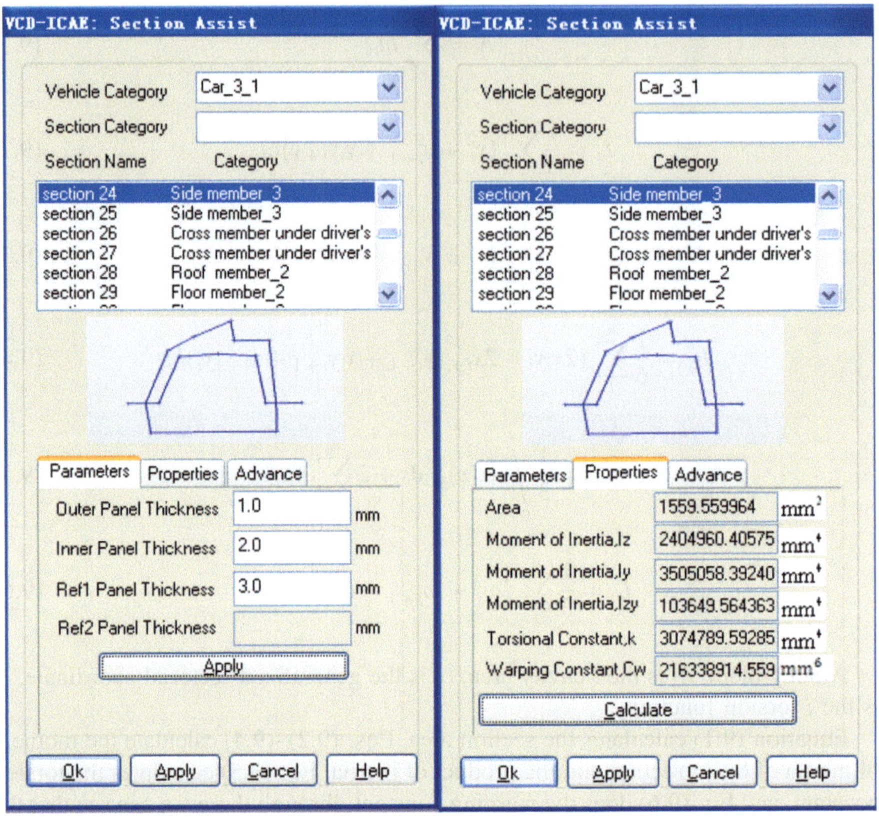

Fig. 9.14 The main interface of the section design model

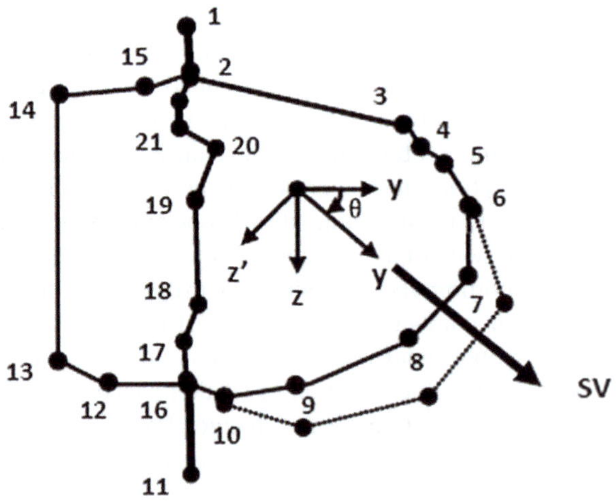

Fig. 9.15 The scale vector control method

Section 8 node in Fig. 9.14, after the confirmation of the rotation angle theta coordinates in a new coordinate system y′oz′

$$\begin{pmatrix} y_s^{'} \\ z_s^{'} \end{pmatrix} = \begin{pmatrix} \cos\theta & \sin\theta \\ -\sin\theta & \cos\theta \end{pmatrix} \begin{pmatrix} y_s \\ z_s \end{pmatrix}_{\text{init}} \tag{9.7}$$

After a given proportion in theta direction vector SV, node 8's coordinates change into

$$\begin{pmatrix} y_s \\ z_s \end{pmatrix}_{\text{new}} = \begin{pmatrix} \cos\theta & -\sin\theta \\ \sin\theta & \cos\theta \end{pmatrix} \begin{pmatrix} y_8^{'}\text{SV} \\ z_8^{'} \end{pmatrix} \tag{9.8}$$

In order to assess potential hot forming materials of lightweight, the optimization model is established without cross section of the beam shape optimization, focusing mainly on the optimization effect of beam cross-section thickness.

Body beam structure is composed of plate (Inner panel), plate (Outer panel), and reinforcing plate (reinforcing panel). Therefore, to determine the thickness of the section, optimized design variables are used. Determination of the concept of body quality is the optimized goal and body bending stiffness and torsional rigidity are the optimized constraints. So, the optimization models:

$$\begin{cases} \text{Design variable: The thickness of the vector } T = (t_1, t_2, \ldots, t_n)^T \\ \text{Optimal object: Min} M(t) = \rho \sum_{j=1}^{m} A_i L_i \\ \text{Constraint condition:} \\ \quad S_{\text{bending}} \geq S_{\text{bending\_low}} \\ \quad S_{\text{torsion}} \geq S_{\text{torsion\_low}} \\ \quad t_i \in (t_{i.\text{low}}, t_{i.\text{up}}) \end{cases}$$

P: density; A_i: cross-section area; L_i: thin-walled beam length; t_i: thickness of plate; $t_{i,\text{low}}$: lower limit of thickness; $t_{i,\text{up}}$: upper limit of thickness; *M:* mass; S_{bending}: bending stiffness; $S_{\text{bending\_low}}$: lower limit of bending stiffness; S_{torsion}: torsional stiffness; $S_{\text{torsion\_low}}$: lower limit of torsional stiffness.

To improve the efficiency of optimization, the sensitivity of calculation optimization variable thickness on the quality of the body, large enough sensitivity of unit are chosen as the optimization object.

Body structure refers to the sensitivity of the change of performance parameters (f_i) in body structure analysis to the change of the body structure design parameters (x_j) are defined as

$$\text{Sen}(f_i, x_j) = \frac{\partial f_i}{\partial x_j} = \frac{f_i(X + \Delta X * e) - f_i(X)}{\Delta x_j} \tag{9.9}$$

X Vector composed of all the body structure design parameters
Δx_j Change of the body structure design parameter values
e Vector sharing same dimension with X and element corresponding to x_j is 1, the others is 0.

In this paper, the optimization goal is M, cross-section design parameters is SV and thickness t; deformation of formula is

$$\mathrm{Sen}(M,t) + \frac{\partial M}{\partial t} \tag{9.10}$$

Aiming at a white body, its typical front anti-collision beam body parts (B1), the threshold beam (B9), A column (B11), before the roof beam (B12), in the roof beams (B15), after the roof beam (B18), the top front side beam (Bl3), after the top side beam (B16), column (B19), C B-pillar upper beam (B20), the B-pillar bottom beam (B21) and the front-seat beams (B22) are used to replace the 22MnB5 hot forming material. Detailed information of each beam is shown in Fig. 9.16, each beam raw material (STEEL) and the replaced the 22MnB5 hot forming materials (HSS) basic properties are shown in Table 9.4.

After replacing the above 12 elements of the beam material with 22MnB5 hot stamping materials, due to the large optimization variables, in order to improve the efficiency of optimization, the concept of the white body model for sensitivity will be analyzed and calculated first. Sensitivity of the calculated results is shown in Fig. 9.28 and the sensitivity value is shown in Table 9.5. It can be seen that the

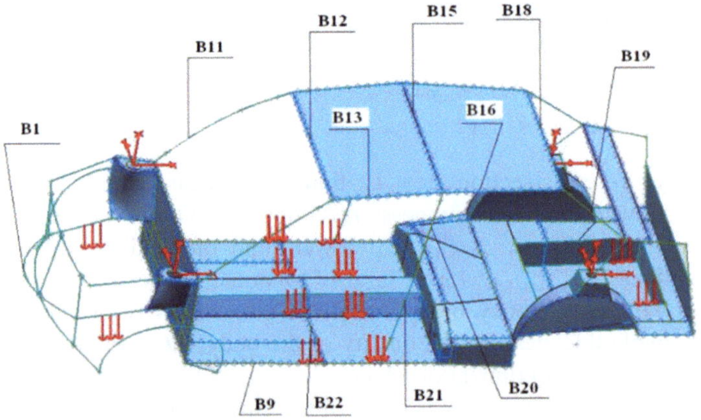

Fig. 9.16 The 12 beams applied by 22MnB5

Table 9.4 Main properties of the two materials

Material	Density (Kg/m^3)	E (Mpa)	YS (Mpa)	TS (Mpa)	ν
Steel	8.0E+03	206940	138	262	0.29
HSS	8.0E+03	209000	1028	1554	0.33

Table 9.5 Sensitivity calculating results of panel thicknesses

Variables	Sensitivity value	Variables	Sensitivity value
T_B1_O	2.128123	T_B19_O	1.250924
T_B1_I	0.766887	T_B19_I	0.882013
T_B11_O	1.937444	T_B19_R1	1.036298
T_B11_I	1.182576	T_B20_O	1.210539
T_B11_R1	1.430408	T_B20_I	1.396874
T_B12_O	1.568472	T_B20_R1	1.138198
T_B12_I	1.781522	T_B20_R2	0.831113
T_B13_O	1.383223	T_B21_O	2.173775
T_B13_I	0.831773	T_B21_I	0.714839
T_B13_R1	1.071244	T_B21_R1	2.086043
T_B15_O	1.307375	T_B21_R2	1.493516
T_B15_I	1.511508	T_B22_O	4.847929
T_B16_O	1.701928	T_B22_I	2.171561
T_B16_I	1.036186	T_B9_O	6.913873
T_B16_R1	1.344283	T_B9_I	5.060672
T_B18_O	1.631824	T_B9_R1	5.154398
T_B18_I	1.751500		

influence on body quality of different beams are different for the selected 12 girders, the thickness of outer plate of threshold beam section (T_B9_O) has the greatest influence on body quality, the sensitivity value is 6.914, the plate with smallest sensitivity is the inner plate of B-pillar bottom beam (T_B21_I), whose sensitivity is only 0.715. In order to improve the efficiency of optimization, only the plate thickness with sensitivity value no less than 1.000 will be selected as the final optimized object; so only 28 plates of the beams are determined as the final optimization objects, which are outer plate of front anti-collision beam (T_B1_O), inner, outer and reinforcing plate 1 of the threshold beam (T_B9_O T_B9_I T_B9_R1), A column of the inner and outer plate and the reinforcing plate 1 (T_B11_O, T_B11_I T_B11_R1), inner and outer plate of front roof beam (T_B12_O, T_B12_I), front side beam in the top plate and the reinforcing plate 1 (T_B13_O, T_B13_R1), both inner and outer plates of the roof beams (T_B15_O, T_B15_I), the reinforcing plate 1 and inner, outer plate of the top side beam (T_B16_O, T_B16_I T_B16_R1), inner and outer plate of rear roof beam (T_B18_O, T_B18_I), outer plate and reinforcing plate of C column (T_B19_O, T_B19_R1), inner and outer plate and the reinforcing plate 1 of B-pillar upper beam (T_B20_O, T_B20_I, T_B20_R1), the outer plate and reinforcement plate 1 and reinforcing plate 2 of the B-pillar lower beam (T_B21_O, T_B21_R1 T_B21_R2) and the inner and outer plate of front-seat beam (T_B22_O, T_B). The T_ before the variable means thickness, _O, _I, _R1 and _R2 stand for the outer plate, inner plate, reinforcing plate 1 and reinforcing plate 2 of beam section, respectively, as shown in Fig. 9.17.

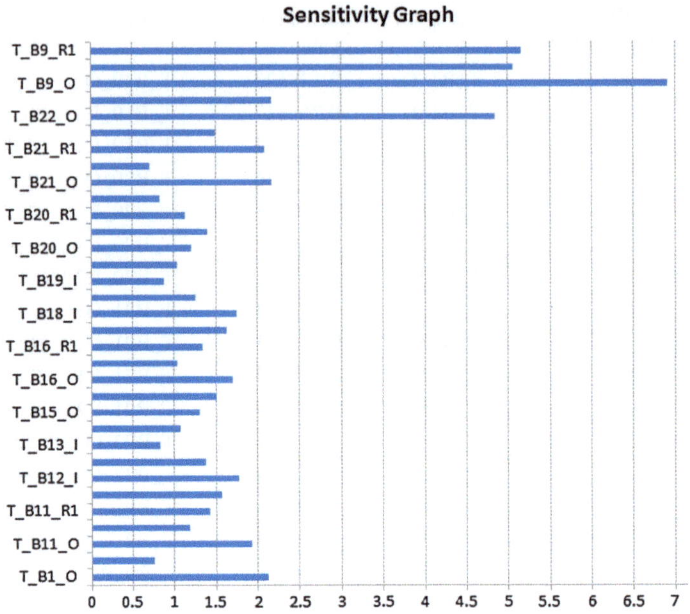

Fig. 9.17 Sensitivity calculating histogram (*O* outer plate of the section, *I* inner plate of the section, *R1* reinforcement plate 1, *R2* reinforcing plate 2)

The above 28 thickness are optimization variables using the bending stiffness and torsional rigidity of the car body model as constraint, lightweight car body as the optimization goal using NX/Nastran solver, to calculate the optimization. The calculation results are shown in Fig. 9.18. Body quality after three iterations reached a minimum value of 386.378 kg compared with the initial body quality 427.940 kg, lost 41.562 kg, lightweight rate was 10.76 %, and the use of 22MnB5 hot forming parts to replace the traditional cold forming can achieve very good effect on lightweight body parts, hot forming technology is to solve the lightweight car body and car security is one of the most effective measures to this contradiction.

The introduction of hot forming material 22MnB5 beam cross-section thickness optimization results point out that 23 objects out of 28 section thickness design variables get thinner; only the top plate and the reinforcing plate inside and outside of the side beam after 1 (T_B16_O, T_B16_I T_B16_R1) as well as the threshold beam plate (T_B9_O) thickness increased; C column plate (T_B19_O) thickness remains unchanged before and after optimization, which showed that the thickness of different parts of the components in the optimization of changing trends are also different. The vast number of car beam thickness decreases during optimization, the biggest drop is 20 %, reducing the optimization goal body quality from 427.94 to 386.378 kg, at the same time optimization constraint stiffness increased from 4357.45 to 4362.115 N/mm, torsional rigidity from 6438.47 N N m/deg increase to 6439.62 m/deg, stiffness of body improved slightly, namely, under the precondition of lightweight car body at the same time makes sure about the safety of the car.

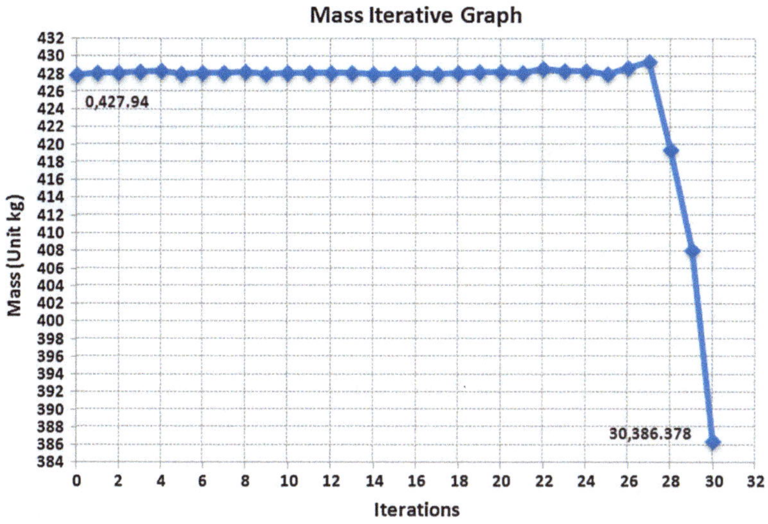

Fig. 9.18 Body mass optimization result

The 22MnB5 hot forming material can realize the lightweight car body and improve the strength of the body, which is an obvious advantage than traditional cold forming. As the trend of modern car requires lightweight design and safety, it definitely has a very broad market prospect.

In order to verify the optimization results, based on the UG NX5 Nastran solver, the optimization model under the working condition of bending displacement nephogram and stress nephogram are analyzed, and the two kinds of cloud picture are compared respectively with that of the original model and compare the two kinds of cloud picture, as shown in Figs. 9.19 and 9.20. It can be seen that the optimized model of displacement and stress are within the scope of the security, and the original model has greatly improved, so the thickness optimization is practical. Constraints on the bending conditions before and after the spring support is a total of bending load of 8852.8 N direction (down). In order to reduce the load point of stress concentration, each load on the three points, the specific method is: in the

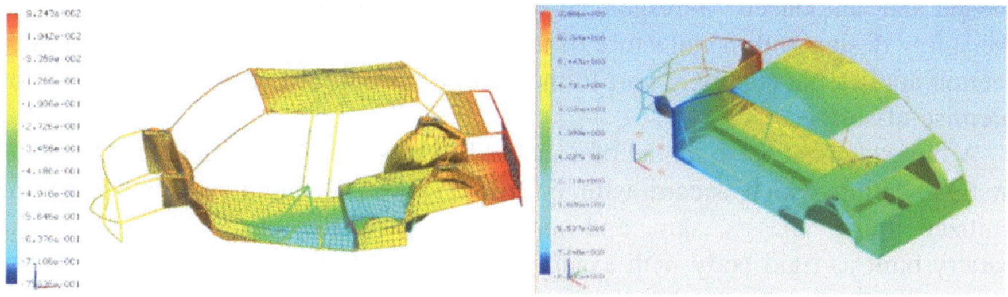

Fig. 9.19 Displacement cloud chart comparison between original and optimal modes

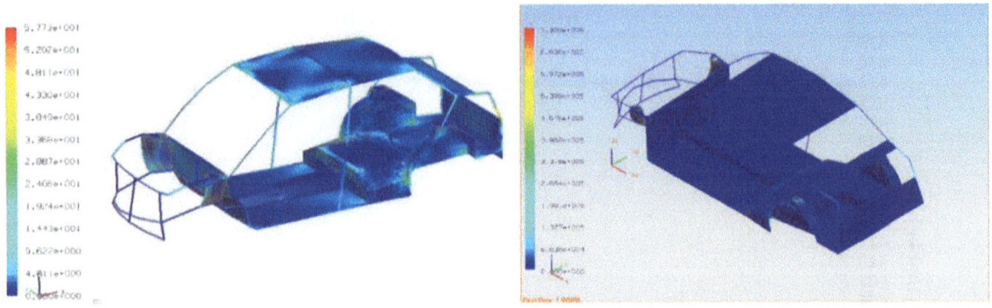

Fig. 9.20 Stress cloud chart comparison between original and optimal modes

front longeron imposed a total of 2100 N engine gravity (1050 N) around each. On the lower beam of front seat, a total passengers gravity of 3140 N is applied, equal to 785 N for each of the four position. On longitudinal beam under the rear seats, a total passengers gravity of 3132.8 N (783.2 N for each of the four position) is applied, and a total baggage gravity of 480 N is applied on the baggage compartment, equal to 240 N on right and left sides.

VCD_ICAE, this section introduces the concept of car design system, and gives the sectional thickness optimization and shape optimization module of the algorithm; based on VCD_ICAE thickness of section optimization module, the preceding research basic material mechanics properties of 22MnB5 introduces some typical parts of white body, and then to optimize the model thickness of section, optimization, according to the results of 28 thickness of section, optimization variable thickness decreases and 23 body quality reduces by 10.76 %; at the same time, body to improve the bending stiffness and torsional stiffness of small, fully shows the hot forming effect of lightweight materials and potential.

9.1.4 The Application of Hot stamping Component in Lightweight Design of Large School Bus

In recent years, the safety of school bus has become the focus of public concern gradually. On March 28, 2012, the 197th executive meeting of the State Council passed and announced "management regulations of school bus safety" to manage the safety design and manufacture of the school bus. Based on the simulation, this section applies hot stamping part to the school bus and tests the light effect on the premise of safety.

A finite element mesh model of a school bus was built based on the CAD model in this section which accord with one domestic real school bus, including the unitized frame, chassis, skin, and some other parts like electrical machine and battery built as rigid body with exact mass (as shown in Fig. 9.21). The center of mass of both element model and real bus is nearly the same. There are almost 186,000 elements and 192,000 nodes in this finite element mesh model.

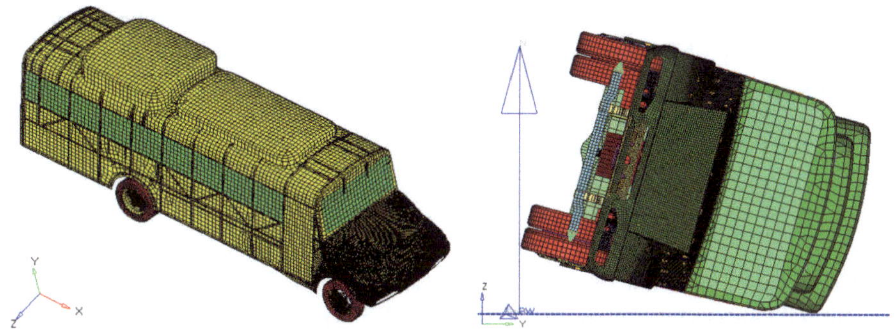

Fig. 9.21 Finite element mesh model of whole school bus (*left*) and the moment of the initial crash (*right*)

Most parts of the mesh model are modeled as deformable body which is meshed with quadrilateral shell elements and a few of triangular shell elements. To make sure if simulation is accurate enough, the length–width ratio of quadrilateral shell elements should be less than 3, warp degree should be less than 20°. Belytschko-T say shell element type should be used whose computational accuracy and speed are especially suitable for large deformation simulation. For the most parts of the bus in this section, it is Q235 low-carbon steel and the one-dimensional rigid body should be used as the connection of parts. The material model should be Cowper–Symonds strain rate model which fits the mild steel materials, so $C = 30(1/\mathrm{s})$, $P = 5$. And survive room is built as rigid body and fastened on floor, based on some related standards, and has no contact with other parts so the result could show whether survive room is invaded by other parts, as shown in Fig. 9.22.

According to ECE R66, in this section, simulation was conducted when school bus first hit the ground and ended while the deformation was over. Based on Table 9.6 and Eq. (9.11), it is calculated that the angular velocity is 2.387 rad/s

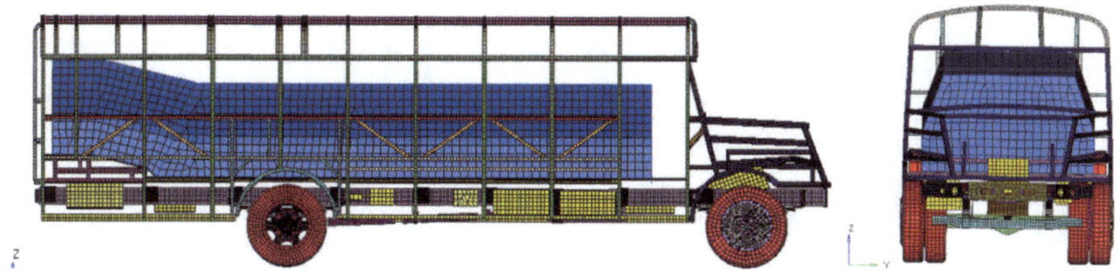

Fig. 9.22 Survive room finite element model

Table 9.6 Relative parameters of school bus (unit: mm)

Name	Height (H)	Height of centroid (h_0)	Centroid offset (t)	Half width (B)	Height of platform (h)	Length of bus (L)
Value	3100	1216	16	1200	800	10040

when the bus is about to touch ground. In the Eq. (9.11), v, ω are the linear and angular velocity of CG″ respectively, m is the mass of school bus and g is acceleration of gravity.

$$\omega = \frac{v}{l} = \frac{\sqrt{2g\left[1 - \sin\left(\arctan\frac{B+t}{h_0} - \arcsin\frac{h}{H}\right)\right]}}{\sqrt{4}(B+t)^2 + h_0^2} \tag{9.11}$$

Numerical simulation of rollover crash is performed using LS-Dyna software. Figure 9.23 shows the simulation result using LS-Dyna and the relative position of survive room and side vertical beams.

As shown in Fig. 9.23, though the right side beams are very close to survive room, survive room is not invaded by side vertical beams. As shown in Table 9.7, the front side vertical beams' deflection are bigger than the rear ones, because at the initial crash time, the skin and frame of battery located on the rear hit the ground first and absorb some crash energy most of which should have been absorbed by the rear side vertical beams. The energy absorbed by front side vertical beams is stronger than rear beams, which leads to a bigger deflection of front side vertical beams.

From the results of rollover crash numerical simulation, side vertical beams and top arch beams absorb the most crash energy and are most liable to get deflection. However the maximum deflection of side vertical beams can directly decide

Fig. 9.23 Numerical simulation result at time 0.12 s in LS-Dyna and relative position of survive room and side vertical beams at time 0.8 s

Table 9.7 Deflection of origin school bus's right side vertical beams (numbered from school bus head to rear) (unit: mm)

1	2	3	4	5	6	7	8	9	Average deformation
943	933	913	894	878	864	841	823	781	874.9

whether side vertical beams invade survive room and harm passengers or not. So side vertical beams and top arch beams are very worthy to optimal design.

According to the analysis of origin school bus rollover crash numerical simulation result and the deformation of vertical beams shown in Fig. 9.23, side vertical beams and top arch beams mainly have bending deformation during the crash. Therefore increasing the flexural rigidity of those beams determined by second moment of area could decrease bending deflection effectively during rollover crash. At the meantime, to reduce the mass of school bus, two new high strength and lightweight structures of side vertical beams and top arch beams are designed in this section. As shown in Fig. 9.24 and Table 9.8, the cross-section shape has changed from rectangle to C type whose flexural rigidity is higher, and also its material use DP980 or hot stamping part instead, the property is shown in Table 9.9.

The rollover crash numerical simulation result of the school bus model with high strength and lightweight structures of side vertical beams and top arch beams are as shown in Fig. 9.25 and Table 9.10. Side vertical beams do not invade survive room and are further from the survive room than the origin, and also deflection of side beams are smaller than the original one, but weight is lighter. Based on Table 9.8 and the number and length of side vertical beams and top arch beams, the new design school bus reduces 94.5 kg in total. Hot stamping materials are not only lighter but also safer.

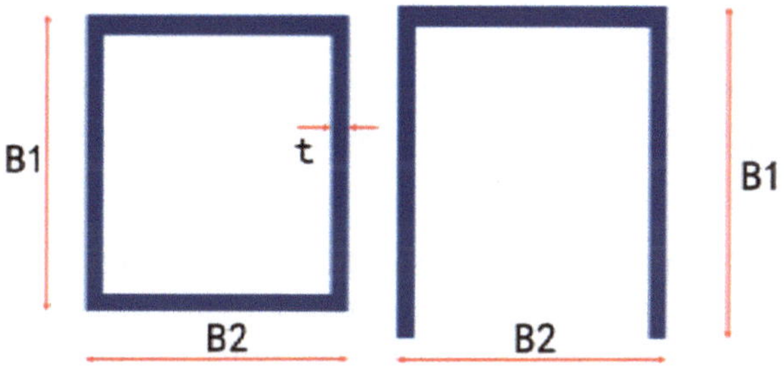

Fig. 9.24 Cross-section optimization of side vertical beams and top arch beams

Table 9.8 Size of cross-section of side vertical beams and top arch beams

	B1 (mm)	B2 (mm)	Thickness (mm)	Second moment of area (kg mm^2)	Mass (%)
Rectangle	50	40	2.0	86690.67	100
C type	58	40	1.6	87589.24	71.07

Table 9.9 Mechanical property of DP980 and hot forming materials

Material	0.2 % yield strength (MPa)	Tensile strength (MPa)
DP980	552	972
Hot stamping	1000 MPa	1500

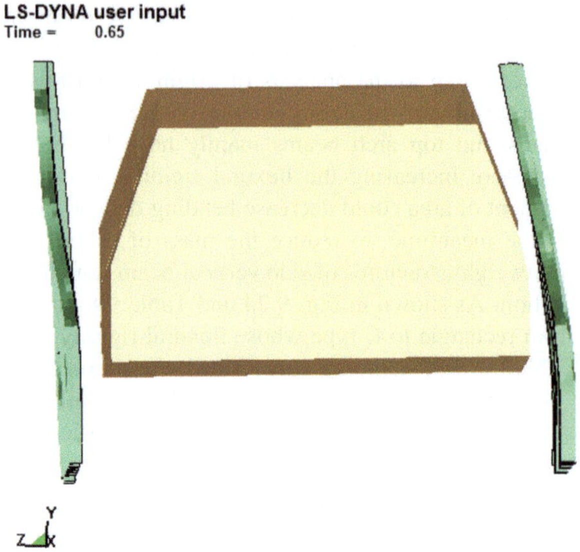

Fig. 9.25 Relative position of survive room and side vertical beams at time 0.65 s

Table 9.10 Comparison between the side vertical beam deflection of the origin and the new design (unit: mm)

No.	1	2	3	4	5	6	7	8	9	Average
Original	943	933	913	894	878	864	841	823	781	874.9
DP980	911	899	879	862	849	840	824	820	796	853
Hot stamped	890	854	831	821	816	801	785	780	768	816

With the above analysis, these two new high strength and lightweight structures of side vertical beams and top arch beams introduced in this section can guarantee the rollover safety effectively and lower the quality of the bus structure at the same time. For electric school bus, safety can be guaranteed, meanwhile, energy consumption can be reduced, which in turn increases the voyage range. But it is not proved to be good in other aspects of safety, which needs further research and supplement.

9.2 The Application of Gradient Hardness Hot stamping Component in Vehicle Bodywork

Hot stamping high-strength steel has characteristics of high strength and hardness, which leads to the shortcomings in crash energy absorption. Designers tend to get the products whose overall intensity is high, but local energy absorption is outstanding, traditional laser welding technology is an example. In hot stamping technology aspect, the third chapter of this book shows that, by controlling the heat transfer between mold and sheet metal, auto parts with variable gradient property can be obtained [30]. Therefore, this section will focus on the research of energy absorption property of typical gradient hardness structure [31, 32].

9.2.1 The Research of Crash Energy Absorption Property of Gradient Hardness Hot stamping Component

In order to research the collision energy absorption property of gradient hardness hot stamping component, a cap-shaped thin-walled part composed of U-shaped gradient hardness hot stamping components was used for crash simulation. Cap-shaped thin-walled parts are main components of the vehicle bodywork, which undertake the main in the process of collision energy absorption, such as front and rear vertical beams, B-pillars, bumpers, B/C pillar stiffener, etc. The geometric dimension of the cap-shaped thin-walled structure described in this section is shown in Fig. 9.26.

The finite element model of the cap-shaped thin-walled structure is shown in Fig. 9.27, in which the BT shell elements are adopted to scatter. This structure is fixed at one end and imposed a rigid plate at the other end, which impacts the structure at a constant speed of 10 m/s along the axis direction. The U-shaped gradient hardness component of the thin-walled structure includes part A and part B corresponding to the quenched hot stamping steel with two kinds of material properties and signed respectively as Mat A and Mat B, whose yield strength are 1400 and 600 MPa, respectively. In Fig. 9.27, LA is on behalf of the length of Mat A and LB is on behalf of the length of Mat B.

The comparative analysis result of the impact energy absorption of thin-walled structures with different proportions of LA and LB is shown in Table 9.11. As it can be seen, along with the proportion of high-strength material increasing, the axial impact energy absorption and specific energy absorption greatly improve, while the increase of peak impact force is not very significant. The peak force will

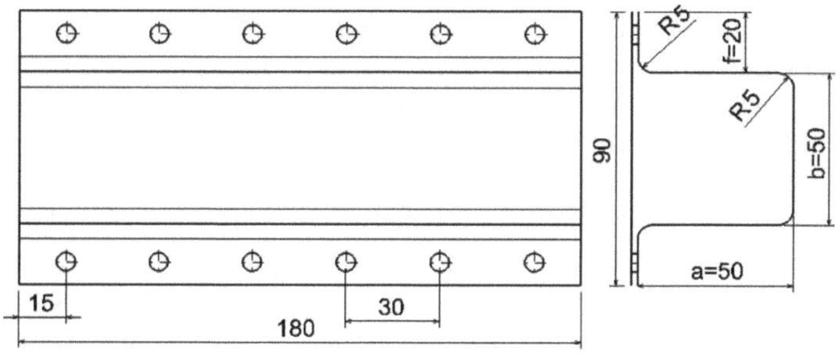

Fig. 9.26 Geometrical configuration of top-hat straight beam and section properties (mm)

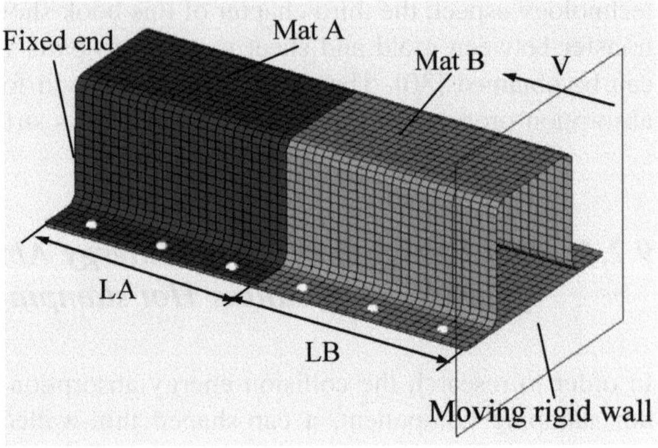

Fig. 9.27 The finite element model of the top-hat thin-wall structure for numerical simulation

Table 9.11 Comparison of different material distribution

Model	LB: LA	Peak force (KN)	Augment (%)	Energy absorption (J)	Augment (%)	Specific energy absorption (KJ/Kg)
1	1:0	181.65	–	6184.0	–	13.4716
	1:1	183.61	1.08	8515.5	37.70	18.5508
	1:2	188.53	3.79	9163.0	48.17	19.9611
	1:4	197.09	8.50	9812.5	58.68	21.3761
	0:1	257.31	41.65	9308.9	50.53	20.2791

not increase substantially until high-strength material is applied completely, while at this time, the energy absorption capacity of the structure will decline. In Table 9.11, the third thin-walled structure has the best impact energy absorption capacity, which is 58.7 % higher than model 1 with single hot stamping material, but the peak force is 23.4 % lower than model 5 with single hot stamping material. It fully shows that hot stamping composite combines the excellent performance of the individual single-phase material, which is suitable to the structure that needs to withstand shock.

Figure 9.28 shows the crash force–displacement curves of the straight beam with different materials distribution. As it can be seen, the level of impact force increases gradually throughout the augment of material strength. From the initial stage of the impact to the moment when compression displacement is up to 40 mm, the impact force curves are almost identical from model 1 to model 3. This is because low-strength material occurs plastic deformation primarily at the beginning of the impact. Until the high-strength material gradually get into plastic deformation stage, the curves become different. The force curve peak of the composite structures does not increase greatly, which can be attributed to the reason mentioned above. From the view of the material distribution throughout the structure, the existence of low strength material is equivalent to introducing the initial defect, which can effectively reduce the peak impact force and lead stable plastic folds of the structure [33]. In the structure of high-strength material, the impact force curve decreases gradually

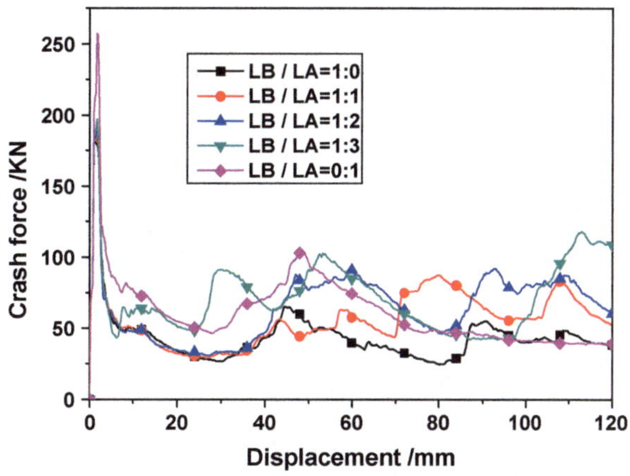

Fig. 9.28 Crash force-displacement curves of different material distribution

after the second peak. This is because the deformation mode of the structure changes and collapse occurs in the subsequent wrinkle process.

It is can be seen from the energy absorbing—displacement curves described in Fig. 9.29 that with the compression displacement increasing, the energy absorption gradually improves. For composite structures, because low-strength material occurs, plastic deformation primarily at the beginning of the impact, energy absorption curves at the initial compression stage remain almost same. In the subsequent stages, the slope of energy absorption curves gradually increase, which marks that high-strength material accesses to plastic deformation.

Therefore, these composite structures with "hard and soft" combination contribute to reducing the peak impact force and improving the energy absorption efficiency. Relative to conventional impact energy absorber, the absorption pattern to guide drapes by setting gaps has a greater advantage in terms of manufacturing costs or crashworthiness performance [34].

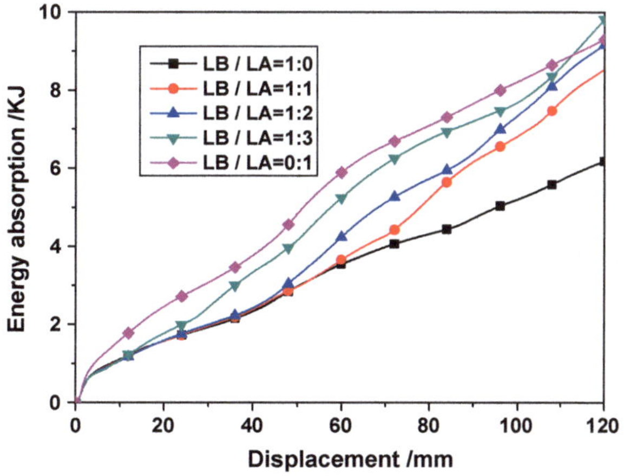

Fig. 9.29 Energy absorption–displacement curves of different material distribution

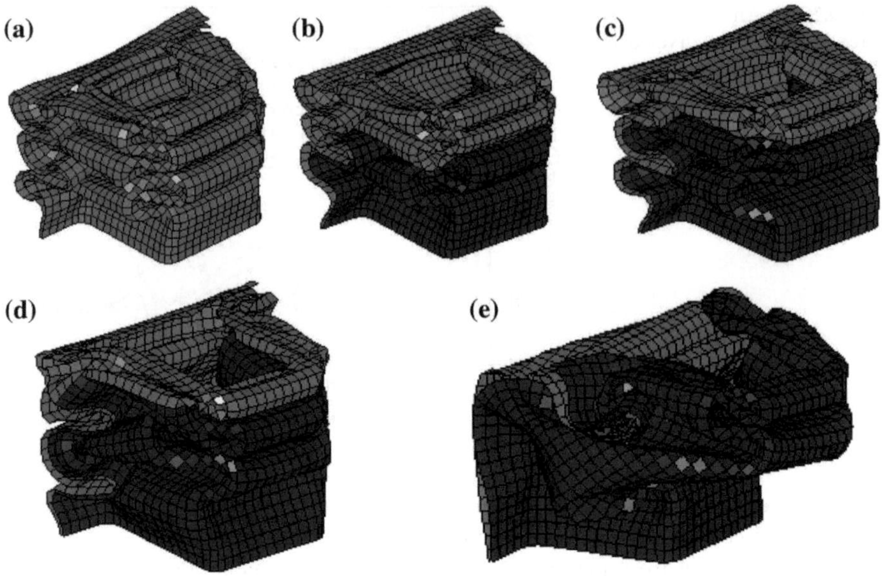

Fig. 9.30 Deformation profiles of composite material thin-walled tubes. **a** Model 1, **b** model 2, **c** model 3, **d** model 4, **e** model 5

Figure 9.30 shows the deformation outlines of five structures with different material combinations. Low-strength material begins to fold at the crash end, then subsequently high-strength material is compressed gradually. During the impact process, every model folds regularly from model 1 to model 4, which is related to the guide function of the precedence buckling of low-strength material. As the proportion increasing of high-strength material, the number of plastic folds becomes from 5 to 6, and the fold wavelength of low-strength material, which is compressed more compact, is shorter than high-strength material. In the model 5 of high-strength material, overall buckling is occurred after the model folds twice at the impact end, which is also the main reason for the decline of energy absorption ability.

In summary, according to the analysis of the crash absorption capacity of cap-shaped thin-walled structures consisting of U-shaped metal composite material, the energy absorption capability can be 58.7 % better and peak force can be 23.4 % lower through reasonable material arrangement of U-shaped metal composite parts compared with the corresponding single hot stamping material. By analyzing the fold outlines of the cap-shaped structures, it is found that the energy absorption of composite material is superior in contrast with single "soft material" or single "hard material". This metal composite material is particularly suited for what needs to withstand collision impact.

9.2.2 The Application of Gradient Hardness Hot stamping B-Pillar in Vehicle Bodywork and Optimization Design

B-pillar is one of the major components of the car side impact performance. To improve the side impact performance, the crash characteristic of these major components must be improved. In order to improve the crashworthiness of B-pillar, designers usually design complex shapes to make it possible to meet crashworthiness requirements, such as to set strength members and so on, which will increase the design difficulty of B-pillar undoubtedly. Therefore, it is a method to change the material gradient property of B-pillar to ensure the design requirements of crashworthiness and simplify the shape of B-pillar. The performance study of gradient hardness hot stamping steel described in Sect. 3.4 provides the best implementation of the gradient material of B-pillar.

Experiment and numerical simulation are two methods to achieve gradient material property design of B-pillar. Based on a typical side collision model of B-pillar, this section uses a B-pillar model with gradient hardness property instead of the traditional B-pillar with single hardness to do crashworthiness and energy absorption performance numerical simulation and optimize its strength distribution. With the goal of intrusion and energy absorption characteristics, response surface model and radial basis function approximative model are applied to achieve the multi-objective optimization design of the B-pillar structure with gradient hardness distribution.

In optimization process, high-strength steel DP1180 applied widely in current market was used as the basic scheme to carry out optimization design. The yield strength of DP1180 is about 826 MPa and tensile strength is 1180 MPa, which can fully guarantee the performance of the car crash and high security factor.

In the section, according to the specific size and quality parameter stipulated in the Chinese New Car Assessment Program (C-NCAP) and the Chinese side impact regulation GB20071-2006, calculation model was established and finite element model was established based on the dynamic display algorithm software LS-DYNA. The vertical length of B-pillar model is: $L = 1180$ mm and the relative position and the collision area between crash block and B-pillar are established according to the regulations. Quadrangle shell elements are used and the size of elements is 5 mm. In addition, the key words *CONTACT_AUTOMATIC_SURFACE_TO_SURFACE is adopted for the contact between B-pillar and crash block. Furthermore, adjacent various materials of the B-pillar share same nodes and the top and bottom edges are constrained. The speed of the crash block is 50 km/h in the total impact time is 0.1 s and its mass is 100 kg. The collision calculation model of gradient strength property B-pillar is shown in Fig. 9.31.

Based on the displacement image after the collision shown in Fig. 9.32 and American NASS regulations, the B-pillar can be divided into three parts according to whether the deformation is bigger than 8 cm or not: namely, the top length of the B-pillar $d1 = 375.3$ mm, the middle length $d2 = 537.5$ mm and the bottom length $d3 = 267.2$ mm.

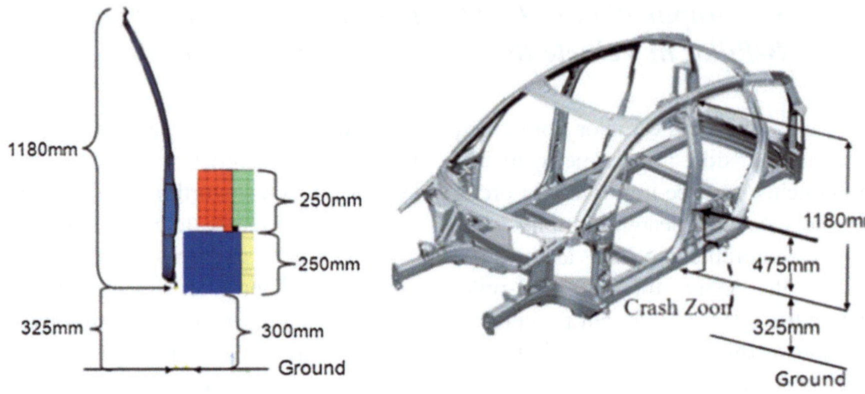

Fig. 9.31 Collision model of gradient strength B-pillars

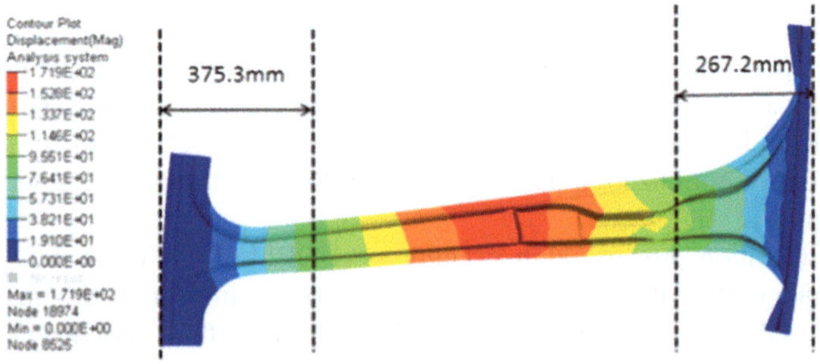

Fig. 9.32 Intrusion distribution of gradient strength B-pillar

As the most important protection structure during a side impact, the inside of B-pillar is usually very close to the protected persons and goods. Therefore, the maximum deformation MaxD of B-pillar during a collision is one of the important indicators. Meanwhile, in order to assess the performance of B-pillar structure comprehensively in a collision, the energy absorption property in a collision need to be evaluated. Energy (EA) is defined as the amount of energy absorbed by the structure in a collision, which can be used as another important crash performance evaluation of gradient B-pillar.

The correctness of the calculation results can be estimated according to whether the total energy is constant or not in the collision. The energy curve is shown in Fig. 9.33. Through the initial speed and quality of the mass block, it can be calculated that the initial total energy is 9.65 kJ. Due to the deformation of B-pillar in the collision, the inner energy increases and the kinetic energy decreases. After the collision, the kinetic energy is reduced to near zero and the inner energy increases similar to the total energy. In the whole collision process, the total energy remains

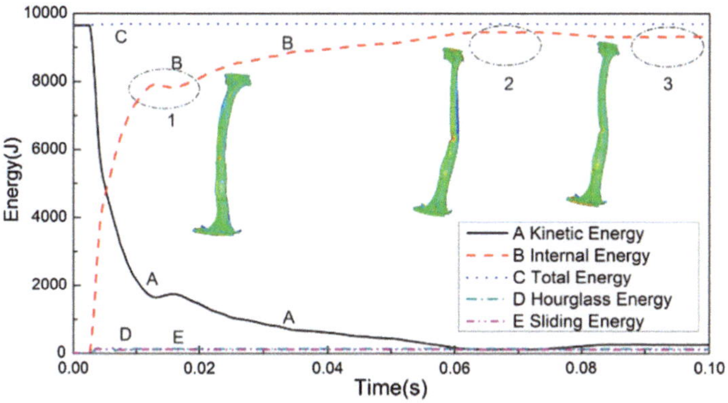

Fig. 9.33 Energy transformation of B-pillar

constant and hourglass energy is less than 7 % of the total energy, which proves that the finite element model is correct and feasible.

According to the district cooling test of plate dies described in Sect. 3.4.2 of this book, the continuous gradient of hot stamping part can be achieved by changing the die temperature [8]. For three parts of strength distribution, four different yield strength values: 592.56, 791.36, 904.55, and 1032.32 MPa are chosen to arrange and combine, corresponding to the die temperatures of 500, 400, 330, and 25 °C respectively, thus there are 64 kinds of different conditions. The stress–strain curves in various temperatures are shown in Fig. 9.34. The MaxD and EA were seen as the evaluation index and side impact finite element analysis were conducted to 64 kinds of gradient hardness B-pillars. The analysis result is shown in Fig. 9.36 and in which, case 1, 22, 43, and 64 show the collision results of single strength B-pillars.

As can be seen, the results of side impact vary due to different strength distributions. For the uniform B-pillars, the MaxD is negatively related to EA, which could imply that the decrease of MaxD and the increase of EA are contradictory.

Fig. 9.34 Tensile curves under different temperatures

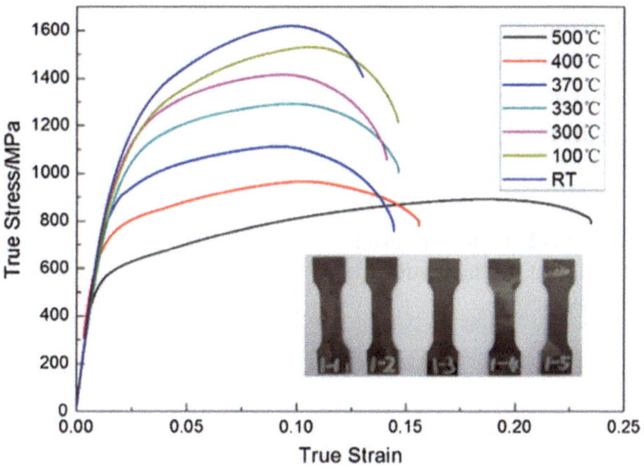

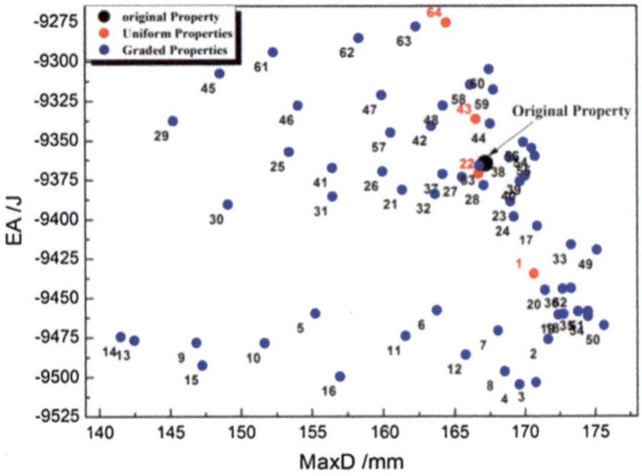

Fig. 9.35 Energy absorption and intrusion of different gradient strength B-pillars

However, for B-pillars with gradient property, it can be found that the MaxD decreases but EA increases as well, which can overcome the contradiction between EA and MaxD. When σ_{middle} is 1032 MPa, which means that the middle strength of B-pillar is higher than or similar to the edge strength, the energy absorption will increase but intrusion will decrease obviously, such as case 14 and 15 seen in Fig. 9.35.

This is due to the large amount of energy absorbed by lower strength material at the ends when the MaxD appears in middle region, which helps decrease the deformation in the middle part and reduce the MaxD.

When σ_{middle} is 592.56 MPa, which means that the middle strength of B-pillar is lower than or similar to the edge strength, the energy absorption and intrusion will both increase, such as case 3 and 4 seen in Fig. 9.35. This is because the lower strength in middle part is too weak to resist the impact and the ends cannot help to absorb energy, which leads to the outstanding increase of intrusion and energy absorption.

Several typical cases are listed in Table 9.12.

As we all know, the main function of B-pillar is to protect passengers from collision, therefore, it is the first priority to enhance crashworthiness followed by the energy absorption performance. In conclusion, when the strength distribution of B-pillar is like $\sigma_{\text{up}} \leq \sigma_{\text{middle}} \geq \sigma_{\text{bottom}}$ and σ_{middle} is large enough, the crashworthiness performance of B-pillar will increase obviously.

Table 9.12 Collision result of different gradient strength B-pillars

No.	σ_{up}	σ_{middle}	σ_{bottom}	MaxD/mm	EA/J
Origin	826	826	826	167.145	9364.63
3	593	593	905	170.776	9503.47
4	593	593	1032	169.586	9504.76
14	593	1032	791	141.493	9474.51
15	593	1032	905	147.253	9492.43

In order to further improve the crashworthiness of B-pillar, yield strength σ_s of the three parts were seen as design variables and to EA and MaxD were seen as objective function based on the strength configuration of case 13, 14, 15 from above researches. The multi-objective design optimization problem of a B-pillar subjected to a side impact load with the two objectives (EA and MaxD) mentioned above can be formulated as Eq. (9.12). The MaxD of original material DP1180 is used as upper baseline and EA is set as lower baseline.

$$\begin{cases} \text{Min MaxD}, -\text{EA} \\ \text{s.t. EA} \geq \text{EA}_{\text{caseori}} = 9364.63\,\text{J} \\ \text{MaxD} \leq \text{MaxD}_{\text{caseori}} = 167.145\,\text{mm} \\ \sigma_{\text{up}} \leq \sigma_{\text{middle}} \geq \sigma_{\text{bottom}} \\ 592.56 \leq \sigma_i \leq 1032.32 \\ (i = \text{up, middle, bottom}) \end{cases} \quad (9.12)$$

To improve the solution efficiency, the four-order response surface method (RSM) approximative models were applied in the optimization process to replace the original finite element models. Firstly, 50 samples among design domain were selected by the Latin Hypercube design method. Second, four-order RSM approximative models were established based on the simulation results of these 50 samples and the accuracy of the RSM was evaluated. Finally, the NSGA-II multi-objective optimization algorithm was used based on the RSM models to get the optimization of the optimal Pareto solution set as shown in Eq. (9.12). The flow chart is shown in Fig. 9.36.

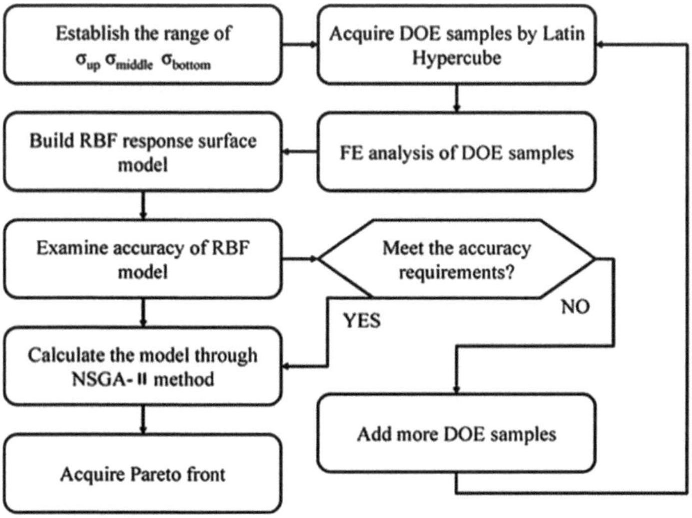

Fig. 9.36 Flow path of multi-objective optimization

Table 9.13 Accuracy verification of four derivative response model

ID	σ_{up}	σ_{middle}	σ_{bottom}	MaxD/mm			EA/J		
				FEA	RSM	Error/%	FEA	RSM	Error/%
1	905	593	905	173.77	173.98	0.12	9458.68	9459.35	0.01
2	905	905	1032	167.50	167.19	0.18	9339.49	9345.22	0.06
3	1032	593	971	175.62	175.59	0.02	9467.49	9473.95	0.07
4	1032	791	791	170.48	171.15	0.39	9354.89	9354.91	0.003
5	1032	905	593	160.51	160.63	0.07	9345.06	9340.87	0.04

In this section, MaxD and EA were seen as optimization goals and the RSM was established based on fourth-order polynomial response surface method. Five samples were selected randomly to test the accuracy of the RSM. The simulation results and relative errors of RSM models and the finite element analysis (FEA) results of these five samples were listed in Table 9.13. As can be seen, the simulation errors of the fourth-order polynomial RSM models are lower than 1 % and the fitting accuracy are excellent.

One of the distinguishing features of the multi-objective optimization method is that there is no global optimal solution but a set of optimal solutions called Pareto set or Pareto front. Based on the Pareto set, designers can determine the final optimization solution according to the specific questions.

Based on the RSM model, the Pareto set was obtained by NSGA-II (Eq. 9.12) as shown in Fig. 9.37. The optimal solution set paves the way to optimal design in strength aspect of gradient property B-pillar for certain cases. For example, when intrusion should be controlled lower than 132.0 mm, point c on Pareto front can meet the requirement and will provide the largest energy absorption. While point a and point b on Pareto front represent the designs that provide the least intrusion and largest energy absorption respectively. Point a represents the case that the strength at top and bottom ends of B-pillar is small and the intrusion is the least. While, point b represents another case that the strength at bottom ends of B pillar is highest

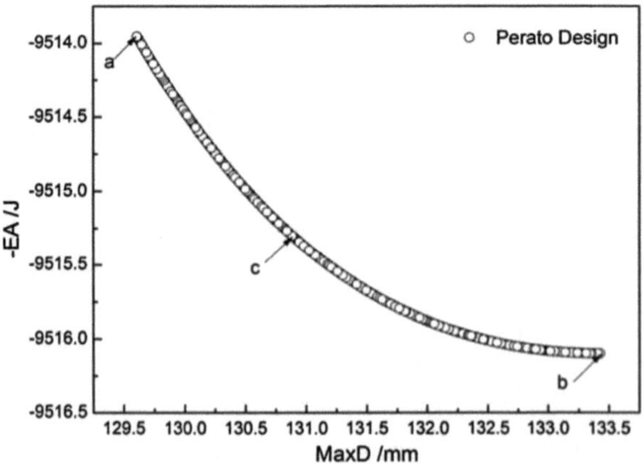

Fig. 9.37 Optimal solution of gradient strength B-pillar based on Pareto front

Table 9.14 Several Pareto optimal solutions

No.	σ_{up}	σ_{middle}	σ_{bottom}	MaxD/mm	EA/J
a	593	1032	593	129.61	9513.95
b	593	1032	654.2	133.43	9516.1
c	593	1032	608.72	130.51	9515.0
ori	826	826	826	166.80	9369.8

and the EA is largest as well. Otherwise, the design with middle strength in bottom end can effectively improve the crashworthiness of B pillar in terms of MaxD and EA. The strength on top of B-pillar should maintain lowest, which should be considered in design (Table 9.14).

In this section, finite element analysis was used to analyze the crashworthiness of gradient hardness B-pillar. By establishing RSM approximate model, multi-objective optimization design was carried out to improve the crashworthiness of B-pillar. It is concluded that

(1) The gradient hardness property of high-strength steel 22MnB5 was studied by way of gradient temperature mold. The relationship between yield strength of high-strength steel and mold temperature was established, which can provide effective guidance for the optimization design based on practice.
(2) The yield strength distribution of gradient hardness B-pillar has a great influence on the collision performance. When the strength in middle is high and on both ends is small, the EA will increase and intrusion will decrease obviously. While, when the strength in middle is small and on both ends is high, the EA intrusion will increase. So, if the distribution of is B-pillar is $\sigma_{up} \leq \sigma_{middle} \geq \sigma_{bottom}$ and σ middle is big enough, the crashworthiness of B-pillar can be improved distinctly. Compared with traditional uniform B-pillar of DP1180 material, the crashworthiness of B-pillar with optimum gradient distribution has been significantly improved.
(3) Using the RSM approximative model and genetic algorithm NSGA–II, the Pareto set of crashworthiness design of gradient hardness B-pillar can be got, which can further improve comprehensive crashworthiness of B-pillar.

It is worth noting that the gradient strength distribution of B-pillar in this section is arranged according to the relevant regulations and collision results. For other collision conditions, the specific distribution should refer to the specific constraint conditions.

9.3 Summary

Based on simulation and experiment, this chapter studied the lightweight design and application of typical hot stamping parts in automotive body. By comparative researches in Sect. 9.1, it is found that hot stamping parts can ensure vehicle safety and achieve lightweight at the same time. From the point of view of energy absorption analysis, Sect. 9.2 studied and designed the gradient hardness property of typical hot stamping thin-walled structure and B-pillar, and revealed the characteristic in the application of impact structures in vehicle body, which is significant. The application and optimization design of corresponding parts in each case and many new concepts, such as technology combination, CAE vehicle collision, and VCD-ICAE system concept car design, have an important guiding function on the eventual application of hot stamping technology in vehicle body.

References

1. Guo R, Hou W, Hu P et al (2012) Research and implementation of beam section optimization for vehicle body light weighting. Automot Eng 1:40–45 (in Chinese)
2. Hongyuan Y (2013) Research on hot forming process improvement of automobile high strength steel and its application. Dalian University of Technology, p 72 (in Chinese)
3. Hu P, Yu H, Ying L et al (2013) Development of hot forming high strength components based on side impact. Automobile Technol 1:57–61 (in Chinese)
4. Runqing G (2012) Research on several fundamental properties of automobile hot forming steel 22MnB5 and its application. Dalian University of Technology, p 79 (in Chinese)
5. Yu Y, Ying L, Hu P et al (2015) Optimization design method of school bus structure lightweight on the rollover safety. Opole, Poland
6. Guozhe S, Ma H, Liu L et al (2013) Optimization of thin-wall straight beam based on crashworthiness and design space. Automobile Technol 7:34–38 (in Chinese)
7. Ying L, Zhang F, Dai M et al (2015) Optimization of hot forming vehicle structure with tailored strength side crash simulation. In: 2015 congress of society of automotive engineering of China, Shanghai, China (in Chinese)
8. Ying L, Dai M, Hu P et al (2014) Strength and Hardness prediction based on cooling rate for hot forming high strength steel. J Jilin Univ (Eng Technol Ed) 6:1716–1722 (in Chinese)
9. Ying L, Yu Y, Dai M et al (2015) Influence of forming-quenching parameters on hardness and strength of high strength steel hot forming product. Trans Mat Heat Treat 1:205–210 (in Chinese)
10. Xiaoqiang H, Yu H, Ying L et al (2013) Research on influence of key process parameters of hot stamping on mechanical properties of parts. China Metal Form Equip Manuf Technol 4:58–62 (in Chinese)
11. Ma N, Shen G, Zhang Z et al (2011) Material performance of hot-forming high strength steel and its application in vehicle body. Chin J Mech Eng 8:60–65 (in Chinese)
12. Sun H, Hu P, Ma N et al (2010) Application of hot forming high strength steel parts on car body in side impact. Chin J Mech Eng 23:252–256
13. Maddever W, Guinehut S (2005) Use of Aluminum foam to increase crash box efficiency. SAE technical paper
14. Sofi F, Kulkarni S, Haarda M et al (2008) A novel energy absorber design technique for an idealized force-deformation performance. SAE technical paper
15. Zhang Z, Liu S, Tang Z (2010) Crashworthiness investigation of kagome honeycomb sandwich cylindrical column under axial crushing loads. Thin-Walled Struct 48:9–18
16. Baoyue L, Hou W, Zhang H et al (2011) Design of multi-objective optimization module in vehicle body structure concept design system. Trans Chin Soc Agric Mach 4:17–21 (in Chinese)
17. Wenbin H, Zhang H, Chi R et al (2009) Development of integrated system for conceptual design of vehicle body structure. Comput Integr Manuf Syst 2:240–244 (in Chinese)

18. Fengchong L, Chen J, Lin J (2005) Research on parameterized modeling and application for car. Comput Integr Manuf Syst 11:183–188 (in Chinese)
19. Hu P, Guo R, Hou W et al (2011) Optimal design for beam section thickness of concept vehicle body. Automobile Technol 10:1–5 (in Chinese)
20. Prokić A (2000) Computer program for determination of geometrical properties of thin-walled beams with open–closed section. Comput Struct 74:705–715
21. Prokić A (1999) Computer program for determination of geometrical properties of thin-walled beams with open profile. Adv Eng Softw 30:109–119
22. Banichuk NV, Ragnedda F, Serra M (2002) Optimum shapes of bar cross-sections. Struct Multi Optim 23:222–232
23. Vinot P, Cogan S, Piranda J (2001) Shape optimization of thin-walled beam-like structures. Thin-Walled Struct 39:611–630
24. Serra M (2005) Optimum design of thin-walled closed cross-sections: a numerical approach. Comput Struct 83:297–302
25. Hou WB, Zhang HZ, Chi RF et al (2009) Development of an intelligent CAE system for auto-body concept design. Int J Automot Technol 10:175–180
26. Donders S, Takahashi Y, Hadjit R et al (2009) A reduced beam and joint concept modeling approach to optimize global vehicle body dynamics. Finite Elem Anal Des 45:439–455
27. Jie X, Lei Y, Zhang J et al (2007) Study on relationship between the geometrical properties of body main sections and stiffness of BIW. Automobile Technol 2:15–18 (in Chinese)
28. Murray NW (1984) Introduction to the theory of thin-walled structures. Clarendon Press, Oxford
29. Yim HJ, Lee SB, Pyun SD (2002) A study on optimum design for thin walled beam structures of vehicles. SAE technical paper
30. Facai L (2014) Research on the influence of model tailored tempering factors on the properties of S-rail beam by experiment. Dalian University of Technology, p 74 (in Chinese)
31. Shouzhao L (2014) Effects of the forming process on the crash performance of S-rail with tailored properties. Dalian University of Technology, p 64 (in Chinese)
32. Yu Y, Ying L, Hu P et al (2014) Optimization of typical structure component of gradient strength hot stamping steel based on front crash simulation. In: 2014 congress of society of automotive engineering of China, Shanghai, China
33. Kalpakjian S (2001) Manufacturing engineering and technology. Pearson Education India
34. Ma N, Hu P, Guo W et al (2009) Feasible methods applied to the design and manufacturing process of hot forming. In: IDDRG2009 conference, Golden. CO USA

Chapter 10
The Optimization Design and Manufacture of Hot Stamping Mold

10.1 The Key Technology of Hot Stamping Mold Design

10.1.1 The Whole Structure of Mold

In general, the structure of cold stamping die mainly includes the following several parts: upper die, lower die, blank holder ring, punch parts, die parts, all kinds of panel boards, all kinds of lifting lugs, directional plate, guide pins, guide column sets, limit device, safety device, baffle plate, waste tank, variety of standard parts, etc. [6].

Cold stamping die is used only for parts forming, hot stamping tool not only for forming, but also for cooling quenching parts always with good mechanical properties and dimensional accuracy. So compared with the traditional cold stamping dies, the main characteristic of hot stamping die on the structure is the cooling system for cooling parts, and other aspects almost like cold stamping die, including the manufacturability of various parts and structural parts. Hot stamping die cooling system is composed of the external power system, mold internal cooling water circulation loop and connected with cooling pipes, etc. The internal cooling circuit design is the key and difficult point for hot stamping die design, also is the core technology of the hot stamping process.

There are many ways of designing cooling water channel in the mold. According to all kinds of literature reports, there are three processing methods for hot stamping mold cooling pipe, mainly include casting, drilling, and inlaying [24] (Table 10.1).

In addition, in the process of hot stamping, when put the steel blank with high temperature in the hot stamping mold, before stamping the area which blank in contact with the mold surface will cool quickly, and phase changes and harden rapidly, so that it can lead to local parts hardening and results in an uneven of forming group composition, huge parts internal thermal stress even causes cracking phenomenon in the forming process. Thus, hot stamping dies must be designed with special loading device to avoid the above potential problems [11].

P. Hu et al., *Hot Stamping Advanced Manufacturing Technology of Lightweight Car Body*, DOI 10.1007/978-981-10-2401-6_10

Table 10.1 The manufacturing methods of cooling channels

Casting	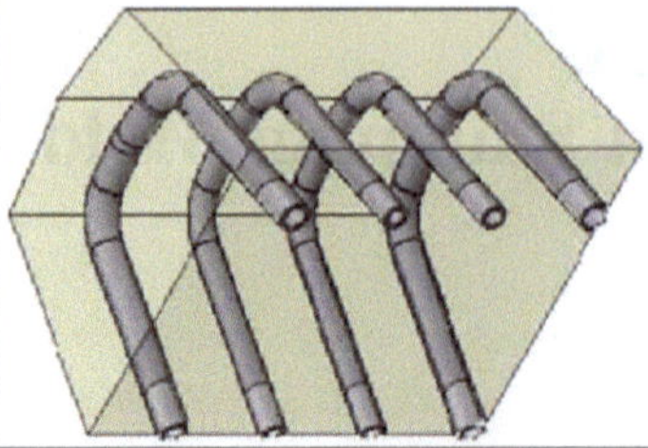	Advantage: cooling uniformity, good sealing, high-mold intensity Disadvantages: difficult to machining, the cooling effect is not good, and is not convenient to maintain
Drilling	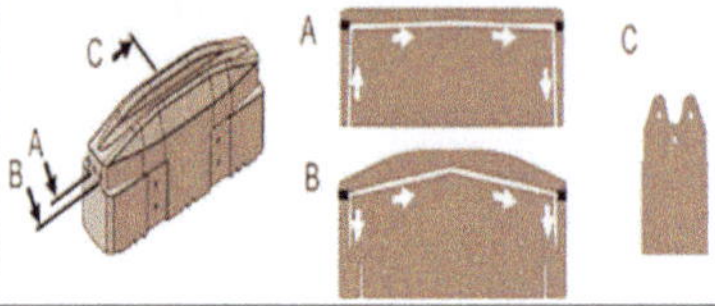	Advantages: good cooling effect, easy processing, good sealing, convenient repair Disadvantages: lack of uniform for curved surface cooling
Inlaying	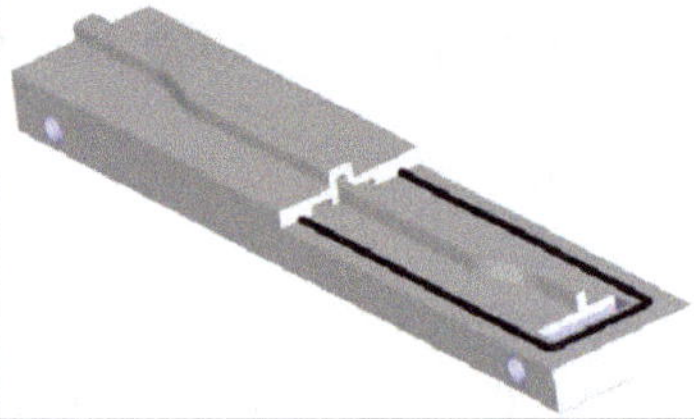	Advantages: good cooling effect, uniform cooling Disadvantages: poor sealing, low mold strength

10.1.2 The Selection of Material

In the process of high strength plate hot stamping, plate through upper austenitizing temperature falls down to room temperature. So when choosing mold materials of hot stamping die we should pay more attention to its thermodynamic performance. Hot stamping die transfers away the heat of parts, and finishes quenching the parts mainly through the contact of working surface and parts. So materials must have good heat conduction, to ensure the rapid heat transfer between steel plate and mold and achieve good cooling function [1]. Second, in the process of high-strength plate hot stamping, stamping sheet strength increased significantly, the structure of the forming die must have sufficient rigidity, surface hardness, and fatigue life. On the other hand, mold is working under the condition of alternate work with severe cold and hot, so the thermal and mechanical properties of mold material must be well [26], in order to guarantee the stability of mold size accuracy at work. Third, good surface hardness not only can resist high temperature and strong heat friction that against the mold, but also protect mold surface from the abrasive wear effect by oxide layer of debris and particles [9, 19]. In addition, due to the need of designing cooling channel in the mold, mold materials also need to have good corrosion resistance to ensure no cooling pipe blockage by cooling medium corrosion [25].

Therefore, when mold material is selected, it is necessary to consider the factors below: the expansionary of mold, the frequent changes of mold surface temperature, the mold plastic deformation failure because of the aggravating wear of the mold by

Table 10.2 The classification of hot work die steels

Sort by property	Sort by chemical component	Grade of steel
Low heat resistance and high toughness die steel	Low alloy hot die work steel	5CrNiMo, 5CrMnMo, 4CrMnSiMoV, 5Cr2NiMov
Middle heat resistance and high hot strength die steel High heat resistance and high wear resistance die steel	Chromium die steel Chromium–tungsten die steel Chromium–tungsten–molybdenum die steel	4Cr5MoSiV, 4Cr5MoSiV1, 4Cr5W2SiV, 4Cr5Mo2MnV1Si, 3Cr2W8 V, 5Cr4W5Mo2 V, 3Cr3Mo3W2 V, 5Cr4Mo3SiMnVA1, 5Cr4W2Mo2VSi, 4Cr3Mo2MnVNbB
Low heat resistance and high toughness die steel High heat resistance and high wear resistance die steel	Chromium die steel Tungsten die steel	8Cr3, 3Cr2W8V
Non-magnetic die steel Aging strengthened die steel	Austenitic die steel Martensite die steel	7Mn10Cr8Ni10Mo3V2, 5Mn15Cr8Ni10Mo3V2, 7Mn15Cr2A13V2WMo, 18Ni(300), 18Ni (250), 18Ni(350)

the mold cavity surface high temperature soften and fatigue failure and cold or hot fatigue, etc. Improper selection of materials not only will cause a fluctuant quality of stamping parts, but also will produce the problem such as die wear, cracking, and scrap. Therefore, we need to select die material and heat treatment process accurately. The selection of mold material in addition to consider the above factors, special attention should be paid to the heating temperature, heat conduction coefficient, cooling speed requirements, and other influences. Generally, specific situation of work and the hot forging steel choice the material of hot stamping die can be regarded as important factors. For example, aluminum, magnesium, and boron alloy plate can use H13 hot stamping die material, such as 3Cr2W8V steel [10, 20].

The relationship between the classification methods of hot work die steels and grade of steel are shown in Table 10.2.

The selection of hot stamping die steel material should be based on different kinds of mold and die stress, process performance object, complexity of structure, selection of equipment, process characteristics, processing batch size to select material. In addition, hot stamping die steel should be used according to the performance requirements, giving full play to the material potential, economic and reasonable principle of material selection.

10.1.3 Surface Engineering of Mold

Mold design includes the design of stamping process and mold structure. Stamping process design is the core of the mold design. And one of the main tasks of stamping process design is that design a reasonable process. Stamping die surface is made on the base of the geometric data of model also includes model repairing, editing, reperforating, flanging, boundary fairing, processing supplement, and so on. This process is called Surface engineering of mold. Therefore, the key to stamping die surface design is surface engineering [3].

In the traditional design, first, mold design personnel always analyze the machinability of parts that according to the product's final configuration and material combination stamping performance. Second, according to the drawing process rule confirm a series of key process elements such as the stamping direction, reperforating, edge unwrapping, blank-holder surface shape, addendum, and so on. Finally, after repeated stamping debugging and design modification it will get a satisfied result. The design process is as shown in Fig. 10.1. If this kind of method which used to design large complex parts such as automotive covering parts is applied, it will cause a waste of time, more difficult to process and higher costs. What is worse is that it is difficult to guarantee the quality of the product parts.

Because of the complexity of manual mold parts design, the industry has begun to seek more advanced and faster way of design. With the rapid development of computer technology, CAD and CAE technology arise at the historic moment and get rapid development. After decades of development, the method of using CAD software design and CAE analysis software for mold design is yet to mature. The current mold design process as shown in Fig. 10.2.

Mold surface engineering is the system implemented by means of software engineering to realize mold surface design. It also can enable users to finish rapid interactive design for model surface with the guidance of system [7]. There are a lot of software which can be used for hot stamping die design and analysis. Such as CATIA developed by the French Dassault company, ABAQUS developed by HKS and Dynaform developed by both the ETA and LSTC company, etc. Due to the rapid development of surface engineering, surface engineering provides great

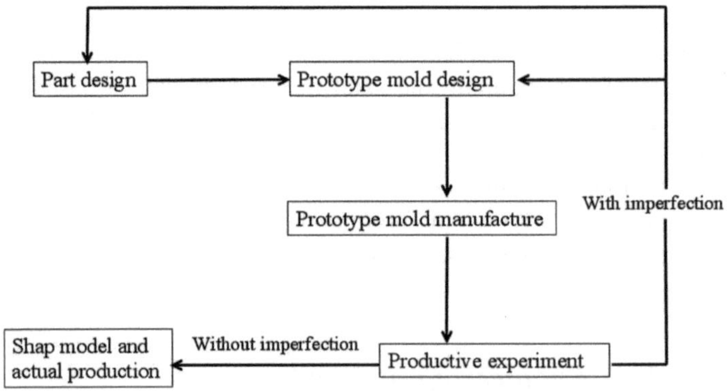

Fig. 10.1 Traditional model design process

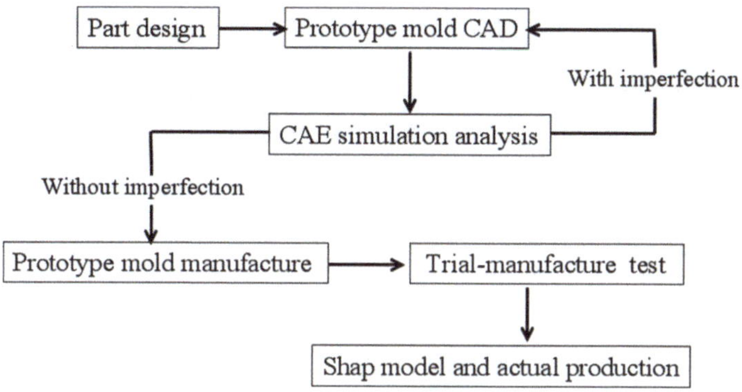

Fig. 10.2 The present model design process

convenience for the hot stamping die design, shorten the design cycle, improve the efficiency and greatly reduces the cost of mold design. Hot stamping die surface structure design parameters that should be considered generally include: radius of punch and die, gaps between punch and die, the size tolerance of punch and die, etc. At the same time, according to material performance, surface requirement and draft parts design the mold surface.

10.1.4 Optimization Design of Mold Cooling System

The hot stamping die with cooling system is an indispensable key component in the hot stamping system. In order to assure the quality of hot stamping products are uniform and reliable. Hot stamping die must be designed and both can meet the cooling requirement of the continuous production and ensure all parts of model can be cooled uniform. In traditional design opposite direction drilling in Fig. 10.4 which was based on practical experience is always be used. Most of model parts cooled are nonuniform and easy to produce stress concentration because this kind of channels cannot ensure the distance between channels and mold surface to be equidistant. But the conformal cooling channel design is a good way to solve the problem of mold cooling intensity and uniformity. The conformal cooling channel design method is in Fig. 10.3. In Fig. 10.3 is the mold cooling temperature field of different channel design type. It is observed that the maximum temperature in

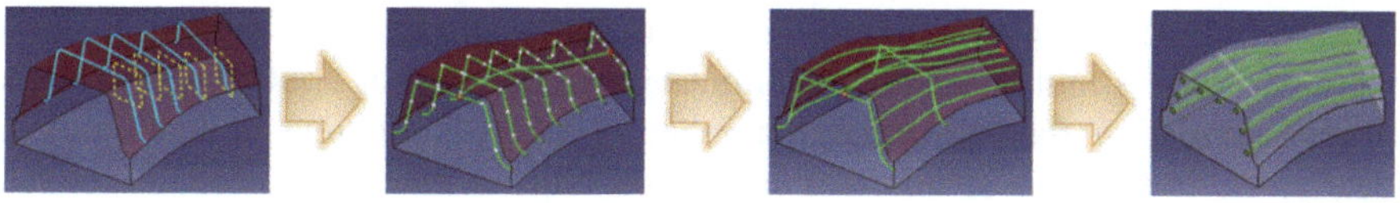

Fig. 10.3 Conformal cooling channel design method

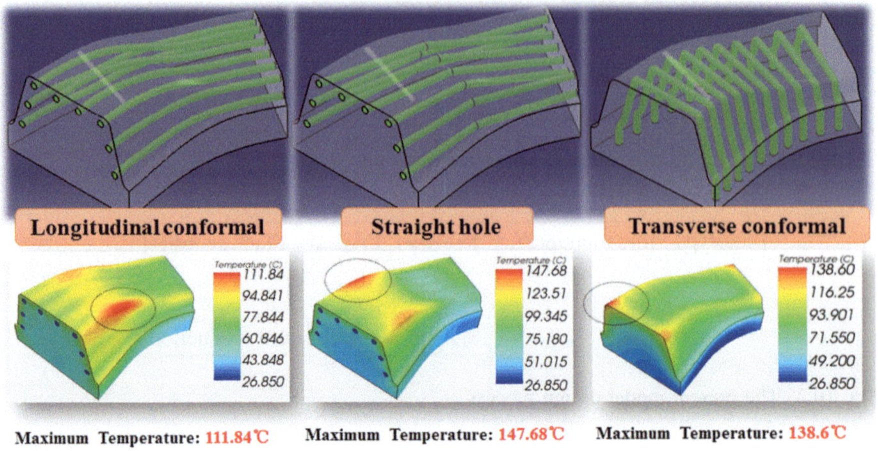

Fig. 10.4 Mold temperature contrast of different conformal cooling channel designs

steady state simulation module is only 111.84 °C when hot stamping die used longitudinal design. It is much better than opposite direction drilling design and transverse conformal design [4, 5, 13].

As the mold shapes become complex, there are some disadvantages in complex shape surface hot stamping dies which designed by the experiential design method. It can only through optimizing significantly improve the performance in the design process. In order to shorten the time and economic cost, it is necessary to use numerical simulation technology. In this case, it would require the digital optimization technique which based on coupling numerical simulation.

10.2 The Optimization of Cooling System in Hot Stamping Dies

10.2.1 Optimization of Subsystem Decomposition

The entire process is divided into subsystems from top to bottom by the simulation and optimization of large complex industrial systems, in which each part can be analyzed separately [17]. But each subsystem should also be considered from an overall level. Simulated environment not only makes it flexible to consider the results of the individual subsystem, but also connects them together for the data transfer and optimization analysis. According to complexity of the hot forming overall system, simulation-based optimization design is not confined to a single process simulation; the entire design necessarily involves different simulation tools. Ideally, the integration of different simulation tools makes it possible to simulate hot forming process in real time.

1. Fast define a parameterized analysis model, subsystems division
2. Fast analyze subsystems
3. Fast optimization of entire process
4. Fast convert between rough and fine analysis
5. Evaluate parameters extracted from the subsystem
6. Overall control of the analysis process
7. Joint analysis of different subsystems and parameters delivery
8. Choose various optimization methods for analysis based on the characteristics of different questions
9. Ensure stability of the long-time and heavily calculated analysis
10. Flexibly add and remove the subsystem to extend the scope of analysis or improve the efficiency of analysis.

10.2.2 Virtual Prototype of the Optimization of Mold Cooling

Hot forming of high-strength steel optimized system is divided into subsystems as follows: (Topology diagram is shown in Fig. 10.5)

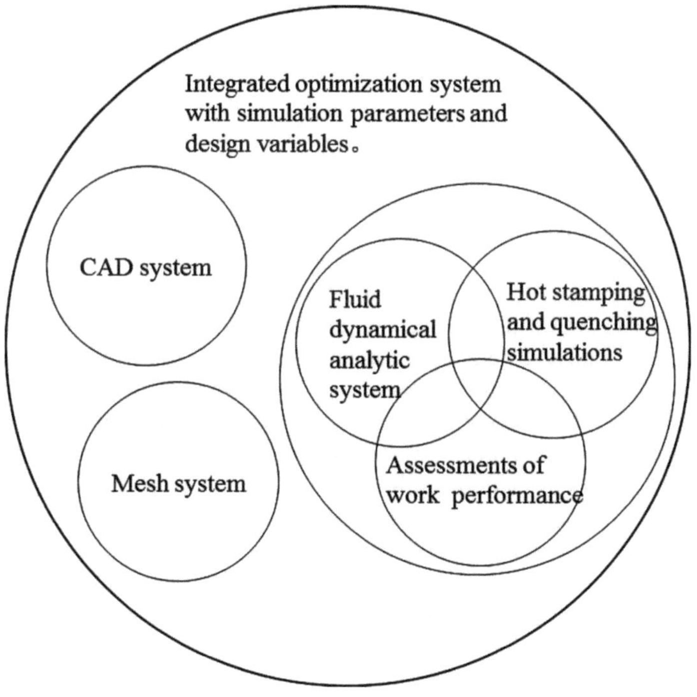

Fig. 10.5 Structure of the optimization of cooling system in hot stamping die

1. CAD modeling system
2. Grid generation system
3. Hot forming and quenching system
4. Fluid dynamic analysis system
5. Product service performance evaluation system
6. Test and process database system
7. Integrated optimization system.

Among them, the core modules include the following parts: product service performance evaluation system, hot forming and quenching system, and fluid dynamics analysis system. Product service performance evaluation system is responsible for the former product service performance evaluation, including crash, fatigue and NVH Dynamics, in which the hot forming products can be applied into the practical automotive engineering. Simulation analysis of hot quenching heat transfer and forming process of sheet metal from the hot forming and quenching system for parts, can realize the accurate prediction and simulation of the experiment. Condition of water-cooled mold with cooling channels and heat transfer during quenching is described in fluid dynamic analysis system. In addition, the CAD system is applied to the design and parameterization of part products, die surface, and water cooling pipe layout. Grid generation system is applied to automatic division of molds, sheet metal FEM meshes, and fluid CFD meshes. Database systems provide important parameters in the simulation process, including material flow curves under different temperatures and different strain rates; contact heat transfer coefficient under different temperature and pressure in the hot forming process; convective heat transfer coefficient under different flow rates, and different flow conditions in a cooling channel; friction coefficient, thermal microstructure conversion and the relationship between temperature and stress. Based on the experimental data, the results can be effectively comprised with the simulations. Integrated optimization system is applied to control the remaining six systems, and to collect their feedback information. The optimal parameters of die surface and water channels are calculated with combined optimization algorithms, which are delivered to CAD system to modify the geometric model.

Figure 10.6 shows the typical design process of linear hot stamping virtual prototyping. The whole process can be divided into two parts: initial design and detailed design, which, according to actual needs, can be divided into more steps for optimization. Effective control system is the key to the whole process. Subsystems illustrated in Fig. 10.5 are combined in the design and analysis of process.

The implementation process of hot stamping virtual prototype is shown in Fig. 10.7. The product models are imported in virtual prototype. First, in order to test if the related performances meet the requirements, some software such as LS-dyna is used to emulate the properties of products including collision. If it does not meet the requirements, the product models should be optimized until the product models meet the requirements. The convergence criteria of this step are closely

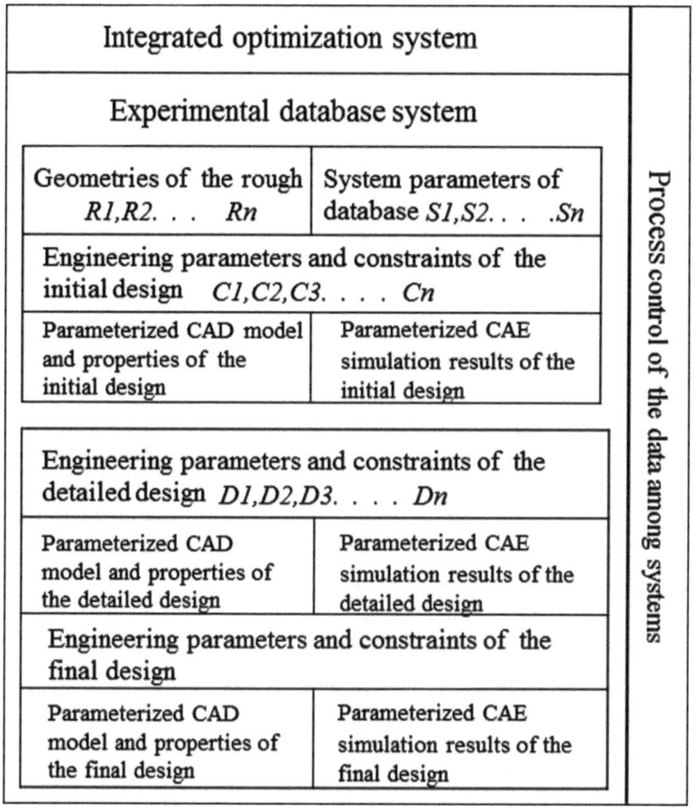

Fig. 10.6 Process schematic of optimization design in hot stamping model cooling system

related to the specific requirements of the model actual structure. According to the different structures, specific processing is accomplished. The initial design modifications must be in accordance with structure which is in the application of the vehicle. Mold surface is designed according to optimization product model when structural geometry has been determined. The size optimization billet is used then coupling analysis simulation method is applied to emulate hot stamping and quenching analysis. Finish overall control with ISIGHT. In order to ensure the sufficient and uniform cooling rate of stamping blank and the uniform type mold surface temperature distribution and smaller temperature fluctuations. Mold surface, hot forming process parameters, and water channel layout are further optimized. The three-dimensional entity model with best cooling channel design is the final output results. At the same time gain hot stamping mold surface and water cooling channel design which both guarantee the hot stamping production requirements of product design and conform to the actual demand.

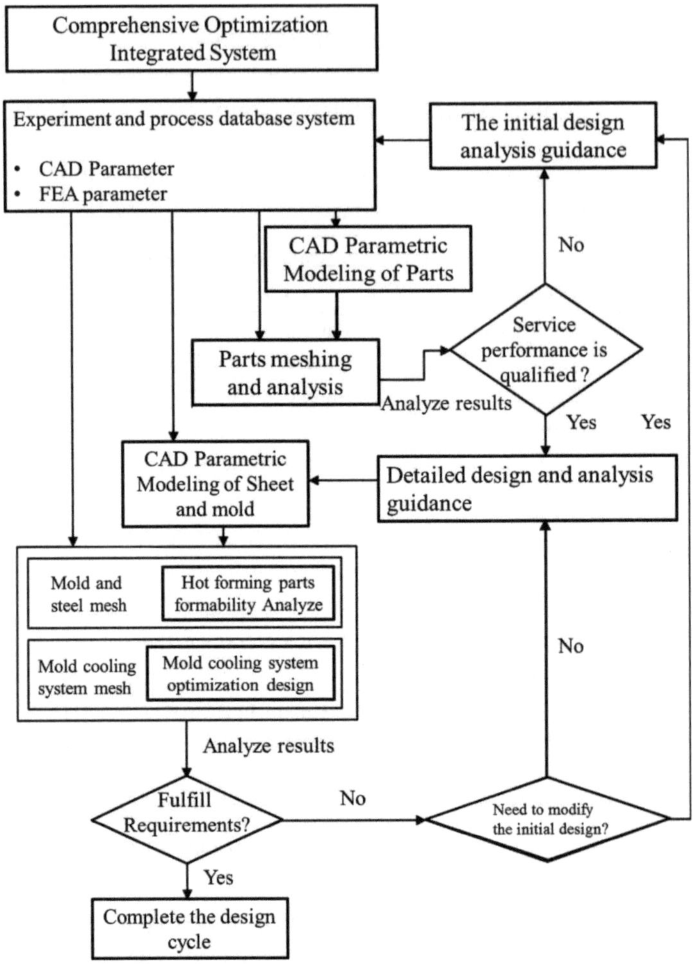

Fig. 10.7 Hot forming mold optimization specific flow chart

10.2.3 Optimizing Core Technology Decomposition

Hot forming mold optimization engineering problems are very complicated. The key technologies involved computer CAD parametric modeling, hot forming and quenching simulation analysis, cold water piping design and optimization.

1. CAD Parametric Modeling Technology

CAD parametric modeling provided strong technical support and platform for optimizing system builders. From design drawing, optimize modified to data output, it can be achieved water-cooled die-face design and pipe size layout optimization based on optimal product shape design. Hot forming cooled mold is a comprehensive structure of complex industrial equipments, including complex mold surface and cooled piping arrangement. If using conventional solid modeling to modify critical parts of the size and shape that is very complicated, and must to manually reconfiguration, it is impossible to achieve the optimization process. Using CAD parametric modeling technology can easily modify the set parameters to achieve the geometry shape and size changes automatically. In the hot forming mold design optimization, based on optimization results can achieve rapid cooling mold design modification by CAD parametric modeling technology, and shorter design cycles.

2. Hot forming and quenching simulation technology

For multi-field coupling hot forming simulation, optimization techniques must achieve multiphysics coupling parameter passing under conditions of fluid–solid coupled, to accurately analyze the actual hot forming production process, then for hot forming production technology and key equipment design optimization provides critical guidance.

3. Water-cooled mold pipeline design and optimization techniques

Design Optimization of cooling pipes shall meet the following basic requirements: first, make sure the water-cooled pipe design is easy processing; second, under the premise of the strength requirements we should ensure maximum sheet quenching rate; third, should ensure uniform temperature distribution of the mold surface, so to ensure a uniform temperature distribution of the sheet; and finally, reduce mold temperature fluctuations, to ensure the fatigue life of the mold, while ensuring the stability of product quality.

10.2.4 Optimization Examples

Figure 10.8 shows the temperature distribution of the middle B-pillar sheet metal and mold before the optimization. This section is the most complex shapes, which has the highest requirements to the optimizing design of the cooling system, so we use this section as an example to optimize water cooling system.

As can be seen in both the sheet or mold before optimization, in convex corner part the cooling effect is bad. The mold temperature is uniform distributed.

Figure 10.9 shows half section of the middle of the B-pillar before optimization, *R* is the pipe radius, *D* is the distance between the pipe walls, *H* is the distance between the pipe wall and the surface of the mold. Figure 10.9 shows the pipeline number. Table 10.3 shows the comparison of RDH before optimization and after optimization.

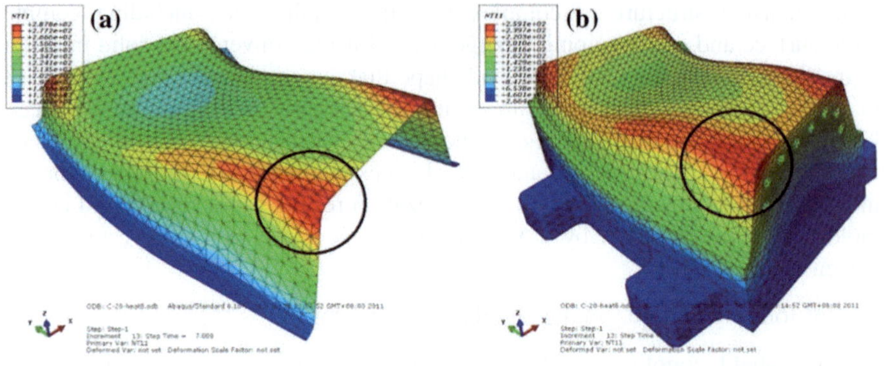

Fig. 10.8 The middle section of "Rainbow car" B-pillar **a** sheet, **b** mold

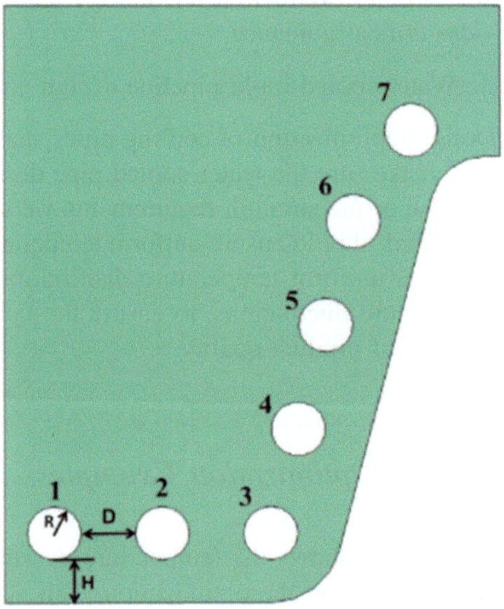

Fig. 10.9 The section of B-pillar mold

Table 10.4 shows the two evaluation criteria before and after optimization that are temperature uniformity and average temperature. "Temperature difference" as the mold surface maximum temperature and minimum temperature difference, which is used to indicate temperature uniformity. "Average temperature" is the average temperatures of the mold surface every point, to measure the cooling effect.

Figure 10.10 shows the effect before optimization and after optimization. As it can be seen, in the protruding position there has the high temperature. Mold surface temperature uniformity is poor. Pipe arrangement is changed, the pipe closer to the original temperature higher position, and take away from the original temperature lower position. The temperature uniformity is improved, and the average

Table 10.3 RDH parameter changes

Parameter	Value before optimization	Value after optimization	Parameter	Value before optimization	Value after optimization
D1	10.0	9.51	H5	8.0	5.31
D2	10.0	11.52	H6	8.0	9.51
D3	10.0	8.88	H7	8.0	11.77
D4	10.0	9.61	R1	5.0	5.25
D5	10.0	8.64	R2	5.0	5.55
D6	10.0	9.76	R3	5.0	5.20
H1	8.0	7.21	R4	5.0	4.97
H2	8.0	5.44	R5	5.0	4.62
H3	8.0	3.0	R6	5.0	6.22
H4	8.0	6.11	R7	5.0	5.45

Table 10.4 The comparison of temperature uniformity and the average temperature of the mold surface before and after optimization

	Temperature difference (°C)	Average temperature (°C)
Before optimization	78.1	211.4
After optimization	15.6	182.0
Change value	62.5	29.4

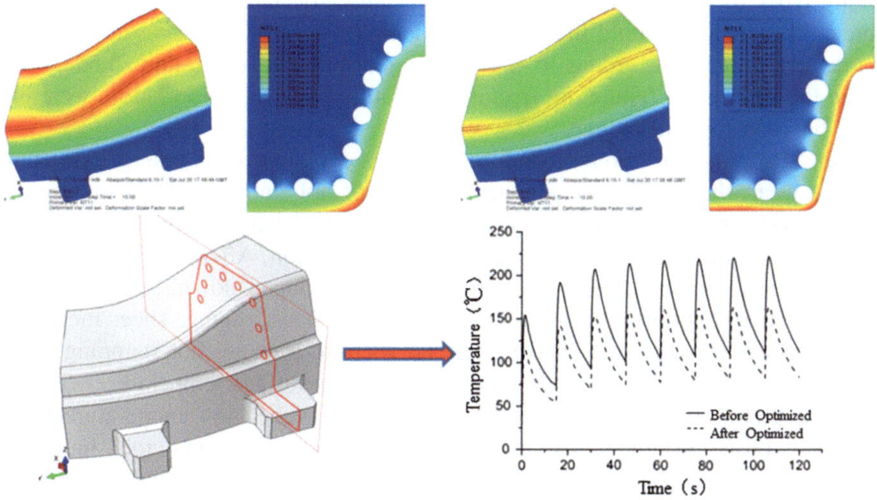

Fig. 10.10 Changes of pipe layout and temperature cloud before and after optimization

temperature is decreased. In the multi-loop simulation, mold surface temperature fluctuation amplitude is decreased, and the overall temperature is decreased.

These results show that the average temperature of the mold decreased significantly after optimization, and uniformity of surface temperature increased significantly. So, the optimization process is effective.

10.3 The Manufacturing of Hot Stamping Mold

10.3.1 Mold Heat Treatment

Mold inserts and parts commonly used heat treatments are normalizing, annealing, tempering, quenching, tempering, carburizing, and nitriding. Hot stamping mold material is usually hot work alloy tool steel, its heat treatment deformation range is associated with the mold size. When the length, width, and height are smaller than 275 mm, heat treatment deformation can be allowed ±0.2 mm; when the length, width, and height within 276 ~ 375 mm, heat treatment deformation can be allowed ±0.3 mm; when the length, width, and height are larger than 375 mm, heat treatment deformation can be allowed ±0.5 mm.

10.3.2 Mold Surface Strengthening Treatment

Many types of mold working conditions vary greatly, failure modes are also different. To extend the mold life, we should look for the most effective mold surface hardening process according to the working conditions and failure modes. The basic forms of mold failure are fracture and fatigue, plastic deformation, wear and tear, bite and thermal fatigue.

Hot stamping die common failure modes are fatigue crack, fatigue fracture, fatigue wear, thermal fatigue, impact fatigue. For hot stamping die with complex work environment, according to the surface shape, the impact of die is not real sense impact, but the result of shock and slip joint action. After sheet contact mold surface, the punch continue going down. With the punch down, sheet and mold surface relative sliding, generating friction. After the stamping, sheet metal and mold surface are separated, and there is no friction. Because hot stamping die at work in contact with the heated blank, die working surface temperature will rise, then the mold is cooled by waterway. Because of cyclic high temperature, mold surface is prone to fatigue cracking, generates impact fatigue and crack failure. In the thermal and mechanical coupling, it is subjected to thermal fatigue affected, and the surface oxide layer accelerates mold wear. Under the influence of thermal stress, during punching the mold surface gradually shows net cracks. And under the impact, the crack grew deeper, the mold surface roughness increases.

The purpose of the mold surface treatment is given new properties on the substrate material. These new features are wear resistance, adhesion resistance, heat-bite resistance, thermal fatigue resistance, fatigue resistance, and corrosion resistance. Mold surface hardening methods are mainly three. The first one is to change the chemical composition of the surface; the second is coated with various coating; third is not changing the chemical composition of the surface [18]. Table 10.5 shows the common methods of mold surface hardening. The table lists from the original chrome to the latest TD treatment method and CVD, PVD [2, 8, 23, 28, 30], and other treatment methods.

Table 10.5 shows that by altering the chemical composition of the surface to achieve the purpose of strengthening. Carburizing, boriding, and other chemical treatment methods can effectively improve die life. Using various deposition methods, the TiC, TiN, VC, NbC carbides coated on the mold surface, forming a layer of hard film. So, the material obtained excellent mechanical properties of high hardness, wear resistance, and corrosion resistance. When the coating layer of TiC, TiN, VC, and other compounds having 2–10 μm, mold surface would have a high hardness and wear resistance, service life increased several times to ten times. It can also help form a lubricating film between the workpiece and the mold surface, played the role of isolation and lubrication. If applied to punch, the life of molds can be increased 4–10 times. Used in plastic mold also had a significant effect. Using these coating processes can make all the mold material, including cemented carbide, gain unprecedented high abrasion resistance, and adhesion. So that those

Table 10.5 Mold surface hardening methods

(A) Strengthening method by changing the chemical composition of the surface	
Carburizing	Gas, solid, liquid carburizing
Nitriding	Nitriding, liquid nitrocarburizing, gas nitrocarburizing
Boriding	Solid, liquid boronizing
Multicomponent permeation	Carbon, nitrogen, sulfur and boron permeation
Ion implantation	Injected chrome plasma into the mold surface by ion implantation machine
(B) Strengthening method by coated with various coating	
Plating	Nickel, chrome, chemical plating, etc.
Chemical vapor deposition (CVD)	Mold surface coating TiN, TiC and other coating
Physical vapor deposition (PVD)	Mold surface coating TiN, TiC and other coating
Plasma vapor deposition (PCVD)	Mold surface coating TiN, TiC and other coating
Carbide deposition (TD)	Mold surface coating VC, TiC, Cr_7, C_3
(C) Strengthening method by does not change the chemical composition of the surface	
Flame hardening	
Laser hardening	CO_2 laser
Machining hardening	Shot peening

molds with high-performance requirements, with difficult machining process, particularly large molds, and those molds worked in harsh conditions, must be subjected to a surface treatment.

10.4 The Thermomechanical Fatigue Test and Life Prediction Simulation of Hot Stamping Die

10.4.1 Fatigue Type of Hot Stamping Die

For fatigue from different angles may be different classifications. Depending on the study object, fatigue can be divided into material fatigue and structural fatigue. Depending on the stress state, fatigue can be divided into uniaxial fatigue and multiaxial fatigue. Uniaxial fatigue is the fatigue that under unidirectional stress cycles, and multiaxial fatigue is under the multidirectional stress, also known as composite fatigue. According to cycles before material fatigue failure, it can be divided into high-cycle fatigue and low cycle fatigue. For high-cycle fatigue, alternating stress of material far below the yield limit of the material, and the number of cycles before fracture is greater than 10^5–10^7. For low-cycle fatigue, material alternating stress is generally close to or exceeds the yield limit. The number of cycles before breakage is generally less than 10^4–10^5 times. According to the amplitude and frequency of loads, it can be divided into constant amplitude fatigue, changing amplitude fatigue, random fatigue. Based on load conditions and working environment, it can be divided into conventional fatigue, mechanical fatigue, thermal fatigue, thermal–mechanical fatigue, corrosion fatigue, contact fatigue, high-temperature fatigue, fretting fatigue, and impact fatigue [15].

1. Impact fatigue [29]

For stamping dies, the most immediate impact load is from the press. Therefore, the impact fatigue has an important role to hot stamping die service performance. Impact toughness test is the most commonly used impact test. Toughness can to some extent reflects the tendency of brittle fracture of the material. In practical engineering applications, we are accustomed to use this indicator to measure the ability of the workpiece to resist destruction under impact loading.

2. Thermal fatigue [16, 22, 27]

Thermal fatigue is a major form of failure for hot die. The more common view is alternating hot and cold stress over the high-temperature strength of the material, resulting in crack initiation and propagation. High-temperature low cycle fatigue tests have shown that high-temperature oxidation to crack initiation and propagation play a significant role. High-temperature strength and oxidation resistance to thermal fatigue properties of hot work die has an effect, but it is difficult to separate the respective roles of the two factors. However, when the mold surface is covered with plating or coating, high-temperature oxidation resistance of the film, high temperature strength of the substrate, the film and the substrate binding fastness will become an independent criterion. It decided to thermal fatigue properties.

3. Thermal fatigue cracks [21, 27]

Hot surfaces fracture caused by thermal effects is an important engineering problem. Thermal cracking, surface flaking caused by thermal cycling is the main failure mode of hot rolls, casting roll, die-casting mold, hot stamping mold. Thermal fatigue crack is a joint action result of variety of pressure, high-temperature oxidation, corrosion, and embrittlement. The thermal stress and impact thermal stress play a leading role.

10.4.2 Thermomechanical Fatigue Test Device

Thermomechanical fatigue test device can simulate the whole process of hot stamping. On the one hand, this experiment can simulate the alternating mechanical stress of mold material suffered under actual operating condition. On the other hand, it can simulate alternating thermal load state of the mold material suffered. And the test device should have high reliability, good stability, and the measured value should be more close to the real situation. Meanwhile, the device should be easy to replace the mold material, easy operation. Based on the above basic principles design the device [14]. The device includes power systems, heating systems, transmission systems, recovery systems and control test systems.

(1) Power system: 40 t hydraulic power source, a pressure sensor, a fixed mold handle, a round table punch pedestal, punch;
(2) Heating system: single-head heating rods, heating temperature controller;
(3) Transmission system: mansard rocker, spring, base, rod, fixed shaft;
(4) Control test system: press counter, press pressure controller, press the automatic shock controller, MX100 temperature test equipment.

According to the above, thermomechanical fatigue test device main body is shown in Fig. 10.11.

10.4.3 Experimental Principle and Content

The principle of thermomechanical fatigue test device is that the punch is in motion because the impact of the press forms a periodic cycle. When the punch accesses to the material block, the punch and material block generated a contact force, then the punch will be pressing the rocker arm downward movement. When the punch presses rocker arm to limit, the press will produce a packing process, which can

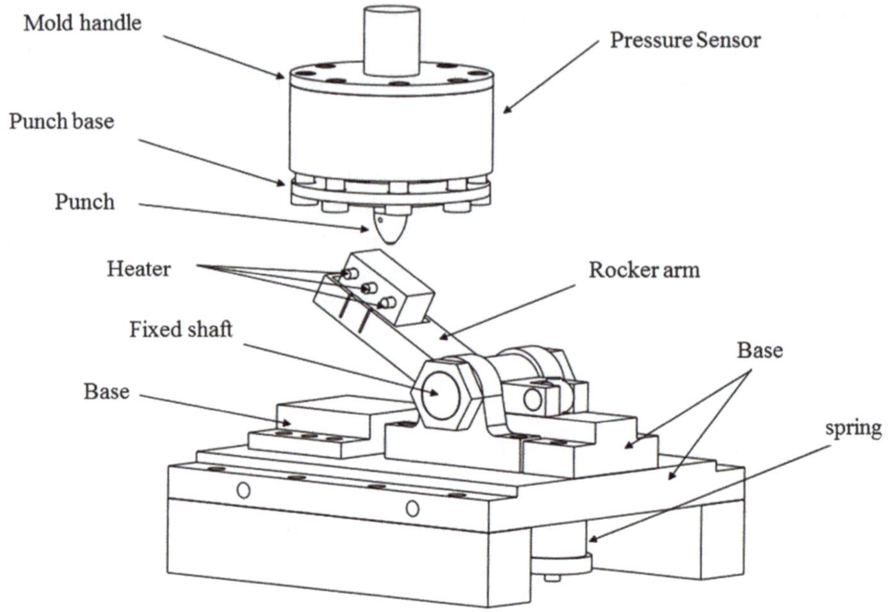

Fig. 10.11 The main body of thermomechanical fatigue test device main body

simulate the packing process in sheet forming. Thermomechanical fatigue test device schematic is shown in Fig. 10.12.

This experiment studies the mold material service performance under thermal shock effects. Due to the high intensity of high-strength steel, the stamping process requires heating the sheet to 810 °C, austenitizing the material, then stamping, and rapid cooling sheet. So the sheet is turning from austenite into martensite. However, in order to achieve uniform martensite and high strength, mold surface temperature must be kept below 200 °C [12]. In order to simulate mold fatigue service

Fig. 10.12 The schematic of thermomechanical fatigue test device

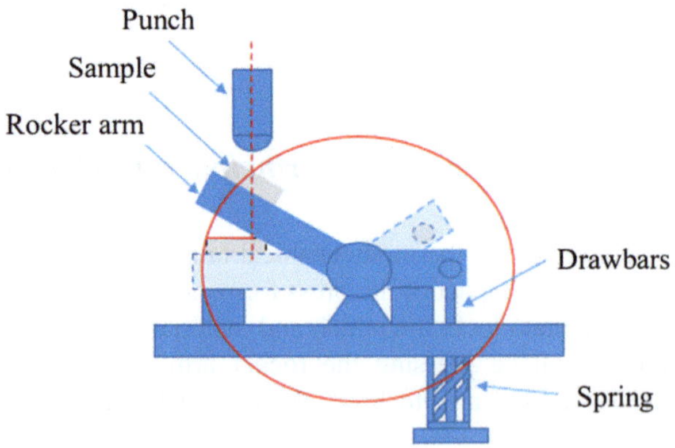

performance at 200 °C working conditions, the test temperature was set to 200 °C. However, many die because the waterway defect does not always keep 200 °C, in order to highlight contrast and simulate the performance of mold in extreme conditions, the comparative temperature was set at 250 °C. Set the number of thermal shocks were 1500, 2500, and 5000. The samples were 45 steel, tool steel H13, 45 steel coated with CrN, steel H13 coated with CrN die.

10.4.4 Experimental Results

In order to study crack characteristics, study on uncoated 45 steel surface cracks at 200 °C working conditions. According to the characteristics of the cycle impact, the growth of cracks in shock, slip, and packing stages are mainly studied. The position of the three stages are shown in Fig. 10.13.

In order to verify the high-cycle fatigue properties, the sample surface crack growth at the conditions of 200 °C and 5000 times impact is studied, as shown in Fig. 10.14. In the electron microscope, it can be seen that there are different degrees of crack growth in the impact—slip—packing stage. According to Hertz contact theory, sliding vice generates shear stress at the material surface and the material surface will generate fatigue cracks. As can be seen, with the increase of stress, the crack growth rate is increased. And the crack growth direction is perpendicular to the direction of the force that meets the high-cycle fatigue characteristics.

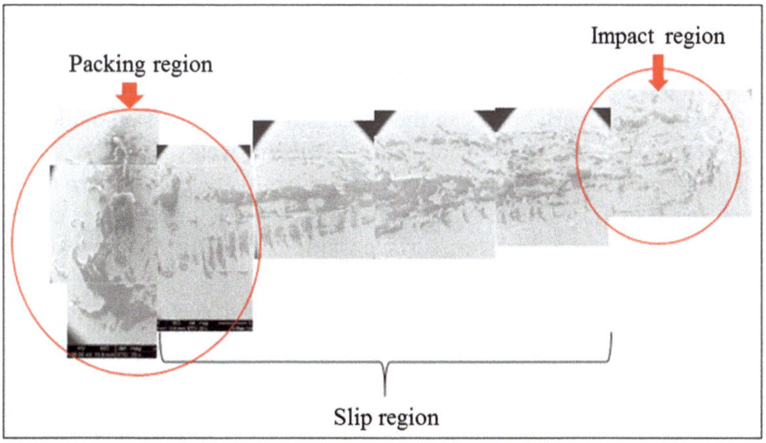

Fig. 10.13 45 steel surface topography in SEM

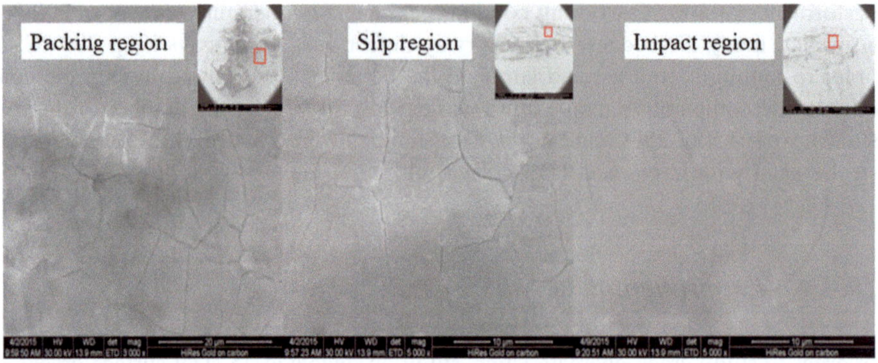

Fig. 10.14 The specimen surface cracks in three regions

The surface scratches topography has different characteristics depending on the working condition, the number of impact, and the base material. Therefore, comparative studies on the impact, slip and packing phase.

1. Comparative analysis of different material fatigue damage at the same temperature and the impact of different times

According to the experimental plan, mainly exploring fatigue characteristics of H13 uncoated steel and 45 uncoated steel sample times in different impact, as shown in Fig. 10.15.

From the horizontal comparison, due to the different forces, the morphologies of three regions have some differences. In the impact stage, it can be seen from the figure, the scale formed on the surface has been shedding. This proves that the adhesive wear occurs during the impact, so the scale sticking off with the punch down slip. In the slip stage, we can see that there exist many scratches. That proves shedding oxide layer participates in the wear and tear at this stage, and adds new types of wear—abrasive wear. That exacerbated the wear of surface, so that the material has shifted in the slip stage, and the number of scales has sharp drop. When the punch in packing position, as pressure increases the surface wear continues to intensify. As the material transfer, surface oxide is difficult to form, and then the substrate fatigue after several cycles.

Vertical contrast, in the low number of cycles, the surface of each stage is rough, and there is a lot of surface fatigues and oxidized defects. As the increase of the cycle, in the impact stage the surface wear is intensified, scratches also began to appear, and the surface oxide shedding serious. In slip and packing region, surface scratches disappear gradually under the combined effect of adhesion and wear slip,

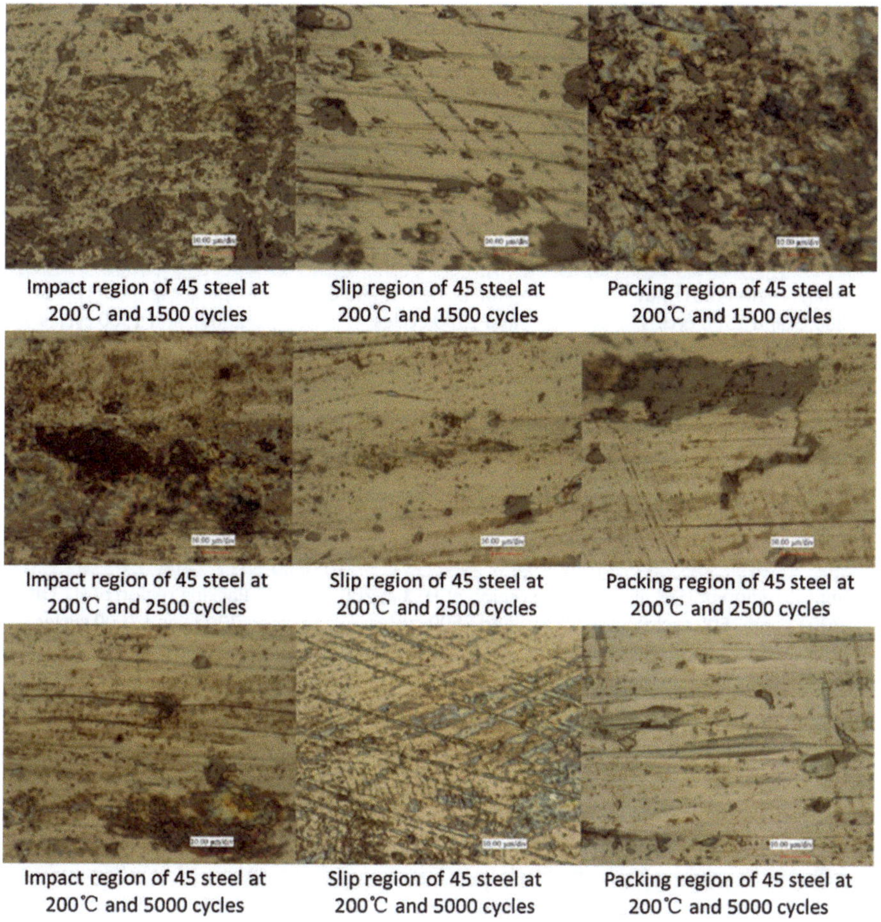

Fig. 10.15 The morphology of 45 steel and H13 steel at different times of impact

and the wear surface becomes smooth. This indicates that the sample reaches a break-in period at around 5000 times, and the fatigue damage is stabilized.

2. Fatigue damage comparison of different materials in the same condition

By comparing the fatigue morphology of different substrate specimens, study on fatigue service performance of different material, and study on service performance of coatings. According to the tests, the sample is shown in Fig. 10.16.

Comparative sample includes uncoated 45 steel, uncoated H13, coated 45 steel, and coated H13. The results show that CrN has a good antiwear friction properties

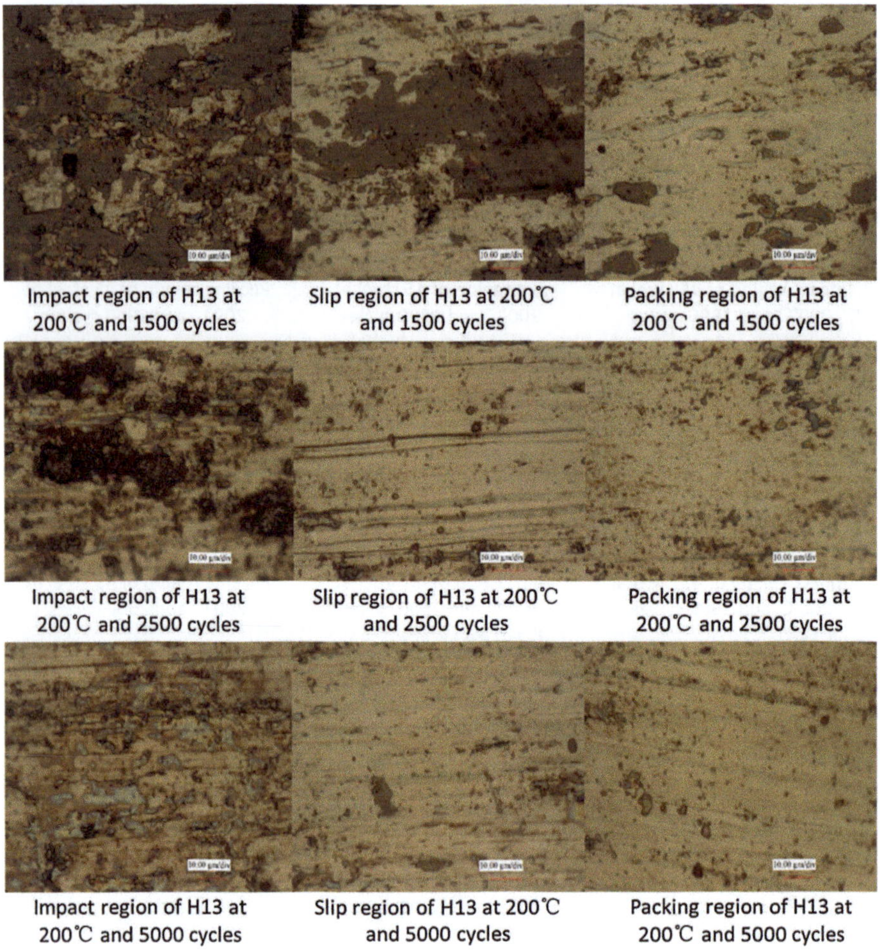

Fig. 10.15 (continued)

compared to uncoated samples. The coating surface has not obvious scratches, which is due to the high hardness, adhesive wear, and abrasive wear resistance and good oxidation resistance of the coating.

3. Under the same impact times, compared to the fatigue damage at different temperatures of each material, as shown in Fig. 10.17.

The figure above shows the oxidation of uncoated 45 steel at 250 °C is higher than 200 °C. In impact, slip, packing stages, the number and area of oxide at 250 °C is higher than 200 °C. At high temperatures, the surface oxide at 250 °C is darker that shows the oxide thickness is higher than oxide at 200 °C. It proved that the oxide generation rate at 250 °C is higher than 200 °C. But from oxide shedding morphology to analyze that there are not significant differences between the two

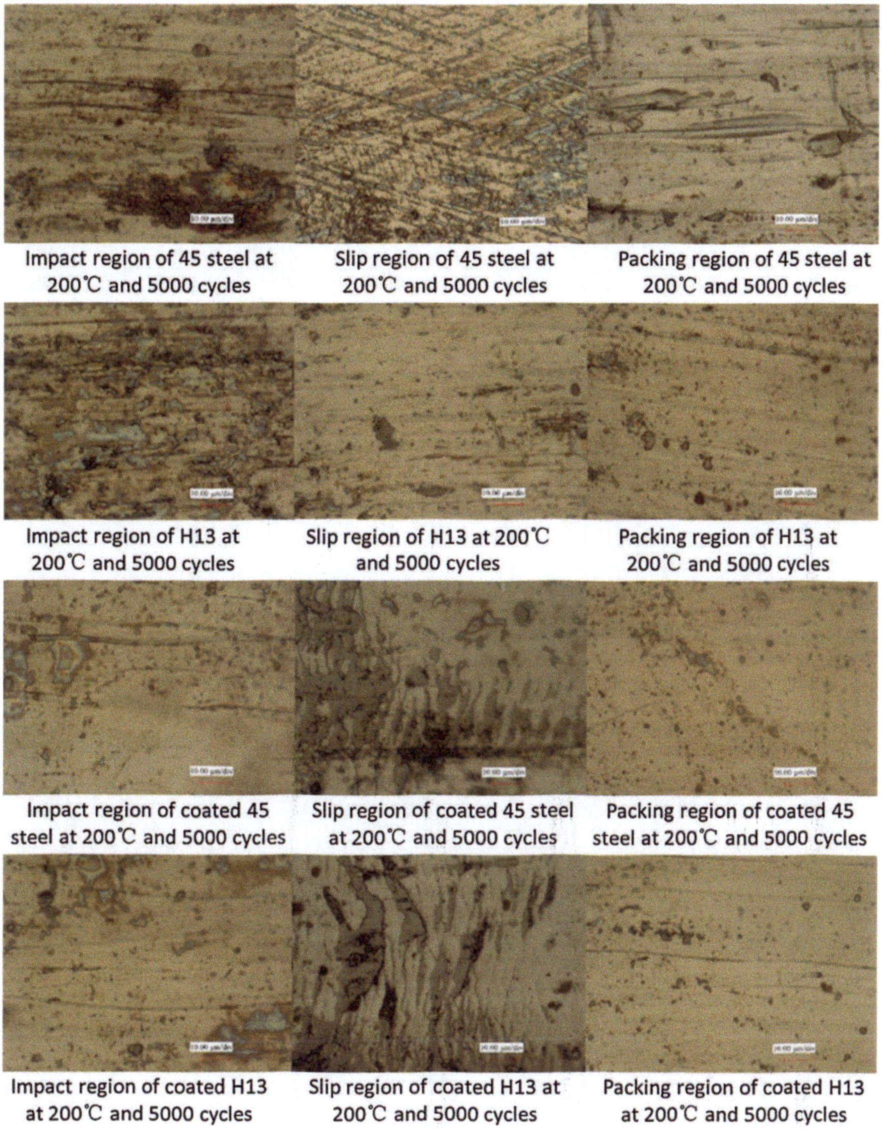

Fig. 10.16 The damage morphology of each sample at 5000 cycles

temperatures, and there were a large area oxide. It can be proved that in the high number of cycles, temperature only affects the oxide growth rate. Temperature difference has little effect on antiwear and oxidation resistance. Compared to 45 Steel, H13 oxide growth is relatively uniform at 250 °C, and oxide loss is less. Compared to 200 °C, morphology at 250 °C have more oxide, that increased wear resistance and oxidation resistance of the material surface to some extent. But

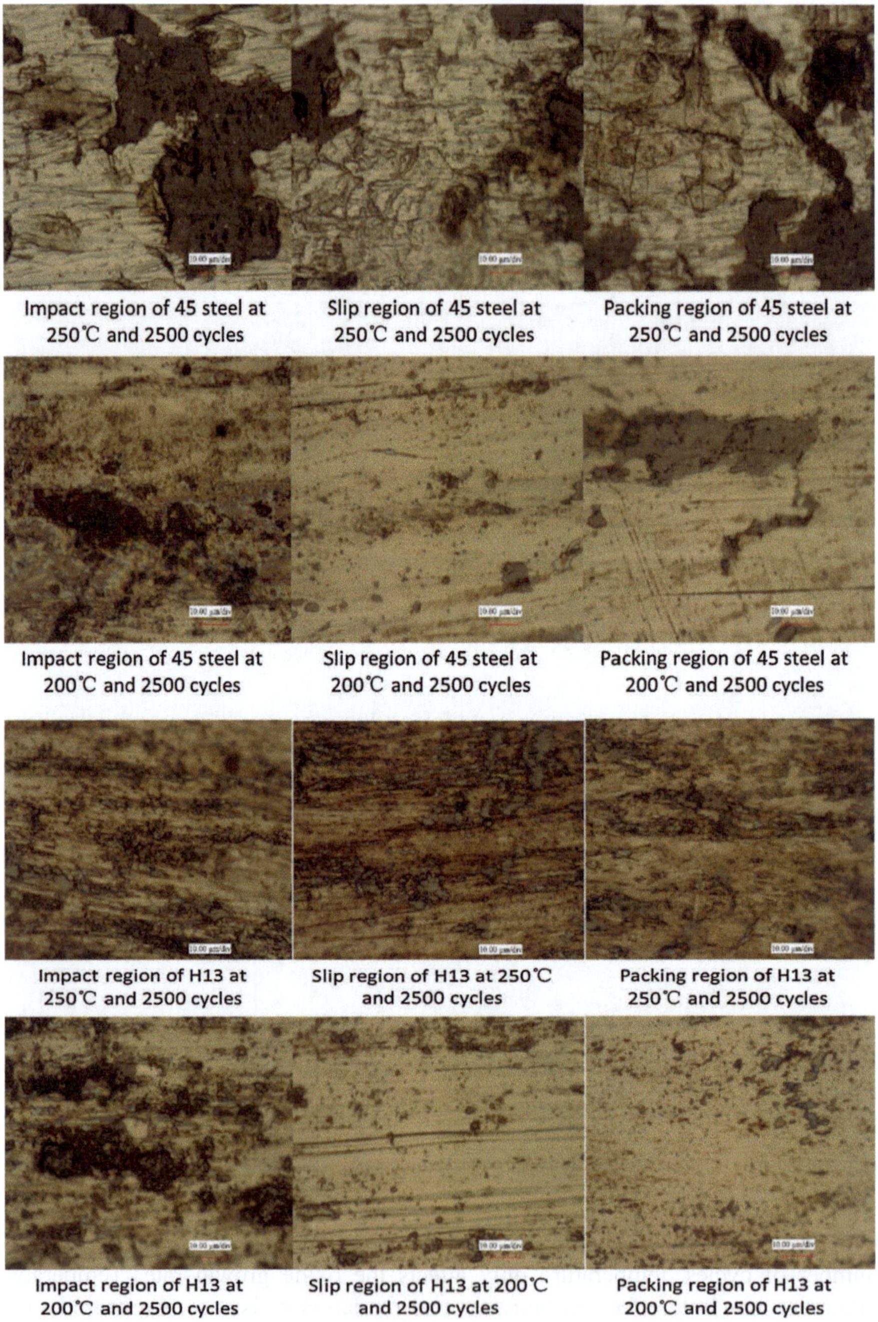

Fig. 10.17 The surface damage of each material at different temperatures

speaking from the surface scratch of the slip stage, the wear resistance was not significantly improved.

As can be seen by comparing the different temperatures, temperature difference has little effect on the sample surface topography. That indicates that at a constant temperature the temperature difference has little effect on the sample fatigue service performance. Therefore it requires a combination of simulation to supplement and verify the above conclusion.

10.4.5 Life Prediction Simulation

1. Finite element simulation model

Finite element model includes the finite element model of sample and the finite element model of coating. In accordance with the actual stress distribution meshed samples model, and arranged load on the corresponding node. By experimenting sample surface scratches, we can accurately obtain the actual load distribution. Because the coating is thin models, so using skin function in Abaqus to establish the model. According to the curve recorded by the displacement sensor, can get punch average speed of 78.5 mm/s. According to the average length of sample surface scratches, the punch moving distance is 38 mm. According to the relationship between load and displacement, the model is divided into separate sections. The relationship of load distribution and time is shown in Fig. 10.18.

2. Simulation result analysis

Contrast to uncoated H13 surface stress at 200 and 250 °C, as shown in Fig. 10.19. As can be seen, the largest force region is packing region. The slip stage has more evenly distributed stress, and stress concentration does not appear. Stress of impact

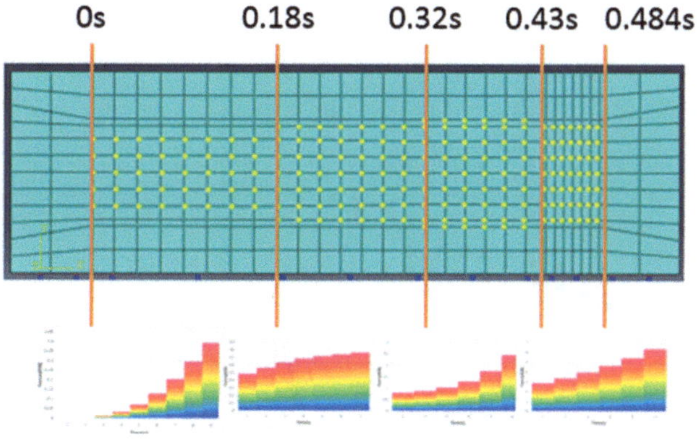

time	force (KN)
0	0
0.047	0
0.078	0
0.125	0.2685
0.156	0.327
0.203	0.4065
0.25	0.5745
0.281	0.7695
0.328	0.924
0.359	0.9465
0.406	1.1505
0.437	2.9325
0.484	6.3045

Fig. 10.18 The relationship of load and time

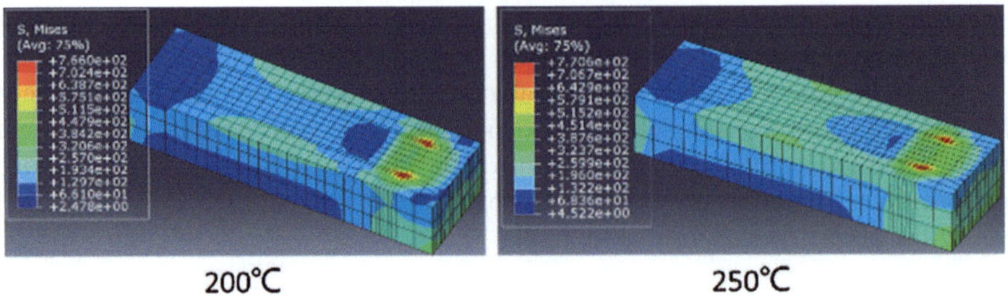

Fig. 10.19 The surface stress of H13 uncoated at 200 and 250 °C

stage is the smallest in three stages, and it is the least prone to fatigue. And the stress is less than the heat-treated material ultimate stress. Therefore, according to the results of the stress, the loads that sample suffered has a high-cycle fatigue characteristics.

The surface stress distribution of each region is shown in Fig. 10.20. As can be seen from the figure in the impact and slip region, the temperature has some effect on the stress of the sample surface. Slip region at 250 °C has a maximum stress, maximum stress is 297.59 Mpa, and stress difference is 117.46 Mpa. Therefore initially identified in a small stress range temperature has a certain influence on the sample surface stress. In the high stress region, temperature difference has little effect to the results.

The surface stress of H13 coating at different temperatures is shown in Fig. 10.21. Through the stress distribution can be seen the coating which can reduce the maximum stress in the surface of the mold to some extent. Its trends are consistent with uncoated H13, packing region remains the largest region of stress, and that most prone to fatigue.

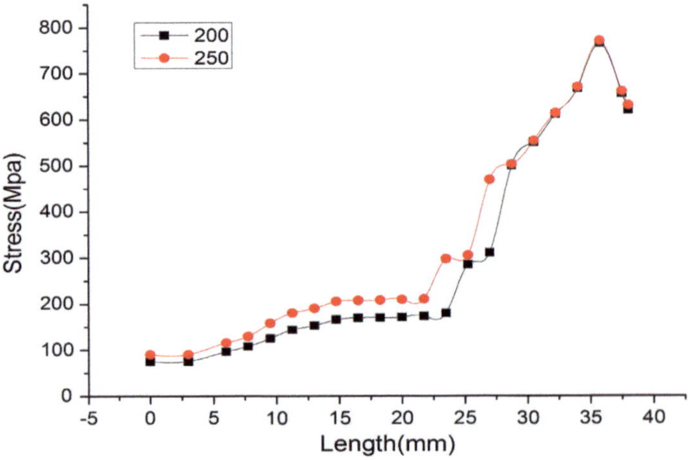

Fig. 10.20 Surface stress of uncoated H13 at different temperatures

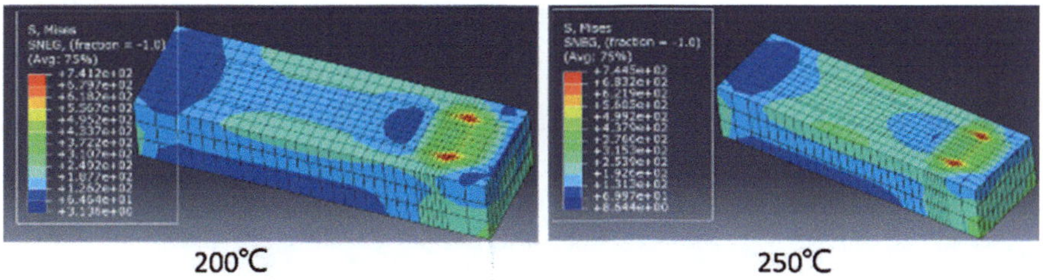

Fig. 10.21 Surface stress of coated H13 at different temperatures

By contrast stress curve, as shown in Fig. 10.22, it can be seen its sensitivity to temperature is mainly concentrated in the low-stress region. The maximum stress is 230.55 Mpa, stress difference is 37.4 Mpa, and compared to uncoated H13 the stress difference is small. In the packing region, at 250 °C H13 uncoated surface stress is slightly higher than coated H13, and its maximum stress is 770.6 Mpa, stress difference is 26.1 Mpa. The maximum stress at 250 °C is 4.6 Mpa higher than 200 °C. It can be concluded that the coating for ease stress difference have a positive effect.

It can be drawn from the above study, whether coated H13 or uncoated H13, at a constant temperature, temperature difference of 50 °C has little effect on stress. The cloud fatigue damage at 200 °C is shown in Fig. 10.23. The main damage and stress occurred in the packing stage, and it showed that the coating can effectively mitigate the impact of stress on the surface damage.

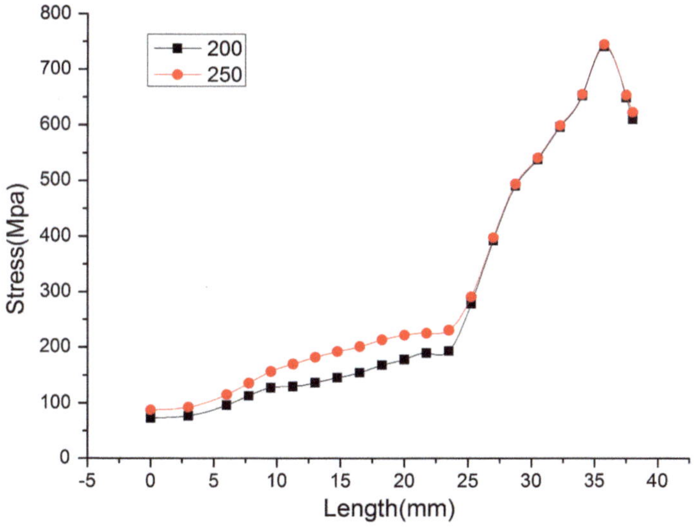

Fig. 10.22 Surface stress of coated H13 at different temperatures

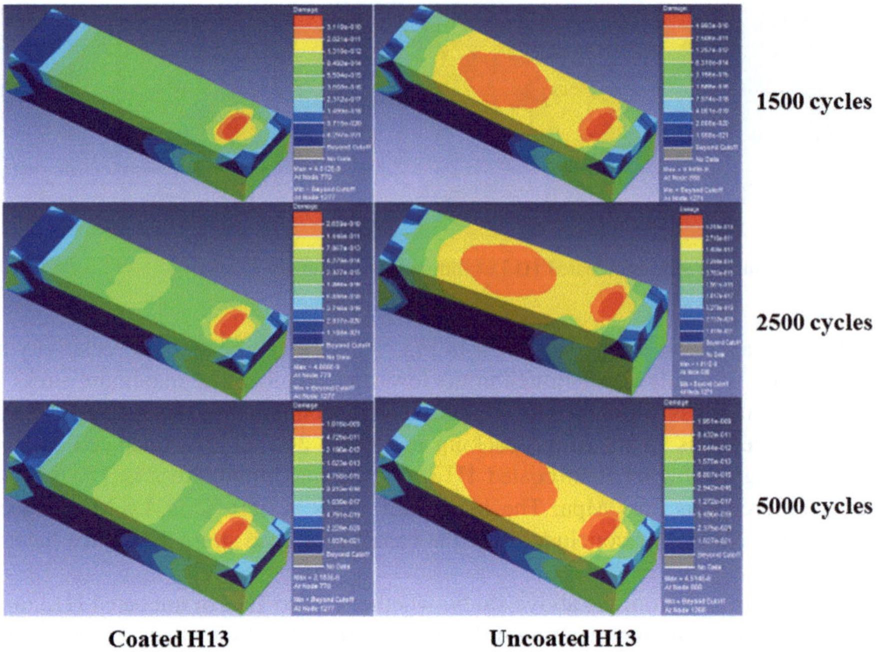

Fig. 10.23 The cloud fatigue damage at 200 °C under different impact times

Surface damage curve is shown in Fig. 10.24. As can be seen from the figure, the damage of H13 coated with PVD is less than the uncoated H13. Compared to uncoated H13, anti-fatigue properties of coated H13 at 1500 cycles improved 40.1 %, at 2500 cycles improved 40.7 %, and at 5000 cycles improved 26.5 %.

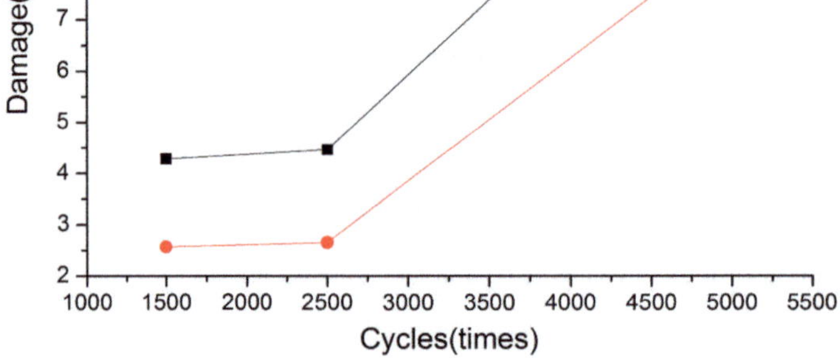

Fig. 10.24 Surface damage curves of uncoated H13 and coated H13 at different time impact

By comparing different temperatures and different impact numbers, study on the mold surfaces damage and crack growth at different conditions. From the stress and damage simulation and experiments can be concluded that at a constant temperature the temperature difference of 50 °C mainly has an impact on low-stress area, to high stress has relatively less affected. Therefore, in the premise that mold surface temperature is constant, extreme temperatures have little effect on the fatigue of the mold. In the above premise, the temperature of the mold is not a major factor in fatigue damage. High hardness and oxidation resistance of coating improved the mechanical properties of the mold surface, enhanced its ability of anti-fatigue damage, thereby increasing the life of molds.

According to the theory of linear fatigue, miner hypothesis suggests that if the material is under constant stress amplitude effect, when the number of stress cycles is N, material damage or fatigue failure will occur completely. If the material is in effect under this stress amplitude, after less than N stress cycles, the material is only part of the damage occurred. It can be seen from Fig. 10.24, the damage did not reach 1 at 5000 cycles. So in the whole mold material only small damage occurred but it is not destroyed. But this does not mean that there is no fatigue, from the surface characteristics of the sample can be seen, the material has occurred fatigue damage characteristics after several cycles. Mold material fatigue life and mold damage is corresponding, and N code software can predict the fatigue life based on the material damage. In 200 °C, fatigue life of each sample prediction results are shown in Fig. 10.25.

As can be seen from the figure, the life of coated H13 is the longest, which is more than 1.3×10^5 cycles. The coated 45 steel life has increased to near the uncoated H13, reaching 1.0×10^5 cycles. Life prediction results are consistent with the results of the stress and damage. Binding test results, mold material after the

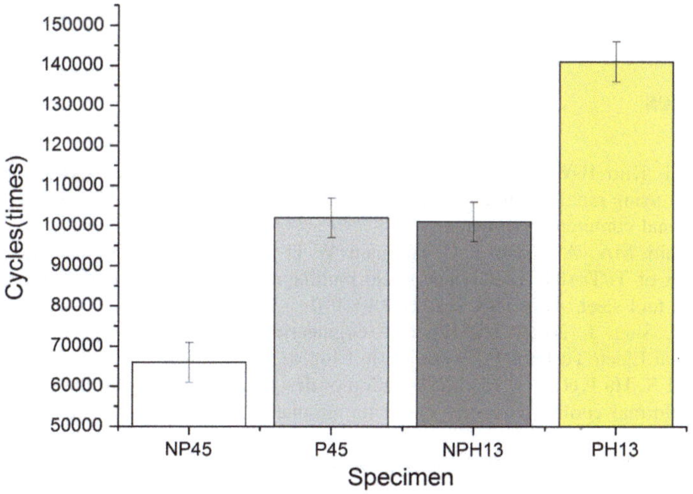

Fig. 10.25 Fatigue life of each sample

surface optimized with PVD coating, the surface hardness, and surface smoothness are higher than the uncoated samples. PVD not only increases the hardness, reducing the roughness of the surface to reduce friction damage, but also its oxidation resistance play a protective effect to the material surface, reduced surface damage. From the view of stress, the coating can experience the same level or even higher levels of stress, so that the material surface deformation slowed, thus prolonging the life of the sample. Simulation results of fatigue life can prove the reliability of the above conclusions, and PVD coating to improve the surface quality of the mold has a significant effect.

10.5 Summary

This chapter makes a deep explanation to hot stamping mold material selection, hot stamping die design, manufacturing. Then describes the optimization design ideas and methods for hot stamping cooling channels, and deeply describes the influence of hot stamping mold cooling performance to forming parts quality. Finally, it discusses the service performance of hot stamping dies, especially thermomechanical fatigue characteristics, fatigue characteristics, fatigue life. Comparison of surface morphology of each sample in different conditions, and different stages, and combined with the fatigue mechanism to contrastively analyze the fatigue characteristics of each sample. Hence it is concluded that the antioxidant and antiwear of coated samples is superior to uncoated samples. In different conditions, uncoated H13 service performance is superior to uncoated 45 steel. The effect of coating to mold material service performance is obvious. Through PVD coating process, performance of 45 steel can be improved which indicates that traditional H13 steel can be replaced by using coated 45 steel.

References

1. Ahn D-G, Kim H-W, Park S-H et al (2010) Manufacture of mould with a high energy efficiency using rapid manufacturing process. In: NUMIFORM 2010, proceedings of the 10th international conference, Korea
2. Al-Bukhaiti MA, Al-Hatab KA, Tillmann W et al (2014) Tribological and mechanical properties of Ti/TiAlN/TiAlCN nanoscale multilayer PVD coatings deposited on AISI H11 hot work tool steel. Appl Surf Sci 318:180–190
3. Benri W, Shen J (2013) The die face engineering for the stamping die of auto cover. J Shanghai Electr Technol 1:22–26+30 (in Chinese)
4. Bin H, Li X, Hu P et al (2015) Investigation of design and manufacture in hot stamping tools with conformal cooling channels based on simulation and 3D-printing technology. Chin J Mech Eng 1–9 (in Chinese)
5. Bin H, Li X, Ying L et al (2015) Optimal design of hot stamping tools with conformal cooling channels. J Jilin Univ (Eng Technol Ed) 1–7 (in Chinese)

6. Chao Z (2010) Designing and optimizing of hot stamping tools of ultra high strength steels. Jilin University, p 90 (in Chinese)
7. Chi D (2009) Several key techniques for CAE-based die-face design system. Jilin University, p 140 (in Chinese)
8. Gali OA, Shafiei M, Hunter JA et al (2014) The tribological behavior of PVD coated work roll surfaces during rolling of aluminum. Surf Coat Technol 260:230–238
9. Ghiotti A, Sgarabotto F, Bruschi S (2013) A novel approach to wear testing in hot stamping of high strength boron steel sheets. Wear 302:1319–1326
10. He B, Ying L, Hu P et al (2012) Design of water cooling tool for hot stamping process of high strength steel. China Metal Forming Equip Manuf Technol 6:62–65 (in Chinese)
11. He B, Ying L, Hu P et al (2014) Investigation of mechanical property and spring back behavior with hot stamping RCP process. Adv Mater Res 1063:186–189
12. Hoffmann H, So H, Steinbeiss H (2007) Design of hot stamping tools with cooling system. CIRP Ann-Manuf Technol 56:269–272
13. Hu P, He B, Ying L (2016) Numerical investigation on cooling performance of hot stamping tool with various channel designs. Appl Therm Eng 96:338–351
14. Hu P, Ying L, He B et al (2015) A kind of hot-work die material heat engine fatigue test device, p 12 (in Chinese)
15. Jianru F, Jiang Q, Han Z et al (2002) Life prediction of hot work die steel under thermomechanical fatigue. J Mech Strength 4:571–574 (in Chinese)
16. Klobčar D, Tušek J, Taljat B (2008) Thermal fatigue of materials for die-casting tooling. Mater Sci Eng A 472:198–207
17. Li Y (2013) The optimization design of cooling system in hot stamping dies. In: Automobile engineering. Dalian University of Technology, Dalian, p 76 (in Chinese)
18. Lingping Y, Dong M, Yang Y (2002) Die surface strengthening technology and its application. Die Mould Manuf 2:50–55 (in Chinese)
19. Nannan S, Wu X, Zhou Q (2011) Effect of surface decarburization layer on thermal fatigue behavior of steel H13. Shanghai Metals 26–29 (in Chinese)
20. Pengcheng X, Chen Y, Ge X et al (2008) Research status and development trends of thermal fatigue property of hot die steels. Heat Treat Met 12:1–6 (in Chinese)
21. Persson A (2004) Simulation and evaluation of thermal fatigue cracking of hot work tool steels. Int J Fatigue 26:1095–1107
22. Persson A, Hogmark S, Bergström J (2005) Thermal fatigue cracking of surface engineered hot work tool steels. Surf Coat Technol 191:216–227
23. Pujante J, Vilaseca M, Casellas D et al (2014) High temperature scratch testing of hard PVD coatings deposited on surface treated tool steel. Surf Coat Technol 254:352–357
24. Schieck F, Hochmuth C, Polster S et al (2011) Modern tool design for component grading incorporating simulation models, efficient tool cooling concepts and tool coating systems. CIRP J Manufact Sci Technol 4:189–199
25. Shizhe S, Yin L, Wu J et al (2005) Corrosion electrochemistry of brass tube in simulated circulating cooling system. J Chem Ind Eng 1:121–125 (in Chinese)
26. Sjöström J, Bergström J (2004) Thermal fatigue testing of chromium martensitic hot-work tool steel after different austenitizing treatments. J Mater Process Technol 153–154:1089–1096
27. Srivastava A, Joshi V, Shivpuri R (2004) Computer modeling and prediction of thermal fatigue cracking in die-casting tooling. Wear 256:38–43
28. Vilaseca M, Pujante J, Ramírez G et al (2013) Investigation into adhesive wear of PVD coated and uncoated hot stamping production tools. Wear 308:148–154
29. Wang L, Su JF, Nie X (2010) Corrosion and tribological properties and impact fatigue behaviors of TiN- and DLC-coated stainless steels in a simulated body fluid environment. Surf Coat Technol 205:1599–1605
30. Wang Y (1997) A study of PVD coatings and die materials for extended die-casting die life. Surf Coat Technol 94–95:60–63

Index

P. Hu et al., *Hot Stamping Advanced Manufacturing Technology of Lightweight Car Body*, DOI 10.1007/978-981-10-2401-6

MIX
Papier aus verantwortungsvollen Quellen
Paper from responsible sources
FSC® C105338

If you have any concerns about our products,
you can contact us on
ProductSafety@springernature.com

In case Publisher is established outside the EU,
the EU authorized representative is:
Springer Nature Customer Service Center GmbH
Europaplatz 3, 69115 Heidelberg, Germany

Printed by Libri Plureos GmbH
in Hamburg, Germany